D1486340

# THE MEDIA, THE PUBLIC AND AGRICULTURAL BIOTECHNOLOGY

# The Media, the Public and Agricultural Biotechnology

*Edited by*

**Dominique Brossard**
*School of Journalism and Mass Communication, University of Wisconsin-Madison, Madison, USA*

**James Shanahan**
*Department of Communication, Cornell University, Ithaca, New York, USA*

*and*

**T. Clint Nesbitt**
*USDA-APHIS Biotechnology Regulatory Services, Riverdale, Maryland, USA*

**CABI is a trading name of CAB International**

CABI Head Office
Nosworthy Way
Wallingford
Oxon OX10 8DE
UK

Tel: +44 (0)1491 832111
Fax: +44 (0)1491 833508
E-mail: cabi@cabi.org
Website: www.cabi.org

CABI North American Office
875 Massachusetts Avenue
7th Floor
Cambridge, MA 02139
USA

Tel: +1 617 395 4056
Fax: +1 617 354 6875
E-mail: cabi-nao@cabi.org





A catalogue record for this book is available from the British Library, London, UK.

A catalogue record for this book is available from the Library of Congress, Washington, DC.

ISBN 978 1 84593 204 6

Typeset by Columns Design Ltd, Reading, UK
Printed and bound in the UK by Biddles Ltd, Kings Lynn

# Contents

# Contributors (by alphabetical order)

**Raju B. Barwale** is the Managing Director of Maharashtra Hybrid Seeds Co. Ltd. (Mahyco). He graduated in Agriculture from G.B. Pant University of Agriculture and Technology, Pantnagar, India. He has been with Mahyco for over 24 years in various capacities. His main emphasis on human resources development has helped Mahyco to sustain and grow in these changing times. He is on the Board of Management of several companies, trusts and trade associations.

*Contact details:* Mahyco, 4th Floor, 78 Veer Nariman Road, Mumbai 400200, India
e-mail: raju.barwale@mahyco.com

**Heinz Bonfadelli** (PhD, University of Zürich, Switzerland) has been a professor at the Institute for Mass Communication and Media Research at the University of Zürich since 1994. His post-doctoral studies were conducted at Stanford University, California, USA. His research fields include media and children, uses and effects of media and public information campaigns.

*Contact details:* Institute of Mass Communication and Media Research (IPMZ)
University of Zürich
Andreasstr. 15
CH-8050 Zürich
Switzerland
e-mail: h.bonfadelli@ipmz.unizh.ch

**Dominique Brossard** is an Assistant Professor of Mass Communication and a faculty affiliate of the UW-Madison Robert and Jean Holtz Center for Science and Technology Studies at the University of Wisconsin-Madison, Wisconsin, USA. Her studies include the dynamics underlying public opinion processes related to controversial science. She earned a Master in Plant Biotechnology at the École Nationale Supérieure d'Agronomie de Toulouse (France) and a PhD in Communication at Cornell University, New York.

*Contact details:* School of Journalism and Mass Communication, University of Wisconsin-Madison, 5168 Vilas Hall, Madison, Wisconsin 53706, USA
e-mail: dbrossard@wisc.edu

**Ildeu de Castro Moreira** is Professor of Physics at the Institute of Physics (Federal University of Rio de Janeiro, Brazil) and at the Graduate Program in Epistemology, History of Science and Technology. He also works in science education and public communication of science. He was coordinator of the World Year of Physics (2005) in Brazil. He is the Director of the Department for the Popularization of Science and Technology, Ministry of Science and Technology, Brazil.

*Contact details:* Department of Physics and Program of History of Science and Techniques and Epistemiology, Federal University of Rio de Janeiro/Praia do Flamengo 200, 80 andar, Flamengo CEP 22210-030, Rio de Janeiro RJ, Brazil
e-mail: icmoreira@uol.com.br

**Madhavi Char** graduated in the humanities and law from the University of Bombay. She has been with Mahyco for 7 years as a Legal Consultant addressing legal matters. Currently, she looks after intellectual property matters and licensing for the organization.

*Contact details: (see entry for R. Barwale)*
e-mail: madhavi.char@mahyco.com

**Urs Dahinden** is a Senior Researcher and Lecturer at the Institute of Mass Communication and Media Research, University of Zürich, Switzerland. His research interests include science and risk communication, political communication and online media.

*Contact details: (see entry for H. Bonfadelli)*
e-mail: u.dahinden@ipmz.unizh.ch

**Sanjay Deshpande** graduated with a degree in Commerce from Marathwada University, Aurangabad, India. He has been with Mahyco for the last 18 years and has played a key role in obtaining government approval for the marketing of transgenic cotton. He currently serves as Assistant General Manager (PR Coordination) at Mahyco.

*Contact details: (see entry for R. Barwale)*
e-mail: sanjay.deshpande@mahyco.com

**Robin Downey** is a PhD Candidate in the Faculty of Communication and Culture at the University of Calgary, Canada. Her doctoral research examines the influence of social movements on technology innovation, and specifically stem cell research.

*Contact details:* Communication Studies, University of Calgary, Calgary, Alberta, Canada, T2N 1N4
e-mail: rdowney@ucalgary.ca

**Sharon Dunwoody** is Evjue-Bascom Professor of Journalism and Mass Communication and an Associate Dean in the Graduate School at the University of Wisconsin-Madison, USA. She studies public understanding of science issues and has been exploring the risk communication domain specifically for some 15 years. Her primary interest there is the role of mediated messages in individuals' knowledge of and judgements about risk. She earned both her undergraduate and PhD degrees from Indiana University, USA, and her MA from Temple University, USA.

*Contact details: (see entry for D. Brossard)*
e-mail: dunwoody@wisc.edu

**Edna Einsiedel** is Professor of Communication Studies at the University of Calgary, Canada. Her research has focused on public representations of science and public participation models for technology assessment. She is a Principal Investigator on a Genome Canada-funded project on Genomics and Knowledge Translation in Health Systems. She serves as Editor of the journal *Public Understanding of Science.*

*Contact details: (see entry for R. Downey)*
e-mail: einsedel@ucalgary.ca

**Margarita Escaler** obtained her undergraduate degree in Biotechnology from Imperial College of Science and Technology, UK and completed her PhD in the field of plant virology at the John Innes Centre, UK. After finishing her studies, she joined ISAAA in August 2000, where she helped establish its knowledge-sharing initiative. She currently coordinates and manages its network of Biotechnology Information Centres in Asia, Latin America and Africa.

*Contact details:* ISAAA c/o, 111 Arthur Road, Windsor, Berkshire, SL4 1RU, UK
e-mail: m.escaler@isaaa.org

**Wendy Fink** is the Manager of Science and Public Policy at the Pew Initiative on Food and Biotechnology (PIFB), a research and education project of the University of Richmond, USA.

*Contact details:* Science and Public Policy, The Pew Initiative on Food and Biotechnology (PIFB), 1331 H Street, Suite 900, Washington, DC 20005, USA
e-mail: wfink@pewagbiotech.org

**Albert C. Gunther** (PhD, Stanford University, USA) is professor in the Department of Life Sciences Communication at the University of Wisconsin-Madison, USA. His research focuses on the psychology of the mass media audience, with particular emphasis on perceptions of media influence on others and biased evaluations of message content. His theoretical work has been applied in contexts ranging from science controversies such as genetically modified foods to health issues such as adolescent smoking adoption.

*Contact details:* Department of Life Sciences Communication, University of Wisconsin-Madison, 440 Henry Mall, Madison, Wisconsin 53706, USA
e-mail: agunther@wisc.edu

**Robert J. Griffin** (PhD, University of Wisconsin-Madison, USA) is full Professor in the Diederich College of Communication at Marquette University in Milwaukee, Wisconsin, USA and Director of its Center for Mass Media Research. His teaching and research interests focus on communication about environmental issues, energy, health, science and risk. He has been principal investigator or co-PI on federally funded research into environmental risks and has served on a National Research Council standing committee concerned with emerging issues in environmental contamination.

*Contact details:* College of Communication, Marquette University, Milwaukee, Wisconsin 53201-1881, USA
e-mail: robert.griffin@marquette.edu

**Michael Huge** (MA, Ohio State University) is a research associate at The Ohio State University School of Communication, and has previously worked for the Ohio Legislative Service Commission.

*Contact details:* School of Communication, The Ohio State University, 3016 Derby Hall, 154 North Oval Mall, Columbus, Ohio 43210-1339, USA
e-mail: huge.8@osu.edu

**Jane Kolodinsky** is Professor and Chair of the Department of Community Development and Applied Economics and Co-director of the Center for Rural Studies at the University of Vermont, USA. She has a BS in Nutrition and a Master of Business Administration, both from Kent State University, USA and a PhD in Consumer Economics from Cornell University, USA. Dr Kolodinsky has studied consumer information issues as they relate to agricultural biotechnology since 1990.

*Contact details:* Department of Community Development and Applied Economics, University of Vermont, 202 Morrill Hall, Burlington, Vermont 05405, USA
e-mail: jane.kolodinsky@uvm.edu

**Janice L. Liebhart** (MS, University of Wisconsin-Madison, USA) is a doctoral candidate in the Department of Life Sciences Communication at the University of Wisconsin-Madison, USA. Her research interests include the psychological processing of media messages about science and the influence of mass media on health and social perception. She has published articles on the hostile media effect in *Public Opinion Quarterly*, *Communication Research* and the *Journal of Communication.*

*Contact details: (see entry for A. Gunther)*
e-mail: jlliebhart@students.wisc.edu

**Martina Leonarz** studied communication science, cultural anthropology and film theory at the University of Zürich. She is a lecturer at the Department of Communication, University of Zürich, Switzerland and works for a private office in the field of technology assessment. Her fields of interests are risk communication, gender studies and media contents.

*Contact details: (see entry for H. Bonfadelli)*
e-mail: martina.leonarz@bhz.ch

**Luisa Massarani**, Dr Sc, is a science journalist. She combines both practical and research activities in science communication. Her research focuses on public perception toward science and on historical and contemporary aspects of science communication, mainly in Brazil and other Latin American countries. Massarani coordinates the Center of Studies of the Museum of Life at House of Oswaldo Cruz/Fiocruz, Brazil). She is also the Latin American coordinator of SciDev.Net (http://www.scidev.net).

*Contact details: (see entry for I. de Castro Moreira)*
e-mail: lunassa@coc.fiocruz.br

**Jennifer Medlock** is a PhD student in the Faculty of Communication and Culture at the University of Calgary, Canada. Her doctoral research focuses on the role of public and stakeholder involvement in science and technology policy development.

*Contact details: (see entry for R. Downey)*
e-mail: jemedlock@ucalgray.ca

**Luke Evuta Mumba** is Professor with the New Partnership for African Development (NEPAD) S & T headquarters in South Africa; formerly an Associate Professor in Genetics and Molecular Biology and former Dean, School of Natural Sciences, University of Zambia (UNZA). Dr Mumba is a member of the ASTM International Technical Committee on Biotechnology (E48), Board of Directors of the Zambia Wildlife Authority and also Chairman of the National AIDS Council – Institutional Biosafety Committee. He is the founding Chairman of the Biotechnology Outreach Society of Zambia.

*Contact details:* NEPAD S & T, SANBIO, CSIR Building 20, Mering Naude Road, Pretoria 0001, Republic of South Africa
e-mail: lemumba2004@yahoo.co.uk

**Mariechel J. Navarro** (PhD, University of the Philippines Los Banos) is Manager of the Knowledge Management Service of the Global Knowledge Center on Crop Biotechnology (KC). The KC, a programme of the International Service for the Acquisition of Agri-biotech Applications (ISAAA), is a science-based information network that includes 15 biotechnology information centres located in Africa, Asia and Latin America.

*Contact details:* ISAAA SEAsia Center, c/o IRRI DAPO Box 7777, Metro Manila, Philippines
e-mail: m.navarro@isaaa.org

**T. Clint Nesbitt** (PhD, Cornell University, USA) studied plant breeding and plant molecular biology. He is currently an outreach and communication specialist with the Biotechnology Regulatory Services of the US Department of Agriculture's Animal and Plant Health Inspection Service. For 4 years, he was an extension associate in the Department of Communication at Cornell University, where he was the principal educator, writer and developer of Cornell Cooperative Extension's Genetically Engineered Organisms Public Issues Education (GEO-PIE) Project.

*Contact details:* US Department of Agriculture, Animal and Plant Health Inspection Service, Biotechnology Regulatory Services, 4700 River Road, Riverdale, Maryland, USA
e-mail: thomas.c.nesbitt@aphis.usda.gov

**Matthew C. Nisbet** (PhD, Cornell University, USA) is Assistant Professor in the School of Communication at American University, USA, specializing in political communication and public opinion. His current research focuses on understanding in generalizable ways the communication dynamics of science-related controversies, exploring the interactions between public opinion, mass media and policy-making. He describes these research topics at http://www.framing-science.blogspot.com.

*Contact details:* School of Communication, American University, 4400 Massachusetts Avenue, NW, Washington, DC, USA
e-mail: nisbetmc@gmail.com

**Hans Peter Peters** is a Senior Research Fellow at the Program Group Humans–Environment–Technology of the Research Center Juelich, Germany, and adjunct professor at the Institute for Media and Communication Studies of the Free University of Berlin. His research deals with public sense-making of science, technological innovations and environmental problems in the context of a media society, and with the involvement of science in public discourses.

*Contact details:* Research Center Juelich, MUT, 52425 Juelich, Germany
e-mail: h.p.peters@fz-juelich.de

**Nick Pidgeon** is Professor of Applied Psychology at Cardiff University, UK. His research looks at how public attitudes and institutional responses form a part of the dynamics of a range of risk controversies, including those of radioactive waste, climate change, GM agriculture and nanotechnologies. Co-author (with B. Turner) of the book *Man-made Disasters* (2nd edn 1997) and (with R. Kasperson and P. Slovic) of *The Social Amplification of Risk* (Cambridge, 2003).

*Contact details:* School of Psychology, Cardiff University, Tower Building, Park Place, Cardiff CF10 3AT, UK
e-mail: n.pidgeon@uea.ac.uk

**Inez Ponce de Leon** is a science communication specialist with the ISAAA Knowledge Center. Her background is in molecular biology and biotechnology, but her interests span nearly everything from forensics to wildlife genetics, from web design to blogging, and from film to literature. She is also a novelist, and writes fiction and poetry when not involved in science writing or speaking engagements.

*Contact details: (see entry for M.J. Navarro)*
e-mail: illustria@gmail.com

**Wouter Poortinga** is an Academic Fellow at the Welsh School of Architecture and the School of Psychology, Cardiff University. Wouter's research focuses on people's responses to various environmental and technological risks. One of his recent topics was on the role of trust in the perception and acceptability of risks. His wider research interests are in studying the interaction between the psychological and social/environmental basis of people's health, well-being and quality of life.

*Contact details:* The Programme of Understanding Risk – Centre for Environmental Risk, School of Environmental Sciences, University of East Anglia, Norwich NR4 7TJ, UK
e-mail: w.poortinga@uea.ac.uk

**Michael Rodemeyer**, JD, is the founding Executive Director of the Pew Initiative on Food and Biotechnology, a non-profit research group, and is currently an adjunct professor at the University of Virginia, USA. Previously, he was the Chief Counsel to the US House Committee on Science and served as the Assistant Director for Environment in the White House Office of Science and Technology policy in the Clinton Administration.

*Contact details: (see entry for W. Fink)*
e-mail: michael@mrodemeyer.com

**Magdalena Sawicka** is a PhD candidate at the Program Group Humans–Environment–Technology of the Research Center Juelich, Germany. Her research deals with cultural roots of attitudes towards technological innovations. In particular, she is analyzing 'concepts of nature' and their relevance for the explanation of cross-cultural differences in public attitudes towards food biotechnology.

*Contact details: (see entry for H.P. Peters)*
e-mail: m.sawicka@fz-juelich.de

**Dietram A. Scheufele** is a Professor with a joint appointment in the School of Journalism and Mass Communication and the Department of Life Sciences Communication at the University of Wisconsin, Madison, USA. He is also a faculty affiliate of the Robert and Jean Holtz Center for Science and Technology Studies. Scheufele's recent work has examined the public opinion dynamics and media coverage surrounding nanotechnology, stem cell research and GMOs.

*Contact details: (see entry for D. Brossard)*
e-mail: scheufele@wisc.edu

**James Shanahan** is an Associate Professor of Communication at Cornell University, USA. His research interests include environmental communication, media effects and public opinion. He obtained a PhD in communication from the University of Massachusetts at Amherst, USA, in 1991. He has authored and edited several books, including *Propaganda without propagandists?* (Cresskill, New Jersey: Hampton Press, 2001), *Television and its Viewers: Cultivation Theory and Research* (Cambridge, Massachusetts: Cambridge University Press, 1999), *Nature Stories: Depictions of the Environment and Their Effects* (Cresskill, New Jersey: Hampton Press, 1999).

*Contact details:* Department of Communication, Cornell University, 314 Kennedy Hall, Ithaca, New York 14853, USA
e-mail: jes30@cornell.edu

**Mahendra K. Sharma** graduated with a doctoral degree in Plant Physiology from the University of Udaipur, India. He has been working with Mahyco in various capacities for the last 25 years. He is currently Managing Director of Mahyco-Monsanto Biotech, an affiliate of Mahyco.

*Contact details: (see entry for R. Barwale)*
e-mail: mahendra.sharma@mahyco.com

**Sonny P. Tababa** completed her undergraduate and Master of Science degrees in agronomy at the University of the Philippines Los Banos. She was a Senior Science Research Specialist at the Philippines Department of Science and Technology's sectoral council in agriculture and natural resources research and development until mid-2000. She currently manages the Biotechnology Information Center at the Southeast Asian Regional Center for Graduate Study and Research in Agriculture, Philippines.

*Contact details:* SEAMEO SEARCA Biotechnology Information Center, College, Laguna 4031, Philippines
e-mail: spt@agri.searca.org

**Usha B. Zehr** did her doctorate in Agronomy at the University of Illinois at Urbana-Champaign, USA. After a period of teaching at Purdue University, USA, she joined Mahyco as Joint Director of Research, a position she has held for the last 10 years. She also serves on the board of management of several companies and trusts and is a member of several committees.

*Contact details: (see entry for R. Barwale)*
e-mail: usha.zehr@mahyco.com

# Preface

This book examines public perception, public deliberation and media presentation of agricultural biotechnology and genetically modified (GM) foods. Evenson and Santaniello (2004) provided an outlook on this topic that was primarily economic and focused on consumer behaviour; our book looks at public opinion data, communication theory and international examples to provide a complementary overview of how the public sees this controversial topic.

The book is derived from the authors' experience with perceptions of agricultural biotechnology at both national and international levels. Thus, one of the main foci of the book is to be grounded in actual case studies. For this, we have called on experts in communicating about GM and on experts in actually bringing GM foods to market, often against great resistance. But we also saw a need for a book that would summarize public opinion. Most studies of agricultural biotechnology have not provided comparative examples of public opinion. Normally, it is simply assumed that there is support in the US *versus* opposition in Europe. This dichotomy tends to ignore the developing world, where there are often many exciting examples of real world field-testing and implementation of GM food.

To preview the book: it is about the communication of agricultural biotechnology. It begins with data on public opinion, moves toward theory that can explain how public opinion is formed and then concludes with cases from around the world.

Part I is a review of how people 'feel' about agricultural biotechnology. The chapters constitute a review of some of the studies of public opinion about agricultural biotechnology, especially genetically modified organisms (GMOs). This section is intended to provide a starting point for individuals interested in descriptive information about opinions about biotechnology. The authors will provide both general data – relating to issues such as opposition or support for biotechnology – as well as focusing on specific cases and stories relevant to each national or regional context.

The first chapter, by Dominique Brossard and James Shanahan, is a review of theoretical concepts useful for approaching the issue of public acceptance of GMOs. Concepts of information, trust in institutions and media factors are the main ideas explored in this chapter, which is intended to provide a broad overview of the field.

The chapter by Wouter Poortinga and Nick Pidgeon, 'Public Perceptions of Agricultural Biotechnology in the UK: the Case of Genetically Modified Foods', begins the series of case studies on particular nations' opposition or support for GM food. This chapter provides an understanding of the nature of public attitudes towards this technology in the UK. This includes a history of the biotechnology controversies in Britain in the late 1990s. Results of a recent British survey on the perceptions of genetically modified (GM) food and crops are presented.

The chapter also touches on the nature of the formalized GM debates that were held in the UK. The chapter proposes 'that people hold a complex set of beliefs about a range of health, environmental and social risks and benefits of GM food and crops'. Labelling of GM food, the liability of the biotechnology industry, trust in risk regulation and trust in various sources of information are also discussed.

In Chapter 3, Hans Peter Peters and Magda Sawicka, in 'German Reactions to Genetic Engineering in Food Production', continue the series of case studies. The description of the German case is presented within the European context and partly in comparison to the USA. The food biotechnology issue is related to the environmental issue, consumer preferences for food and food production, perception of food quality, health concerns, the public image of large companies and criticism of the (former) agricultural policy of the EU. The authors also address the role of environmental and consumer organizations and the likely impact of the industry's strategies on public responses to genetic engineering.

In the second part of the chapter, the authors analyse German public reactions to (food) biotechnology based on an analysis of surveys. The third part of the chapter discusses the media and their potential impact on the structure of public discourse between (political) actors, the framing of the issue and their effect on individual attitudes. The discussion is based on content analyses of media coverage of biotechnology and analyses of media reception and media effects. In their conclusion, the authors argue that attitudes towards food biotechnology are not primarily the result of individual or societal risk–benefit calculations, but are deeply rooted in culture.

In Chapter 4, Heinz Bonfadelli, Urs Dahinden and Martina Leonarz describe and analyse the controversy about red and green biotechnology in Switzerland, a controversy that was strongly influenced by an ongoing, intensive, public debate about biotechnology because of the specific Swiss political system with its mechanisms of direct democracy. The impact of the political debate on media coverage is analysed on the basis of a longitudinal content analysis. The public frames of 'red' and 'green' biotechnology are discussed differently, the analysis being based on qualitative group-discussion data.

The last part of Chapter 4 deals with media effects on public opinion. This analysis is based on quantitative data of three Eurobarometer surveys, carried out in Switzerland in 1997, 2000 and 2002/2003. The data allow comparisons between different applications of biotechnology (red and green), concerning dimensions like benefits and risks, moral concerns and support. Attitudes concerning red and green biotechnology are tracked over time and linked to media coverage.

In Chapter 5, Wendy Fink and Michael Rodemeyer, of the Pew Initiative on Food and Biotechnology (PIFB), examine consumer understanding and acceptance of agricultural biotechnology in the USA. Their chapter places consumer opinion of agricultural biotechnology products in context with the factors that may affect them; it can help explain why these products have received varying levels of acceptance around the world.

This chapter examines several factors affecting consumer attitudes, including: (i) the economic, political and social context into which GM foods were introduced; (ii) the level of awareness and knowledge of the topic of agricultural biotechnology; (iii) the end use and potential benefits of a product; and (iv) consumer confidence and trust in the regulatory system. Additionally, this chapter looks at options for increasing consumer acceptance of new technologies about which the public is troubled or uncomfortable.

In Chapter 6, Jane Kolodinsky examines biotechnology from an American consumer information perspective. She looks at how consumer information plays a vital role in the ongoing debates about biotechnology. She examines differences in results obtained from different groups and tries to bring an overall perspective as to how consumers in the USA get information about biotechnology. She concludes that the collective results are confusing to consumers, while the voluntary labelling of GM foods is seen as at least a partial victory for consumer information.

In Chapter 7, Luisa Massarani and Ildeu de Castro Moreira examine 'What do Brazilians think about Transgenics'. In Latin America, few studies on GMOs have been conducted, so this chapter provides useful new information. The chapter reviews the key results of the main surveys and studies – both quantitative and qualitative – held in Brazil on what people think about transgenics. The first survey on transgenic crops was conducted by IBOPE, the Brazilian Institute for Opinion Public and Statistics, and showed that most Brazilians preferred non-transgenic food.

The authors discuss how results vary slightly for various demographic factors. The authors then present results gathered though various qualitative case studies, one among young people in Rio de Janeiro high school students, another though a Citizens Jury held in 2001. The last case study involves an anthropological study held among farmers in the south of the country – a region in which the controversy surrounding transgenics is especially important due to the illegal growth of transgenic soy from seeds smuggled from Argentina.

While social scientists have easily documented the *state* of public opinion about biotechnology, biotechnology also provides an interesting laboratory in which theories of opinion formation and the policy process can be tested. The chapters in Part II focus on empirical tests of theories of opinion formation using biotechnology as a case study.

In Chapter 8, Matthew C. Nisbet and Michael Huge argue that power in policymaking revolves in part around the ability to control media attention to an issue while simultaneously framing an issue in favourable terms. These two characteristics of media coverage both reflect and shape where an issue is decided, by whom and with what outcomes. In understanding this process, a number of studies have observed cyclical waves in media attention and historical shifts in how an issue is framed, linking these features to policy decisions. Yet, there has been little theoretical specification and testing of the social mechanisms that drive these cycles. With this in mind, this chapter outlines a model for understanding 'mediated issue development'.

Using data from a content analysis of 25 years of coverage at *The New York Times* and *Washington Post*, the model is applied and tested against the issue of plant biotechnology. The authors conclude with comparisons of other issues such as the Human Genome Project and intelligent design. Understanding, however, why plant biotechnology remains at low levels of controversy in the USA compared to the rest of the world remains the object of considerable curiosity, and the focus of this study posits several explanations.

In Chapter 9, Dietram A. Scheufele examines 'Opinion Climates,

Spirals of Silence and Biotechnology: Public Opinion as a Heuristic for Scientific Decision-making'. The author argues that the fact that citizens perceive themselves to be uninformed about the issue does not mean that they will not make decisions or judgements about agricultural biotechnology and the policy issues surrounding it. Rather, people's opinions are influenced by a range of factors other than information, such as ideological predispositions, the way mass media present the issue and – most importantly – their perceptions of what the climate of opinion in the country looks like.

This chapter examines opinions of others as one of the key shortcuts that people use when making decisions about biotechnology. It compares the competing influences of information and heuristic cues about public opinion. The research presented in this chapter relies on a theoretical model of opinion formation in societies called the 'spiral of silence'. The spiral of silence model assumes that people are very aware of the opinions of people around them and adjust their behaviours (and opinions) to majority trends, and also very aware of the fear of being on the losing side of a public debate.

In Chapter 10, Albert C. Gunther and Janice L. Liebhart examine 'The Hostile Media Effect and Opinions about Agricultural Biotechnology'. The chapter opens with a selection of anecdotes illustrating strong sentiments among some groups on biotechnology-related issues – specifically GM foods. The chapter then gives a broad overview of the hostile media effect (HME) phenomenon, which is the notion that partisan groups tend to perceive media coverage of an issue as biased against their own viewpoint.

Issues addressed in the chapter include: (i) partisanship; (ii) asymmetry between groups; (iii) processing mechanisms; (iv) public opinion outcomes; and (v) relation to other theories – describing how the HME may fit into existing communication and public opinion theories, including agenda-setting and spiral of silence, among others.

In Chapter 11, Sharon Dunwoody and Robert J. Griffin discuss 'Risk Communication, Risk Beliefs and Democracy'. Scholars of risk communication are borrowing and, in some cases, developing theory to inform practice. In this chapter, Dunwoody and Griffin describe a set of conceptual frameworks for risk communication decision-making and relate those frameworks to agricultural biotechnology issues.

Prominent among the borrowed are risk perception frameworks such as Slovic's psychometric paradigm, as well as information-seeking and processing frameworks such as Eagly and Chaiken's Heuristic–Systematic Model. The authors also describe a recent Model of Risk Information-seeking and Processing that is specific to risk communication settings.

There is a range of opinion about technology around the world, ranging from enthusiastic acceptance to outright rejection. While social science theory can explain some of these differences, case studies provide access to the specific events, product developments and the policy processes that have arisen to deal with them. The chapters in Part III therefore provide examples of communication strategies designed to increase understanding of agricultural biotechnology and on practical communication strategies tested in international settings.

In Chapter 12, T. Clint Nesbitt of USDA-APHIS analyses the GEO-PIE Project, a web-based outreach at Cornell University, USA. In 2000, researchers at Cornell University acquired funding to create the Genetically Engineered Organisms Public Issues Education (GEO-PIE) Project. As explained by the author, consistently with research findings, the GEO-PIE project analysed in this chapter was developed with a conscious effort to be a neutral, non-partisan purveyor of up-to-date information on agricultural biotechnology. In addition to summarizing the activities of the GEO-PIE Project, this chapter reviews several methods used to evaluate the impacts of the project's outreach efforts.

In Chapter 13, Jennifer Medlock, Robin Downey and Edna Einsiedel of the University of Calgary, Canada, examine 'Consensus Conferences as a Communications Tool'. Using the frame of Participatory Technology Assessment (PTA), they discuss the evolution of traditional, expert-oriented technology assessment to more participatory, inclusive modes of assessment that aim to involve various 'publics' (citizens, consumers and other stakeholders) in debates about technology policy development.

They then discuss how PTA employs a family of consultation methods with the common feature of being based on a model of discussion and deliberation. Next, they discuss the increasing use of deliberative models of consultation, with specific reference to agricultural biotechnology examples. Following this, they discuss how these communication practices influence the trajectory of technology as well as policy development.

In Chapter 14, Mariechel J. Navarro, Margarita Escaler, Inez Ponce de Leon and Sonny P. Tababa discuss 'The Bt Maize Experience in the Philippines: a Multi-stakeholder Convergence'. From 1996 to 2002, the debate in the Philippines about Bt Maize saw a plethora of stakeholders, which even included the religious community, all trying to win the hearts and minds of the public with the government assigned to assess the technology. Activist groups uprooted field trials and lobbied for a moratorium on genetically modified (GM) crops. Catholic priests and nuns pleaded with local government units to refrain from giving support to GM activities in the community.

Filipino scientists battled it out with various groups that linked the Bt maize technology to multinational greed and 'playing God'. Addressing the different concerns of such a diverse group of stakeholders became a real challenge, but was critical to the eventual commercial approval of Bt maize in the country. This chapter provides a case study of the eventual acceptance of Bt maize in this contentious environment.

In Chapter 15, Luke Evuta Mumba examines 'Food Aid Crisis and Communication about GM Foods: Experience from Southern Africa'. The aim of this chapter is to review the food aid crisis of 2001/2002 in southern Africa and to examine how communication about GM foods was handled. First, this chapter gives the background and magnitude of the food crisis faced by the SADC countries in the 2001–2002 period and the humanitarian efforts which were taken by the international community to address the problem.

Second, it reviews the reaction of the SADC countries to the offer of GM food aid by the international community. The reasons behind the decisions taken by these countries to either accept or reject GM food aid are also discussed. Third, the chapter identifies both local and international key players in the GM food aid debate and the roles they played in trying to assist national governments in decision-making.

Fourth, the chapter reviews the current thinking of SADC countries on GM food aid and the initiatives taken individually and collectively to improve levels of acceptance of GM technology and access to agricultural biotechnology in general, since the famine of 2001–2002. Finally, the crucial question: 'What is to be done?' is posed in the last section.

In Chapter 16, Raju Barwale and his colleagues of the Maharashtra Hybris Seeds Co. Ltd. (Mahyco), India, provide an industry perspective on commercializing a GM crop in Asia. Bt cotton became the first transgenic agricultural crop approved for large-scale release in India. When the work was inititated in early 1993, the regulatory system was purely on paper and no precedence of any approval was in place.

The system had to be tested out as events went along. There were no proven processes for the conducting of various studies, nor institutions identified for conducting such studies. Working out the coordination of the various institutions, ministries of the government (namely, Agriculture, Science and Technology, Health, Environment and Forests) was required. The actual process of taking the evaluation forward was yet to be established.

The chapter describes how the path for regulatory approval of the transgenic crops in India was followed. This includes a focus on biosafety aspects, technical aspects, IP issues and communication problems.

## Reference

Evenson, R. and Santaniello, V. (2004) *Consumer Acceptance of Genetically Modified Foods.* CAB International, Wallingford, Oxon.

# I Public Opinion about Agricultural Biotechnology around the World

1

# Perspectives on Communication about Agricultural Biotechnology

D. Brossard[1*] and J. Shanahan[2**]

[1]School of Journalism and Mass Communication, University of Wisconsin, Madison, Wisconsin, USA; [2]Department of Communication, Cornell University, Ithaca, New York, USA;
e-mail: *dbrossard@wisc.edu; **jes30@cornell.edu

## Introduction

It may be observed that there are few hard and fast principles that have been offered in the way of 'best practices' for communication about agricultural biotechnology. The Cartagena Protocol on Biosafety (Secretariat of the Convention on Biological Diversity, 2000), for instance, urges parties to 'promote and facilitate public awareness' (p. 18) and also to 'consult the public' and 'inform the public' about activities related to the protocol and to the products of biotechnology (Article 23).

But how this is accomplished varies enormously from country to country; it is clear from the welter of different extant examples and approaches that no cookie-cutter approach will suffice for developing an approach to understanding how to communicate about biotechnology, much less understanding public opinion about it.

Especially lacking is an approach that links together the main variables involved in communication about biotechnology. While there are comprehensive reviews of some of these variables individually, there are few to our knowledge that discuss the accumulated research and link together findings into a comprehensive model that answers the following question: how do communicators and policy-makers develop information and communication strategies that meet commonly agreed-upon scientific, ethical and practical standards? To begin to answer this question, we will first review some of the research, with illustrative examples from the developing world and from US and European research.

The components of public communication about biotechnology comprise: (i) knowledge/awareness; (ii) trust; (iii) mediated discourse; and (iv) risk communication.

## Knowledge/awareness

Many discussions about the difficulties related to communicating around agricultural biotechnology centre on knowledge and/or awareness of biotechnology. It is assumed, often from a common-sense perspective, that increased knowledge about biotechnology will lead to changes in public acceptance of that technology. However, this assertion has been questioned frequently in studies that include empirical evidence, leading to conclusions that attempts to change public knowledge are both unsuccessful and not needed.

The 'knowledge deficit' model is often posited as the default method for technical communicators, who assume that lay audiences simply lack specific knowledge that will allow them to understand a technology such as genetic modification (Hilgartner, 1990). Most recent reports have debunked the usefulness of this notion in communicating about biotechnology. Hansen *et al.* (2003) argue:

> It has gradually become clear that it is neither helpful nor accurate to view the expert–lay discrepancy as a clash of objective expert knowledge and subjective lay distortions. Instead, what is needed is a better understanding of the way in which consumers look at, or perceive, risk. We need to know precisely how and why this differs from the approach of the experts.
>
> (Hansen *et al.*, 2003, p. 111)

There is a significant literature within the science studies field about the appropriateness of addressing knowledge gaps in science communication strategies. Here we can identify two important components of the knowledge deficit issue that are relevant to policy-makers. From the standpoint of communication effectiveness, knowledge has not been shown to be a major factor, nor is it the only factor in predicting variables such as support for biotechnology. There is not much of a basis for assuming that increased knowledge of technical facts will lead to greater support for biotechnology. Variables such as trust in institutions are seen to be more predictive.

However, from an ethical standpoint, we have noted that the international community has already identified knowledge and informedness as a key goal of biosafety programmes. While knowledge itself is not sufficient to motivate support for biotechnology, and may even occasionally motivate rejection of the technology, it must be considered as a valuable goal in and of itself. To this end, policymakers must consider knowledge as an outcome of a successful programme on communication about biotechnology, but not necessarily as a tool to increase support.

## Measures of knowledge of biotechnology

While there are some studies that incorporate some measures of how well the public understands biotechnology, there are relatively few datasets that provide comprehensive measures, over time, of how different audiences actually understand biotechnology.

### *US and Europe: knowledge about biotechnology*

Perhaps the most widely cited piece of research comparing knowledge levels cross-culturally is the work of Gaskell *et al.* (1999), comparing differences in US and European support for agricultural biotechnology. They found that Europeans were better informed about biotechnology in a textbook sense, but that Americans were more supportive, confounding the expectation that understanding the principles of biotechnology would lead to greater acceptance. Indeed, their study revealed no consistent cross-cultural predictors, though they implied that media coverage and its frequency played some role.

However, Hallman *et al.* (2003), analyzing more recent data, found that Americans were better informed than Europeans. Hallman also noted the poor correlation between objective performance on knowledge indicators and people's self-reported knowledge. In general, Hallman interpreted the data to mean that Americans have both low awareness and objective knowledge about issues related to biotechnology.

The Pew Foundation also published reports on Americans' knowledge about biotechnology (http://www.pewagbiotech.org/research/gmfood/survey3-01.pdf). They found that knowledge about genetically modified (or biotechnology used in) foods remained low, and had not increased over the previous 2 years. In 2001, 44% had heard a great deal or somewhat about genetically modified (GM) foods; today, that proportion is 34%, a ten-point decline. Similarly, 45% had heard 'a great deal' or 'something' about biotechnology use in food production; today, that figure is 36%, a nine-point decline. There were similar increases in the numbers who had heard 'not too much' or 'nothing at all' (eight points for biotechnology, 11 points for GM foods).

Priest *et al.* (2003), using the European data, have argued that general knowledge is a poor predictor of a country's support for biotechnology. However, other authors such as Pardo and Calvo (2002), have questioned the validity of these most-used measures of science knowledge, which they claim are seriously under-conceptualized.

Despite the confusions about whether knowledge is important or can be measured well, there are estimates of the percentage of the public that is, by some standard, well informed. Moreover, some studies do indicate that support for biotechnology can be at least moderately correlated to knowledge. Pardo *et al.* (2002) estimated that about 20% of Europeans were well informed in the sense of being aware of biotechnology and having some basic terms in their vocabulary. We can assume that US citizens would show a similar percentage. The highest percentages of informedness were in Northern European countries such as the Netherlands, Denmark and Sweden. Further, their path modelling suggested both direct and indirect effects of knowledge on variables such as attitudes about biotechnology and perceptions of benefit from GM technology.

In a 2003 report about biotechnology in New York State (http://www.geo-pie.cornell.edu/media/media.html#opinion), supporters of the technology were also the most aware. Opponents reported the next greatest level of awareness, and those who were undecided were the least aware. Hallman *et al.* (2003) also found that increased actual knowledge was related to increased approval. However, one cannot reach the conclusion that increasing awareness or knowledge would lead to greater support; it is just as likely that knowledge is an outcome and not a cause of support.

Some overviews indicate that knowledge and awareness do not tend to change much. A trend analysis (Shanahan *et al.*, 2001) showed that a variety of US public opinion polls that had been conducted showed variance in awareness that was mostly due to question wording (see Fig. 1.1). Even though media coverage had increased during the early parts of the study period, awareness levels seemed relatively immune to their influence. Declines seen toward the end of the data series may be due to declines in media coverage, but they are minimal in magnitude.

Figure 1.2 depicts a model of public opinion that shows levels of support, knowledge/awareness and attitudes about biotechnology. It is derived from our research in New York State, though we can apply it to developing world cases as well. The figure depicts responses to questions about biotechnology posed on scales of 1 to 10. Responses above 5 reflect higher levels of support and knowledge. In our data, both opponents and supporters tend to have higher levels of informedness. Only the 'undecided' are less aware. This suggests that attempts to increase knowledge and awareness in our market in New

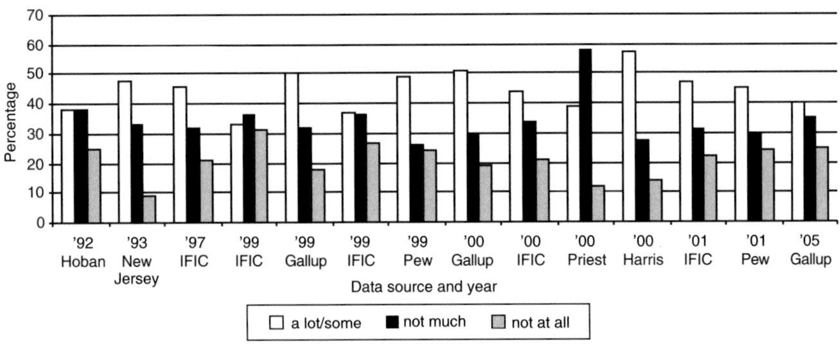


**Fig. 1.1.** Levels of awareness of GMOs by year and data sources in the USA (from Shanahan *et al.*, 2001).

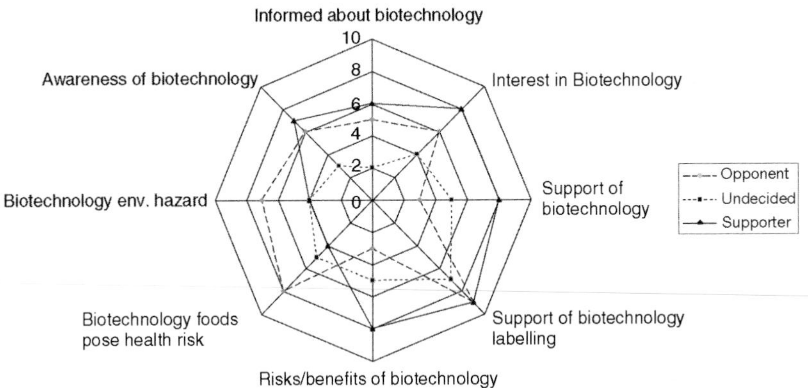


**Fig. 1.2.** Opinion structure for three groups in New York State, USA.

York State could simply have the effect of heightening a polarized dialogue. This has been confirmed in some attempted dialogue formats that bring together polarized opponents.

### *Knowledge and awareness in the developing world*

Not surprisingly, less research has been carried out on knowledge in developing countries. However, there is some information. ISAAA completed a series of country reports on five countries in South East Asia (Indonesia, Vietnam, Philippines, Thailand and Malaysia, http://www.searca.org/~bic/resources/downloads.htm#policy). This series of reports provides some benchmark data on awareness levels in these developing countries. Table 1.1 provides awareness data for the five countries.

**Table 1.1.** Interest in/awareness of biotechnology in Asia.

| Country | Level of awareness[1] |
|---|---|
| Indonesia | 4.93 |
| Thailand | 5.12 |
| Vietnam | 5.13 |
| Philippines | 5.20 |
| Malaysia | 5.33 |

[1] 1, none; 7, highest.

These numbers do not give a sense of actual knowledge – they are perceived interest/awareness. As Hallman showed, these two are not necessarily strongly related. Few studies are available that would report on textbook knowledge in the Eurobarometer sense. ISAAA reached similar conclusions about these five countries: that there was 'moderate' interest in biotechnology. However, the lack of other data makes comparisons difficult, and it is never clear what these self-report estimates really mean in terms of actual knowledge.

Results from other studies provide a pastiche of findings. Zhong *et al.* (2002) found that 'The majority of Chinese consumers have little knowledge of GM foods. Even in a provincial capital city like Nanjing, more than 50% of urban residents had not heard of GM foods at all; of those who had heard of GM foods, only about 25% thought that they knew something about GM foods.'

Similarly, in Brazil, some studies have shown lesser awareness than in developed countries. Kamaldeen and Powell (2000) found that only 37% of the Brazilian public were aware of genetically modified food issues, compared to ratings of 66–95% in other developed countries. Even though issues of biotechnology have now confronted selected developing world countries such as Brazil or Zambia, we can assume that awareness and knowledge rates are lower overall in these countries, but that they are on the increase in the cities. China, India and Brazil are often presented as examples of developing countries where interest and support for GMOs may be higher.

Africa presents a different picture. Apart from South Africa, there are few data at all about awareness and knowledge of biotechnology. Knowledge and awareness among elites are the main concern, as communication media to reach the wider public are often not well developed. One study in West Africa (Alhassan, 2002) examined public awareness of biotechnology among NGOs and media institutions. Even in these most active and informed sectors, awareness/knowledge levels were running at about 50%. That is, half of the NGOs claimed to be well informed about biotechnology, but only about a quarter of the press institutions interviewed had any knowledge of biotechnology. Clearly, in this region, awareness is at its lowest level worldwide.

This review of the literature leads us to conclude that knowledge and awareness should be approached as dependent variables. Awareness in particular is driven by media coverage, at least in urban environments. Knowledge itself, when measured factually, is much more difficult to change. Nevertheless, from an ethical standpoint, in concurrence with international agreements, to increase knowledge is an axiomatic good. But there is no available evidence that would show how increasing knowledge would lead to other desirable outcomes. Indeed, there is not even a clear standard for measuring informedness about biotechnology. As the technology itself remains controversial, there could be significant disagreement about what would constitute the correct suite of facts to transmit to different publics.

## Trust

As with knowledge, there are no consistent conceptualizations of trust in the literature. Some conceptualize trust in institutions as a volatile attitude likely to be influenced by contextual factors and localized incidents. We believe that trust in science has to be considered at two different levels. On the one hand, trust in science can be conceived as a strongly held attitude fostered by cultural contexts and education. On the other hand, trust in science might be a more volatile attitude, likely to be impacted by scientific misconducts and/or errors that influence society at certain points in time, and by media coverage of specific events.

In our research in the USA (Brossard and Nisbet, 2007), we have used three questions to capture strongly held trust toward scientific institutions, which we termed 'deference to scientific authority'. We asked our respondents (*n*, 775) in New York State to rate their agreement with the three following statements:

- Scientists know best what is good for the public.
- It is important for scientists to get research done even if they displease the public by doing it.
- Scientists should do what they do best, even if they have to persuade people that it is right.

We found that these three questions provided a consistent indicator of underlying willingness of an individual to support scientific activity even in the face of public controversy. Our data showed that this variable predicted both awareness and informedness about biotechnology. With this conceptualization, trust precedes awareness and information, because people develop conceptions of broad-scale institutions such as science prior to dealing with specific issues such as biotechnology.

In our data, deference to scientific authority presupposes a readiness to encounter specific technologies such as biotechnology with openness and interest. As Evenson and Santaniello (2004) note: 'Differences in consumer perceptions are partly related to the degree of confidence in existing food safety regulation systems' (p. xi). While such readiness need not immediately lead to acceptance, it is almost always positively related to support.

Other studies have argued that trust is more primary than information. Hansen *et al.* (2003) argue that:

> ... trust enables individuals and organizations to act without knowledge. It is therefore very useful in modern technology-dependent societies ... there is a very real sense in which, because trust is a substitute for knowledge, information does not build it (as is often argued by advocates of the deficit model) but rather makes it redundant. Hence, in one way, trust cannot be effectively created and maintained through information campaigns. Instead, the food sector must aim to improve its trustworthiness by being more socially responsive.
>
> (Hansen *et al.*, p. 119)

This thought captures our view very well. The effect of increasing information is valuable for its own sake, but is best treated as an outcome. For policy-makers wishing to increase public acceptance of a technology such as GM, trust comes first.

Some other studies confirm this view as well. In instances where trust has been compared directly to knowledge as an antecedent of variables such as fear of science or support for biotechnology, trust invariably has stronger relationships. Brossard and Shanahan (2003), for instance, showed that trust in activists was related to greater fear of biotechnological science, while trust in science was related to less fear. Indeed, the relations were strong enough that fear could essentially be conceived as synonymous with trust issues.

Siegrist (2000) also argued that trust directly influences risk perception. His model conceived trust as a replacement for knowledge, incorporates multiple indicators of trust and shows a strong direct relation to perception of risk and benefit, which in turn directly affect acceptance.

Frewer (2003) noted that trust can be looked at both in terms of macro-social trust in institutions and interpersonal source credibility. The social variety seems to be more important in food issues. Bonny

(2003) emphasized that mistrust of institutions and corporations was a major factor in the rejection of GM in France, where rejectionist tendencies appear strongest.

Hornig Priest *et al.* (2003) suggested that 'trust gaps' are the best predictors of national support for biotechnology. They argued that differences between trust in environmental groups and science/technology groups are most strongly correlated to support for biotechnology. They turn away from typical 'knowledge gaps,' arguing that trust functions as an engine of public opinion formation. Thus, rather than focusing on communication of knowledge to individuals, they suggest that trust emerges as a cultural level variable that should receive primary consideration in understanding public opinion about biotechnology.

It may seem tautological to equate trust with acceptance of GM, but the research is clear on two counts: that trust varies nationally and seems to be related to acceptance of technology. To develop our model further it is necessary to show which variables may be conceived of as precursors to trust. Because it is obviously not enough to urge policymakers to 'increase trust', we examine other factors that may be conceived as being closely related to trust.

## Mediated Discourse

As Hagendijk (2004, p. 47) noted, '"distrust" as an explanation is itself in need of explanation'. One common assertion is that public trust is constructed in mediated discourses about science issues. Scientists often believe that activist organizations use the media to foment discord about otherwise non-controversial issues. At the same time, how science and technology institutions use the media to promote science and technology has also received some attention. However, as with most of the variables that we look at, the actual empirical relation of media coverage and use of biotechnology to attitudes is not so straightforward.

Much of the research focuses on the USA/Europe dichotomy. The working hypothesis in the early research was that differences in US and European media coverage might have played some role in the differences in advocacy for the issue. However, the picture is not quite so simple. Both areas have shown times when media coverage has been problematic. For instance, in the USA, early coverage of biotechnology was quite positive, but then issues such as the Starlink controversy or the scare about Monarch butterflies being killed by Bt (*Bacillus thuringiensis*) plants raised concerns. In Europe, coverage seems to have been more consistently critical.

Our view is that coverage of biotechnology needs to be looked at from an issue-cycle perspective. Even though biotechnology is a very important scientific development, news media will not always devote prominent attention to it; other issues will take their place on the agenda as the press and the public consider them worthy of attention. Invariably, thus, the public is asked to consider issues cyclically. In order to understand how public opinion is affected, we need to consider the beginning, middle and end, leading to the 'outcome' of each cycle of attention.

Issue cycles have been studied since the 1970s, when Downs (1972) proposed that the environmental issues then popular could not hold public attention for long. He proposed that public attention would fade for the technically and politically complicated issues. His hypothesis has generally been upheld, though it has not often been studied within the context of biotechnology.

Figure 1.3 depicts a simplified model of how media cover issues, and how they track through the policy system (for a more complete discussion of how an issue is dealt with in the public–media–policy system, see Nisbet and Huge, Chapter 8, this volume).

Four distinct phases can be seen in the model. An issue normally becomes a problem through some signalling, focusing or triggering events. This leads to a period of 'alarmed discovery', as the press notifies the public of a potentially threatening issue. If the issue attracts enough attention, this leads to a period of mobilization during

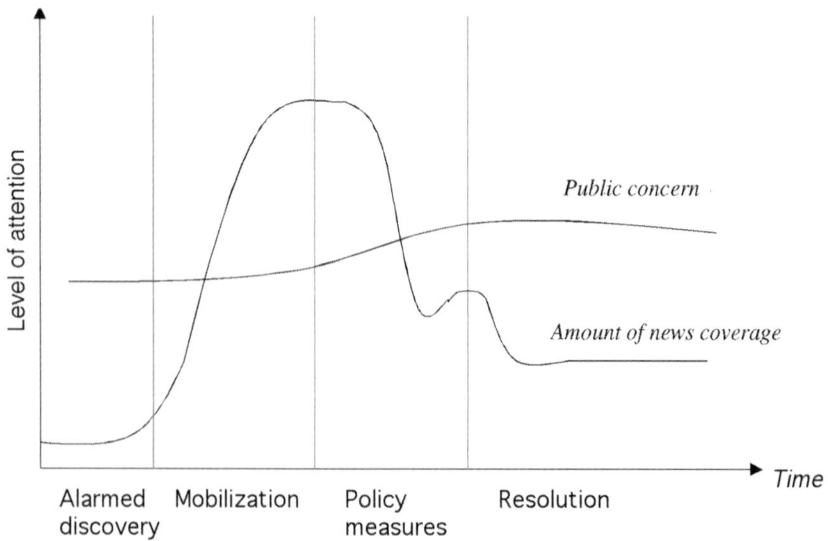


**Fig. 1.3.** Issue attention cycle.

which press and political attention are normally focused on the hazards associated with the risk. During this time, it is not uncommon for media and other claims-makers to overestimate the actual dangers associated with a particular risk.

However, the cyclical nature of press coverage, public attention and policy-making means that attention will begin to turn elsewhere. Also, by this time in the cycle, other institutions have had a chance to temper the most outrageous claims, to assess the certainty associated with the science and to perform their own public relations activities. This is the period during which policy discussions and measures are often being taken. By the end of this period, some resolution will have been reached, even if only to do nothing about the particular problem.

To understand the impact of media coverage on public opinion, it is necessary to have some understanding of how the story is constructed across the cycle. For instance, in the USA, the agricultural biotechnology controversies of 1998–2000 were eventually resolved within the food system policy arena. Overall, public opinion did not change greatly over this time, reflecting the model posed above. While some restrictive regulations were put in place, the technology itself was not irreparably damaged, and trust in food system institutions was never damaged as an outcome of the policy process.

In Europe the situation was entirely different, because previous cycles of news coverage of issues such as mad cow disease had resulted in damage to public trust. Thus, the appearance of new cycles of public interest in agricultural biotechnology were tilted from the start toward a negative outcome, explaining the difference in news coverage effects for the USA and Europe.

This highlights the fact that public trust can be an outcome of a cycle of events in which news coverage plays an important framing role, though not necessarily the decisive role. Many problematic issues never rise to public attention; policy failures to deal with such issues cannot, by definition, affect trust if they are not covered in the media. For those problems that are covered, public trust will be damaged if the institutions fail to deal successfully with the problem. This is one of the main taproots of support for European opposition.

Abbott *et al.* (2001) are among the few to have studied the issue cycle in relation to biotechnology. They studied the influence of ‘triggering events’ on media coverage. In the USA, these events were a Cornell study on Monarch butterflies and a controversy over Starlink maize entering the food supply. Since these events, media coverage has not been as frequent, as biotechnology moves through the policy and market arena. These triggering events moved coverage out of the arena of ‘benign science’ and into controversy over moral, political or social issues. During such issue phases, scientists themselves are only part of the mix, whereas coverage during the down parts of the cycle tends to be more scientifically focused.

Nevertheless, the frequency of news coverage itself is likely to be slightly damaging to public acceptance of biotechnology. As a general rule, if levels of public opinion do respond, they will slowly track news coverage, usually in a negative direction. Their residual level, in the absence of news coverage, will vary by country, though many countries show a general split in opinion. It is likely that these differences in opinion about biotechnology are not as drastic as they appear to be, and mostly reflective of non-specific attitudes that are formulated in response to questionnaires.

We can review several studies that have tied media exposure to concern about biotechnology. Though they sometimes show conflicting results, the overview can be instructive. Gaskell *et al.* (1999) argued that greater attention to biotechnology went together with greater public concern about the issue. Their assertion would be consistent with agenda-setting theory, which argues that news coverage frequency is related to issue salience for publics. However, they did not find that the content of the coverage was related to public perception. This also would be consistent with some versions of agenda-setting theory. However, it does not provide a longitudinal account of media influence on public opinion for this issue.

Another method is to look at the people's reported media use, to see whether there is any relation to attitudes about biotechnology. As we have highlighted before, Brossard and Nisbet (2007) found that trust in institutions, as well as attitudes toward science, are largely motivated by 'deference to scientific authority', which is conceived as a strongly held attitude unlikely to be shaped by localized events. These variables are not affected by media use. In this model, attention to newspapers is associated with somewhat greater knowledge, which in turn is related to higher support. Trust, in these findings, is unrelated to media attention.

Thus, while we generally sustain that trust opens people up to specific information and awareness about biotechnology (as seen in Fig. 1.3), this is not always assessable, especially in cross-sectional surveys. The media measures that are used do not always adequately capture the influence of cumulative media exposure.

A final method is to examine the relationship of attention to certain types of media. Brossard and Nisbet's model shows that news exposure is related to knowledge. They concluded that the impact of the news media on attitudes is limited, and is mediated by informal learning processes. Knowledge and trust both promote support for agricultural biotechnology, but the impacts of trust are stronger. They identified two routes to opinion formation, one stemming from deferent views of science (being willing to follow scientists' advice) and another through mass media and knowledge. Deference to scientific authority promotes trust and positive attitudes, and probably less willingness to turn to media for possibly threatening information. Effects of news media are greatest for those without strong personality-level trust/distrust of science.

Besley and Shanahan (2005) showed that attention to science news specifically was more important. In no case, however, are there dramatic effects of media exposure or attention. Shanahan *et al.* (2001) concluded that media coverage did have some small impacts. A trend series of NSF questions about GMOs showed this decline, as seen in Fig. 1.4, though the decline was not massive. Thus, although the upticks in media coverage were sizable, public opinion did not react at the same magnitude.

We believe that public trust or distrust can be built or destroyed most effectively toward the end of the issue cycle, as portrayed in Fig. 1.3. During the alarmed discovery and mobilization periods, publics are likely to suspend trust. The attempts of technical communicators to present balanced information are least likely to be successful during this period. While institutions must respond factually to criticisms, threats and even to exaggerations or lies, communicators cannot expect to build trust in this period. Neither is this a good time to do science education.

During the policy measures phase, communication efforts need to be more focused on the technical matters of how a balanced approach can be taken, making sure that all viewpoints – including those of accurate science – are heard in the debate. But strong results cannot be expected at the level of public opinion. Communication efforts to inform must also be tempered with attention to presenting a balanced image.

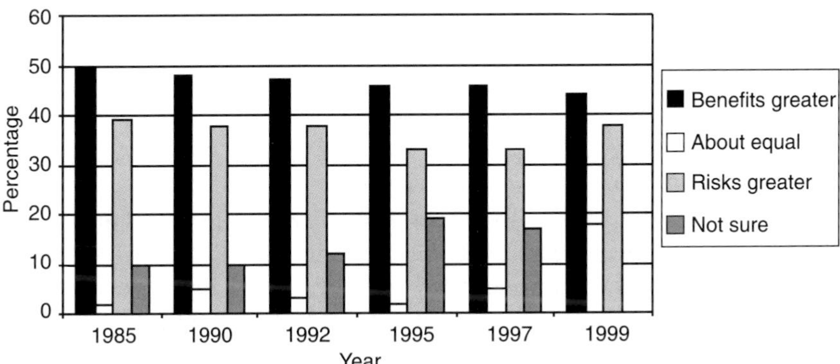


**Fig. 1.4.** Perceptions of risks and benefits from GMO foods.

It is during and after the resolution phase that trust-building can be most effective. Although there is less coverage during these phases, it is more likely to be depoliticized, positive and accurate. Thus, rather than putting one's bets on the hare, we put our bets on the tortoise, who has the advantage of waiting until a propitious time to deliver positive messages and build trust.

## Risk Communication

'Risk communication' is the accepted form of communication with lay audiences about scientific and technical information. It is based on accurate scientific assessment of risks associated with technology, and should present quantitative information suitable to allow audiences to assess risks. Risk communication is intended to influence risk perception, because risk perception is believed to influence audiences and attitudes with respect to the product.

In biotechnology, risk communication and perception present some interesting difficulties. Many existing communication programmes about GMOs emphasize that there are few, if any, demonstrated risks of genetic modification. Working from a standard risk communication perspective, these programmes assume that audiences simply lack correct information and, when informed adequately about low risk levels, they would adopt the technology.

Gaskell *et al.* (2004) have recently shown that risk perception does play a role, but that failure to perceive benefit may play an opposite and stronger role. Their research in Europe shows that some consumers don't see the advantage of GMOs, or assume that the advantages would accrue only to biotechnology companies. These factors, combined with distrust of those industries and food safety institutions, lead to rejection of the technology.

Sjoberg (2002) noted that the standard 'psychometric model' of risk perception looks at two, possibly three factors: dread, novelty and trust. Trust we have already examined above. Dread is the visceral fear associated with the technology; novelty is its perceived newness. In examining a battery of questions on agricultural biotechnology, he also found that 'tampering with nature' was an additional explanatory factor, which was indeed more powerful than the dread and novelty factors. Whether or not people fear biotechnology or whether they feel that it is a relatively new risk, their feelings that it is tampering with nature are more predictive of risk perception. This feeling has been manipulated extensively in the 'playing God' message of NGOs.

Based on these considerations, and our discussion of other variables, the following risk communication principles are relevant to agricultural biotechnology. First, we know that some factors are associated with lower perception of risk (from Covello, http://www.centerforriskcommunication.org). These are risks that are:

- voluntary;
- under the individual's control;
- offering clear benefits;
- natural;
- familiar; and
- affecting everyone equally.

Agricultural biotechnology, in its current form, does not fare well on some of these criteria. Thus, lessening risk perception needs to address some of these issues specifically.

In developing specific messages, messages about agricultural technology need most urgently to focus on benefits. The biotechnology industry realized this in its 'Why Biotech?' campaign (http://www.whybiotech.org), which has placed benefit to developing agriculture as a prime theme.

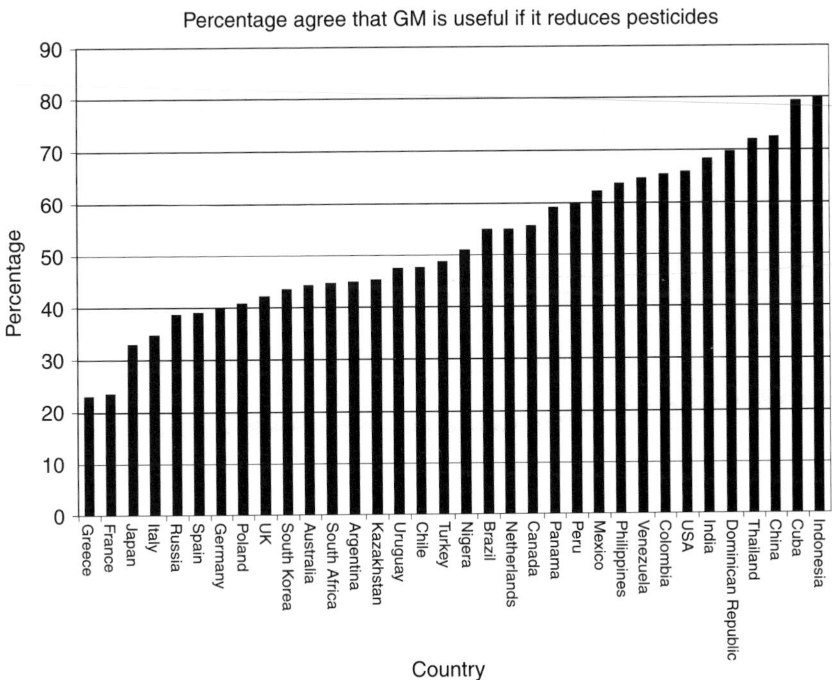


**Fig. 1.5.** Global attitudes toward GM crops (based on data from Environics International Environmental Monitor, 2000).

## Attitudes Around the World: their Impact on Use

Most public opinion research on agricultural biotechnology has focused on consumer attitudes about GM. There is surprisingly wide variance in acceptance in different countries. Figure 1.5 shows results of a survey conducted by Environics, a Canadian communication research firm. When asked whether they would support use of GM for the purposes of reducing chemical use in agriculture, support of respondents ranged from a low of around 20% in Greece, France and Japan to a high of 80% in Indonesia and China.

Given the variance in consumer attitude, it is useful to ask to what extent consumer attitudes shape the actual potential for use of GM in a country. There is no clear relationship between acceptance of the technology from a public opinion standpoint and how much GM is used within a country. There are many countries on the list that do not have extensive GM plantings, as regulatory frameworks are only just beginning to be put in place. Therefore, assessing the impact of attitudes on use and tolerance is a more speculative enterprise.

Of course, there are macro-economic effects on production of GM crops, as some governments fear the loss of exports to anti-GM European states. However, effects of attitudes extend to the individual consumer level. Rousu *et al.* (2004) found that negative information about GM from activist groups reduced consumers' willingness to purchase GM products. However, they highlighted that better access to information from parties perceived to be reliable and disinterested third parties could negate this influence.

This underlines the need for active communication perceived to be balanced, from actors who do not stand to profit from GM activity. Moon and Balasubramaniam (2003) found that consumers in the UK were willing to pay more for non-GM breakfast cereal than were consumers in the USA, consistent with the more sceptical attitudes in the UK. The willingness was directly linked to risk and benefit perception. Other factors, such as the higher prices of GM seeds, mean a confused marketplace in many countries that has not yet taken off.

## References

Abbott, E., Lucht, T., Jensen, J. and Jordan-Conde, Z. (2001) Riding the hoopla: an analysis of mass media coverage of GMOs in Britain and the USA: 1997–2000. *Presentation to Association for Education in Journalism and Mass Communication*, Washington, DC, 5–8 August 2001.

Alhassan, W. (2002) *Agricultural Biotechnology Application in West Africa.* IITA, Ibadan, Nigeria.

Besley, J. and Shanahan, J. (2005) Exploring the relationship between media attention and exposure and support for agricultural biotechnology. *Science Communication* 4, 347–367.

Bonny, S. (2003) Why are most Europeans opposed to GMOs? Factors explaining rejection in France and Europe. *Electronic Journal of Biotechnology* 6 (1), 50–71.

Brossard, D. and Nisbet, M. (2007) Deference to scientific authority among a low-information public: Understanding US opinion on agricultural biotechnology. *International Journal of Public Opinion Research* 19(1).

Brossard, D. and Shanahan, J. (2003) Do citizens want to have their say? Media, agricultural biotechnology, and authoritarian views of democratic processes in science. *Mass Communication and Society* 3 (6), 291–312.

Cartagena Protocol on Biosafety (2000) http://www.biodiv.org/biosafety/

Downs, A. (1972) Up and down with ecology: The issue attention cycle. *The Public Interest* 28, 38–51.

Evenson, R. and Santaniello, V. (2004) *Consumer Acceptance of Genetically Modified Foods.* CABI Publishing, Cambridge, Massachusetts.

Frewer, L. (2003) Societal issues and public attitudes towards genetically modified foods. *Trends in Food Science and Technology* 14 (58), 319–332.

Gaskell, G., Bauer, M., Durant, J. and Allum, N. (1999) Worlds apart? The reception of genetically modified foods in Europe and the US. *Science* 288, 5472.

Gaskell, G., Allum, N., Wagner, W., Kronberger, N., Torgersen, H., Hampel, J. and Bardes, J. (2004) GM foods and the misperception of risk perception. *Risk Analysis* 24 (1), 185–194.

Hagendijk, R.P. (2004) The public understanding of science and public participation in regulated worlds. *Minerva* 42 (1), 41–59.

Hallman, W.K., Hebden, W.C., Aquino, H.L., Cuite, C.L. and Lang, J.T. (2003) *Public Perceptions of Genetically Modified Foods: a National Study of American Knowledge and Opinion.* (Publication number RR-1003-004). Food Policy Institute, Cook College, Rutgers – The State University of New Jersey, New Brunswick, New Jersey.

Hansen, J., Holm, L., Frewer, L., Robinson, P. and Sandoe, P. (2003) Beyond the knowledge deficit: recent research into lay and expert attitudes to food risks. *Appetite* 41 (2), 111–121.

Hilgartner, S. (1990) The dominant view of popularization – conceptual problems, political uses. *Social Studies of Science* 20 (3), 519–539.

Hornig Priest, S., Bonfadelli, H. and Rusanen, M. (2003) The 'Trust Gap' hypothesis: predicting support for biotechnology across national cultures as a function of trust in actors. *Risk Analysis* 23, 751–766.

Kamaldeen, S. and Powell, D. (2000) *Public Perceptions of Biotechnology,* Food Safety Network Technical Report No. 17. Department of Plant Agriculture, University of Guelph , Canada.

Moon, W. and Balasubramanian, S.K. (2003) Willingness to pay for non-biotech foods in the US and UK. *Journal of Consumer Affairs* 37 (2), 317–339.

Nisbet, M.C. and Lewenstein, B.V. (2002) Biotechnology and the American media – the policy process and the elite press, 1970 to 1999. *Science Communication* 23 (4), 359–391.

Pardo, R. and Calvo, F. (2002) Attitudes toward science among the European public: a methodological analysis. *Public Understanding of Science* 11 (2), 155–195.

Priest, S.H., Bonfadelli, H. and Rusanen, M. (2003) The 'trust gap' hypothesis: predicting support for biotechnology across national cultures as a function of trust in actors. *Risk Analysis* 23 (4), 751–766.

Rousu, M.C., Huffman, W.E., Shogren, J.F. and Tegene, A. (2004) Estimating the public value of conflicting information: The case of genetically modified foods. *Land Economics* 80 (1), 125–135.

Secretariat of the Convention on Biological Diversity (2000) *Cartagena Protocol on Biosafety to the Convention on Biological Diversity: Text and Annexes.* Secretariat of the Convention on Biological Diversity, Montreal, Canada.

Shanahan, J., Scheufele, D. and Lee, E. (2001) The polls-trends – Attitudes about agricultural biotechnology and genetically modified organisms. *Public Opinion Quarterly* 65 (2), 267–281.

Siegrist, M. (2000) The influence of trust and perceptions of risks and benefits on the acceptance of gene technology. *Risk Analysis* 20 (2), 195–203.

Sjoberg, L. (2002) Attitudes toward technology and risk: Going beyond what is immediately given. *Policy Sciences* 4, 379–400.

Zhong, F., Marchant, M.A., Ding, Y. and Lu, K. (2002) GM foods: a Nanjing case study of Chinese consumers' awareness and potential attitudes. *AgBioForum* 5 (4), 136–144 (available at http://www.agbioforum.org).

2

# Public Perceptions of Agricultural Biotechnology in the UK: the Case of Genetically Modified Food

W. Poortinga[1] and N. Pidgeon[2]

[1] *The Programme on Understanding Risk, Centre for Environmental Risk, School of Environmental Sciences, University of East Anglia, Norwich, UK;* [2] *School of Psychology, Cardiff University, Cardiff, UK*

## Introduction

During the past 10 years, agricultural biotechnology has been a source of considerable public controversy in both continental Europe and the UK. It has, accordingly, been an issue of high public policy importance. This chapter aims to provide an understanding of the nature of public attitudes towards this technology in the UK. We start with a brief overview of the history of the biotechnology controversy that peaked in the late 1990s. Subsequently, we discuss the results of a recent British survey on the perceptions of genetically modified (GM) food and crops.

The study was designed to get a general overview of British attitudes towards agricultural biotechnology, as well as the general public's levels of awareness, understanding and perceived value of a public debate on the commercialization of agricultural biotechnology that was held during the summer of 2003 (for more details see Horlick-Jones *et al.*, 2004; Poortinga and Pidgeon, 2004a; Pidgeon *et al.*, 2005).

Although public opinion in European countries is often assumed to be overwhelmingly hostile towards GM, we argue that people hold a complex set of beliefs about a range of health, environmental and social risks and benefits of GM food and crops. We also report results illuminating a number of other important issues in the governance of agricultural biotechnology, such as the labelling of GM food, the liability of the biotechnology industry, trust in risk regulation and trust in various sources of information.

## The GM Food Controversy

Genetically Modified (GM) Food has become one of the UK's most contentious risk issues in recent years. Although there has always been some public unease about this application of modern biotechnology, the period between 1996 and 1999 can be seen as the 'watershed years', when a number of successive events dramatically changed public opinion about GM food (Gaskell *et al.*, 2001). As research on the social amplification of risk perceptions has shown, such changes rarely come about for any singular reason, but more typically arise from the interaction between a series of contributory events (Kasperson *et al.*, 1988; Pidgeon *et al.*, 2003) which then go on to form the historical and societal context to an issue's public profile.

The mid-1990s saw the launch of the first recognizable GM product in the UK, which was initially a relative success. Tomato paste made from *Flavr SavrTM* tomatoes initially sold surprisingly well, even if it was clearly labelled as a genetically modified product (Robinson, 1997; Harvey, 1999). Many consumers were willing to try the product, probably because of its price and claims that this tomato paste tasted better.

Second, in autumn 1996 the first shipment of Monsanto's *Roundup Ready®* soya beans arrived in Europe (see, e.g. Lassen *et al.*, 2002). Across Europe, NGOs used the transport of mixed GM and non-GM soya to launch a high-profile campaign focusing on the labelling of GM products. The campaign was aimed at raising public awareness of the fact that, at that time, no regulation was in place for (supposedly non-GM) products containing GM ingredients.

The heavy-handed response of the biotechnology company that segregation was not possible helped the mobilization of opinion against GM food (Simmons and Weldon, 2000). The introduction of non-segregated products can be considered as one of the worst strategies to force consumers to accept GM products, as it directly violates the basic consumer right to choose the products they want to consume. Many interpreted this response as the biotechnology industry showing disdain for consumers (Harvey, 1999).

Third, 'Dolly' the sheep was presented to the world in February 1997. The creation of the first cloned animal by the Roslin Institute in Edinburgh, UK, sparked an intensive public and media debate about the ethics of biotechnology. The cloning of Dolly the sheep also demonstrated the enormous scientific potential of modern biotechnology (for an in-depth account see Einsiedel *et al.*, 2002).

Fourth, one of the key events in the GM food controversy has become known as the *Pusztai affair*. In August 1998, Arpad Pusztai (at that time a researcher at the Rowett Research Institute in Aberdeen,

UK) claimed on UK television that his rat experiments had shown that eating GM potatoes could lead to intestinal changes (Ewen and Pusztai, 1999). Although the (at the time) unreviewed work was subsequently portrayed as being 'unscientific', it focused the attention on the potential human health risks of GM food (Simmons and Weldon, 2000).

Finally, the late 1990s also saw the British government's initiative on a scientific programme of nationwide farm-scale evaluations of selected GM herbicide-tolerant crops, in order to evaluate the impacts of their cultivation regimes upon farmland biodiversity. Further significant media amplification of this issue followed in 1998/1999, when anti-GM campaigners were arrested for damaging a number of the evaluation sites.

The subsequent adverse reactions of the UK public to GM food over this period of time cannot be understood without also considering beliefs about the government's handling of the BSE (or 'mad cow disease') crisis. For many British people, the BSE affair represented a pivotal issue of food safety, which reached a crisis point in March 1996 when the UK government announced discovery of a possible link between the disease in animals and humans.

BSE does not just cause a harrowing incurable disease for humans (Creutzfeldt-Jakob-Disease/CJD): since the late 1980s, when it was first discovered that BSE was prevalent amongst British cows, the government had repeatedly claimed that British beef was safe to eat. Even if the extensive Phillips inquiry (Phillips *et al.*, 2000) concluded that the Government did not lie to the public, the BSE crisis showed a great mismatch between what government institutions were supposed to do and what they actually did in the eyes of the public (Jasanoff, 1997).

As a consequence, BSE is often considered as: (i) a warning of the fallibility of expertise; (ii) ilustrative of the potential unrecognized risks of modern farming techniques; and (iii) a failure to regulate new technology (Grove-White *et al.*, 1997; Jasanoff, 1997; Durant *et al.*, 1998).

Another contributing factor to the GM controversy can be found in the regulatory framework of biotechnology at the time that the controversy emerged (Simmons and Weldon, 2000; Gaskell *et al.*, 2001). In contrast to many other European countries, the UK regulated biotechnology products on a case-by-case basis. As a result, there was no mechanism available for taking into account wider concerns about the impacts of GM food (Simmons and Weldon, 2000).

Increasing Europeanization and Globalization may also have contributed to difficulties in dealing effectively with GM products (Levidow *et al.*, 1999; Grabner *et al.*, 2001). Since 1996, the regulation of GM products in Europe has been multi-layered at the national, European and the wider international levels. The common market brought the EU to harmonize their rules and regulation of biotechnology. Harmonization appeared difficult to achieve, as regulatory styles and the history of the GM debate varied profoundly across Europe (Robinson, 1997; Grabner *et al.*, 2001). The harmonization of the common market made it easier for the biotechnology industry to introduce new GM products across Europe, without taking into account public sensitivities in the different European countries. At the same time, a set of new global trade rules, known as the World Trade Agreement, made it difficult for the EU and individual member states to halt imports of GM products (see, e.g. Grabner *et al.*, 2001).

By the end of the 1990s the landscape of public attitudes towards biotechnology – and GM food in particular – had dramatically changed. Public opinion had turned strongly against GM food, and consumer and NGO pressure had forced supermarkets to remove GM products from their shelves (Simmons and Weldon, 2000). Moreover, trust in risk regulation – especially in the area of food production – was at an all-time low.

What has become clear from this controversy is that public acceptance is of critical importance for the uptake of a new technology such as GM food. Consumers have become a decisive factor for policy-making, both at the national as well as the supranational levels (Grabner *et al.*, 2001). Moreover, the events show that losing trust, a fate that befell the UK government over the BSE crisis, may have far-reaching consequences. Perhaps, not surprisingly, in the light of their previous experiences, people have become suspicious about the ability of the government to regulate a complex new technology such as GM food (see, e.g. Grove-White *et al.*, 1997; Jasanoff, 1997).

## Public Perceptions of GM Food in the UK

In the current section we discuss the results of a major survey on the perceptions of GM food and crops (*n*, 1363) conducted in Britain in the summer of 2003 (for technical details of the survey administration see Appendix A). This was just after the UK government-sponsored *GM Nation?*, a public debate on the commercialization of genetically modified crops – a unique experiment in the governance of technological innovation and risk (see PDSB, 2003; Horlick-Jones *et al.*, 2004), and a number of years after the GM food controversy had peaked in the UK in the late 1990s.

In this section we report on general and specific attitudes towards agricultural biotechnology in the context of other important personal and social issues. We show that many individuals hold essentially ambivalent attitudes towards GM food and crops rather than being simply 'for' or 'against'. We also discuss the importance of trust in risk regulation for the acceptability of GM food and crops. Other issues that are discussed in this section relate to: (i) the labelling of GM food; (ii) the liability of the biotechnology industry for any damage caused by GM products; and (iii) trust in various information sources.

## Importance of GM food: the role of context

When studying a controversial risk issue, such as GM food, it is important to consider that it may not be the only one people are interested in. Controversial issues surface in a society that already has to deal with numerous other issues with which they have to 'compete' for attention. To put the issue of GM food into context, respondents were asked to indicate the importance value of twenty *personal* (P) and *social* (S) issues, as well as four other contemporary risk cases (climate change, radiation from mobile phones, radioactive waste and genetic testing). Table 2.1 shows the proportion of respondents that found the issues 'very important' to them personally. What is clear here is that GM food is relatively unimportant compared to most of the other personal and social issues.

Perhaps, not surprisingly, personal issues (P) are considered more important than social issues (S). In general, issues that touch people's personal life (such as health, partner and family, personal safety and education) are prioritized over wider social (risk) issues such as tackling world poverty and human rights, population growth and climate change (see Table 2.1). Even among the included social issues, GM food is considered as one of the least important: a mere 21% of the respondents indicated that GM food is an issue that is 'very important' to them personally. Only religion (categorized as a personal issue) was considered very important by fewer people (19%).

Although these results seem to suggest that GM food is not particularly salient in people's day-to-day lives, people are nevertheless interested in this issue when asked directly. While 30% of respondents said that they were not very interested, and 10% that they are not at all interested, a majority (56%) said that they were fairly or very interested in the issue of GM food. Similarly, nearly half of the respondents (48%) *disagreed* with the statement 'I am not that bothered about GM food'. In line with these results, more than half of the respondents (51%) indicated that they were concerned about genetically modified food. Overall, the results indicate that, while it is not an issue that features

**Table 2.1.** The importance of various personal (P) and social (S) issues (from UEA/MORI GM Food Survey, 2003; weighted dataset: *n*, 1363).

| Issue surveyed | Very important (%) |
|---|---|
| Your health (P) | 87 |
| Partner and family (P) | 85 |
| Law and order (S) | 80 |
| Personal safety (P) | 77 |
| Education (S) | 75 |
| Being independent (P) | 69 |
| Your privacy (P) | 65 |
| Terrorism (S) | 63 |
| Environmental protection (S) | 59 |
| Having a comfortable life (P) | 58 |
| Personal finance (P) | 56 |
| Social relations/friends (P) | 56 |
| Radioactive waste | 53 |
| Animal welfare (S) | 49 |
| The economy (S) | 46 |
| Excitement/fun (P) | 40 |
| Work (P) | 40 |
| Tackling world poverty (S) | 37 |
| Tackling human rights (S) | 33 |
| Population growth (S) | 29 |
| Genetic testing | 29 |
| Climate change | 28 |
| Radiation from mobile phones | 26 |
| GM food (S) | 21 |
| Religion (P) | 19 |

prominently in daily life (compared to other more personal concerns), GM food is still an issue capable of eliciting levels of attention, depending upon the particular context (see also Zwick, 2005).

## General attitudes towards GM food

The survey included items tapping general attitudes towards GM food (that is positive through to negative beliefs). There is growing evidence that people's initial 'affective' response is an important part of the way in which lay perceptions of risk issues are constructed (see, e.g. Finucane *et al.*, 2000; Loewenstein *et al.*, 2001; Langford, 2002; Slovic *et al.*, 2004). People's general orientation towards an issue – whether it is seen as 'good' or 'bad' – may function as a key filter influencing the way subsequent information is processed, such as perceptions of potential benefits, communications about the issue from others or even trust in risk managers (see, e.g. Poortinga and Pidgeon, 2004b, 2005).

Our own results suggest that, regarding such measures, a proportion of the British public feel particularly uncomfortable about developments in agricultural biotechnology. For example, 40% of respondents said that they felt negatively about GM food, whilst only 15% said that they felt positively. A similar pattern emerged when people were asked whether GM food was a good or a bad thing: a similar proportion (40%) said it was a 'bad thing' and 14% said it was a 'good thing'. Perhaps, more interestingly, these two questions also show that many people do not seem to have clear views on the issue: a sizeable minority could be found in the middle of the scales used (35 and 40%, respectively). As we go on to discuss, these results may have important implications for the understanding the dynamics of public opinion on this issue.

The results with regard to behavioural intentions seem to be in line with the overall 'affective' responses towards GM: a sizeable minority (28%) would be happy to eat GM food, whilst almost half of the sample (56%) disagreed with this statement, and about one in five (23%) neither agreed nor disagreed with the statement (see Table 2.2). Similarly, 50% of the British population said that they would try to avoid purchasing GM food products. In comparison, only 22% said that they would not try to avoid purchasing GM food products. One out of four neither agreed nor disagreed with this statement (see Table 2.2). These results can be interpreted in two different ways. First, it shows that most people surveyed disliked the idea of consuming GM products. However, from a different perspective, the results also suggest that a sizeable minority would not *actively* avoid purchasing GM products.

The current study also contained an item used previously by

**Table 2.2.** Behavioural intentions regarding GM foods (from UEA/MORI GM Food Survey, 2003; weighted dataset: *n*, 1363).

| Behavioural intention | Strongly disagree (%) | Tend to disagree (%) | Neither/ nor (%) | Tend to agree (%) | Strongly agree (%) | Don't know (%) |
|---|---|---|---|---|---|---|
| I personally would be happy to eat GM food[a] | 28 | 18 | 23 | 22 | 6 | 3 |
| I would try to avoid purchasing GM food products | 5 | 17 | 25 | 21 | 29 | 2 |

[a] Statement adapted from *GM Nation?* feedback form (see PDSB, 2003).

MORI to track public support and opposition over time. Table 2.3 shows that support for GM food has fallen since the issue first emerged in the media spotlight in 1996. In 1996, close to one in three (31%) supported GM food. Support for GM food had weakened to 22% by 1998 and had fallen even further, to 14%, by 2003. At the same time, opposition towards GM food grew from 50% in 1996 to 58% in 1998. In 2003, however, opposition towards GM food dropped from 56% in February to 46% in June, and was found to be only 36% in this study (August 2003).

This decreased opposition to GM food does not mean that people have necessarily become more favourable towards GM food. In fact, support for GM food has remained relatively stable since 1998. Rather, people have become more undecided about GM food. Whereas in 1996 and 1998 about one in six (16 and 15%, respectively) neither supported nor opposed GM food, in February 2003 one in four, and in June 2003 about one in three, appeared to be undecided. In the present study it was found that about two in five (39%) neither

**Table 2.3.** Responses to the question: 'How strongly, if at all, would you say you support or oppose genetically modified food?'.

| Support/opposition | 1996[a] (%) | 1998[b] (%) | Feb. 2003[c] (%) | June 2003[d] (%) | Aug. 2003[e] (%) |
|---|---|---|---|---|---|
| Strongly support | 6 | 6 | 3 | 3 | 3 |
| Tend to support | 25 | 16 | 11 | 11 | 11 |
| Neither support nor oppose | 16 | 15 | 25 | 33 | 39 |
| Tend to oppose | 24 | 21 | 26 | 21 | 19 |
| Strongly oppose | 26 | 37 | 30 | 25 | 17 |
| Don't know | 3 | 5 | 5 | 7 | 11 |
| Support | 31 | 22 | 14 | 14 | 13 |
| Oppose | 50 | 58 | 56 | 46 | 36 |
| Net | –16 | –36 | –42 | –32 | –23 |

Please note the question wording was slightly different in 1996 and 1998 to 2003. In 96/98 the question wording was as follows: 'Thinking of genetically modified food or food derived from genetic engineering, what is your opinion towards the development and introduction of such food? Would you say: (i) support it to a great extent; (ii) support it slightly; (iii) neither support nor oppose it; (iv) oppose it slightly; (v) oppose it to a great extent; or (vi) don't know.'

[a] MORI/Greenpeace: 1003 aged 15+, telephone, UK, 13–15 December 1996.

[b] MORI/GeneWatch: 950 aged 15+, face-to-face, in home, UK, 6–8 June 1998.

[c] MORI Environment Tracker/MORI Environment Research Bulletin: 2141 aged 15+, face-to-face, in home, UK, 6–10 February 2003.

[d] MORI Environment Tracker/MORI Environment Research Bulletin: 1958 aged 15+, face-to-face, in home, UK, 19–24 June 2003.

[e] UEA/MORI GM food Survey: 1363 aged 15+, face-to-face, in home, UK, 19 July and 9 September 2003.

supported nor opposed GM food. These results show that, since the peak of the GM controversy in the late 1990s, there has been a trend towards *less polarization*.

Overall, these results paint an interesting picture of recent British public opinion on GM food. Although it is often thought that there is widespread opposition to GM food in the UK, this latest study suggests many individuals now have less clear attitudes than before. In summary, it can be concluded that, whilst public opinion is clearly skewed towards the negative, a substantial proportion can be found 'in the middle'.

## Perceived attributes of GM food

There are a variety of reasons why GM food might have become relatively negatively evaluated. A part of this is clearly the result of the social and historical context of controversy described above. However, the 'psychometric' work pioneered by Slovic and colleagues on nuclear power and other important risk issues (for reviews see Pidgeon *et al.*, 1992; Slovic, 2000) has uncovered a number of generic factors. Risks that are unfamiliar, personally uncontrollable and hold the potential for catastrophic consequences are generally considered less acceptable. Some of these characteristics are also associated with GM food.

Table 2.4 shows that three-quarters think that GM food has unknown consequences, just over one-half feel that GM food poses risks to future generations and that nearly one-half of the respondents do not feel able to control any risks to themselves associated with GM food. In several studies, moral objections were also found to be strong predictors of opposition to genetic engineering (see, e.g. Durant *et al.*, 1998; Gaskell, 2000). In particular, some people view genetic engineering as 'tampering with nature', or as 'playing God'. Congruent with this account, we found that half of our respondents (50%) felt that genetic modification interferes with nature in an unacceptable way.

## Perceived risks and benefits of GM food and crops

In order to further explore public attitudes towards agricultural biotechnology, people were presented with a number of statements about specific risks and benefits, this time in relation to both GM food *and* crops (see Table 2.5). Previous psychometric work has shown that both of these dimensions are important determinants of the acceptability of a risk issue (see Slovic, 2000). Most of the questions were adapted for the

**Table 2.4.** Perceived attributes of GM food (from UEA/MORI GM Food Survey, 2003; weighted dataset: *n*, 1363).

| Perceived attribute | Strongly disagree (%) | Tend to disagree (%) | Neither/ nor (%) | Tend to agree (%) | Strongly agree (%) | Don't know (%) |
|---|---|---|---|---|---|---|
| GM food has unknown consequences | 1 | 5 | 15 | 40 | 34 | 4 |
| GM food poses risks to future generations | 1 | 9 | 27 | 32 | 22 | 9 |
| I feel able to control any risks to myself associated with GM food | 20 | 25 | 19 | 23 | 6 | 6 |
| The risks from GM food are unfair because they fall unevenly on particular groups in British society | 5 | 14 | 33 | 25 | 8 | 15 |
| I feel that genetic modification interferes with nature in an unacceptable way | 4 | 13 | 28 | 25 | 25 | 5 |

present survey from a questionnaire that had been administered as part of the 2003 British *GMNation?* public debate on agricultural biotechnology (see Pidgeon *et al.*, 2005).

Table 2.5 shows that an overwhelming majority (85%) thought that 'we don't know enough about the long-term effects of GM food on our health'. In line with these findings, only 9% agreed with the statement: 'GM crops are safer than traditional crops because they have been more thoroughly tested', whereas almost one-half of the respondents (48%) disagreed. About one in three (32%) neither agreed nor disagreed with this latter statement.

Next to uncertainties about the health impacts of GM food, it emerged that the greater part of the general public (63%) was concerned about the 'potential negative impact of GM crops on the environment', while only 10% were not concerned about the environmental risks of GM food and crops. A large majority (68%) agreed with the statement: 'I am worried that if GM crops are introduced it will be very difficult to ensure that other crops are GM free.' In comparison, only 7% disagreed with this statement. As with the former statements, about one in five neither agreed nor disagreed with the statement: 'I am worried that if GM crops are introduced it will be very difficult to ensure that other crops are GM free.'

**Table 2.5.** Specific risks and specific benefits of GM food (from UEA/MORI GM Food Survey, 2003; weighted dataset: *n*, 1363).

| | Strongly disagree (%) | Tend to disagree (%) | Neither/ nor (%) | Tend to agree (%) | Strongly agree (%) | Don't know (%) |
|---|---|---|---|---|---|---|
| Specific risks | | | | | | |
| I don't think we know enough about the long-term effects of GM food on our health[a] | 1 | 3 | 8 | 33 | 52 | 2 |
| GM crops are safer than traditional crops because they have been more thoroughly tested | 21 | 27 | 32 | 8 | 1 | 9 |
| I am concerned about the potential negative impact of GM crops on the environment[a] | 2 | 8 | 22 | 37 | 26 | 5 |
| I am worried that if GM crops are introduced it will be very difficult to ensure that other crops are GM free[a] | 2 | 5 | 18 | 35 | 33 | 6 |
| GM food will make farmers dependent on big companies that have patents on GM crops | 2 | 5 | 26 | 36 | 20 | 10 |
| I am worried that this new technology is being driven more by profit than by the public interest[a] | 1 | 6 | 13 | 38 | 37 | 4 |
| I think GM crops would mainly benefit the producers and not ordinary people[a] | 1 | 8 | 21 | 33 | 31 | 4 |
| Specific benefits | | | | | | |
| I believe GM crops could help to provide cheaper food for consumers in the UK[a] | 9 | 14 | 23 | 39 | 6 | 8 |
| I think that some GM crops could benefit the environment by requiring less pesticides and chemical fertilizers than traditional crops[a] | 9 | 11 | 26 | 37 | 7 | 9 |
| I believe that GM crops could improve the prospects of British farmers by helping them to compete with farmers around the world[a] | 10 | 16 | 32 | 26 | 5 | 11 |
| I believe that GM crops could benefit people in developing countries[a] | 7 | 10 | 19 | 38 | 18 | 6 |
| I believe that some GM non-food crops could have useful medical benefits[a] | 5 | 6 | 34 | 34 | 6 | 15 |

[a] Statement adapted from the *GM Nation?* public debate feedback form (see PDSB, 2003).

It appeared that a majority (56%) agreed, while only 7% disagreed that 'GM food could make farmers dependent on big companies that have patents on GM crops'. About one-quarter of the British public (26%) neither agreed nor disagreed that this was a concern to them personally. Finally, as shown in Table 3.5, three out of four are worried that 'this new technology is being driven more by profit than by the public interest'. In accordance with these findings, 64% agreed that GM crops would mainly benefit the producers and not ordinary people, whilst only 9% of the sample disagreed.

The results presented above clearly show that a range of concerns exists about the risks of GM food and crops. However, it appears that, at the same time, a substantial proportion of our sample appreciate the various (potential) benefits of GM food and crops. Almost one-half of the sample (45%) believed that GM crops 'could help to provide cheaper food for consumers in the UK'. In comparison, 23% disagreed with this statement. People also acknowledged some potential benefits of GM crops for the environment. About 44% agreed, whereas 20% disagreed, that 'some GM crops could benefit the environment by requiring less pesticides and chemical fertilizers than traditional crops'.

People's responses to the statement 'I believe that GM crops could improve the prospects of British farmers by helping them compete with farmers around the world' were fairly equally distributed: 31% agreed, 26% disagreed and 32% neither agreed nor disagreed with the statement. However, a clear majority (56%) did feel that 'GM food crops benefit people in developing countries'. In contrast, only 17% disagreed with this statement.

Finally, a sizeable minority (41%) believed that 'some GM non-food crops could have useful medical benefits', whereas a mere 11% doubted this assertion. It is worth nothing that a relatively large number of people (34%) neither agreed nor disagreed with the latter statement, while 15% had no opinion or did not respond to the latter question. Once again, and congruent with the general attitudinal measures reported earlier for GM food, this suggests that a large proportion of the general public has no clear opinion about the potential medical benefits of GM crops.

A Principal Components Analysis (PCA) was conducted in order to examine whether people's responses to the 12 specific attitude statements could be summarized by a number of underlying dimensions. Because the statement 'GM crops are safer than traditional crops because they have been more thoroughly tested' loaded highly on multiple factors in an initial exploratory PCA, this item was omitted from further analyses.

The PCA produced two clearly interpretable dimensions. The first dimension reflected the perceived risks of GM food and crops, whereas the second dimension reflected the perceived benefits of GM food and crops. Both dimensions were internally consistent, with a Cronbach's α of 0.82 and 0.81 for the *perceived risk* and the *perceived benefit* dimension, respectively. The existence of the two independent factors means that different levels of perceived risk co-exist with different levels of perceived benefits. In the following section we will argue that this has important implications for the way we conceptualize public attitudes towards GM food.

## Ambivalent attitudes

For a long time it had been thought that people are either 'for' or 'against' GM food. However, recent qualitative studies suggest that the

**Table 2.6.** Factor loadings after Varimax rotation.

| | Factor[a] | |
|---|---|---|
| Statement | 1 | 2 |
| I am concerned about the potential negative impact of GM food crops on the environment | **0.66** | –0.32 |
| GM food will make farmers dependent on big companies that have patents on genetically modified crops | **0.61** | 0.06 |
| I am worried that if GM crops are introduced it will be very difficult to ensure that other crops are GM free | **0.76** | –0.08 |
| I am worried that this new technology is being driven more by profit than by the public interest | **0.78** | –0.13 |
| I don't think we know enough about the long-term effects of GM food on our health | **0.76** | –0.09 |
| I think GM crops would mainly benefit the producers and not ordinary people | **0.71** | –0.34 |
| I believe GM crops could help to provide cheaper food for consumers in the UK | –0.12 | **0.74** |
| I believe that GM crops could benefit people in developing countries | –0.17 | **0.77** |
| I think that some GM crops could benefit the environment by requiring less pesticides and chemical fertilizers than traditional crops | –0.17 | **0.74** |
| I believe that GM crops could improve the prospects of British farmers by helping them to compete with farmers around the world | –0.25 | **0.75** |
| I believe that some GM non-food crops could have useful medical benefits | 0.07 | **0.66** |
| Eigenvalue | 3.21 | 2.95 |
| Explained variance | 29.2 | 26.8 |
| Average agreement | 3.99 | 3.27 |
| Cronbach's α | 0.82 | 0.81 |

[a] The scales ranged from 1, 'totally disagree' to 5, 'totally agree'. Factor loadings higher than 0.50 are in bold. Factor interpretations: 1, perceived risks; 2, perceived benefits.

general public have more complex, nuanced and often conflicting views on agricultural biotechnology (Grove-White *et al.*, 1997; Marris *et al.*, 2001). In these studies, many people expressed elaborated arguments both for and against agricultural biotechnology, suggesting that they were essentially *ambivalent* about the issue.

Pidgeon *et al.* (2005) argue that such ambivalence can arise in a number of ways. For example, people's cognitive and affective responses to a certain facet of biotechnology might be in conflict; or they may believe agricultural biotechnology could indeed bring useful environmental benefits but do not trust corporations or the authorities to invest in delivering those particular benefits (as opposed to attributes, such as a longer shelf-life for products, which more directly benefit producers). Equally, both perceived benefits and risks may be viewed as highly significant, and possibly difficult to trade off directly: for example, environmental or economic benefits set against a perceived long-term health risk to one's family from eating GM food.

An interesting question arises here as to whether attitudinal ambivalence can be captured in more quantitative terms. As rightly pointed out by Marris *et al.* (2001), many quantitative studies have used bipolar dimensions to measure public attitudes towards biotechnology. A problem with such measures is that positive and negative attitudes are mutually exclusive. That is, the endorsement of positive beliefs is constrained to imply the rejection of negative beliefs (Cacioppo *et al.*, 1997).

In mainstream social psychology research, Conner and Sparks (2002) make a useful distinction between *direct* and *indirect* measures of ambivalence. As suggested by its name, direct measures ask people directly to what extent they have mixed feelings about a certain object or issue. Others have questioned the validity of direct measures, as they are more open to extraneous influences (Bassili, 1996, cited by Conner *et al.*, 2003). Indirect measures of ambivalence, by contrast, combine two independent (and opposing) attitude ratings in some form of index expressing both the respective balance between the ratings and their strength (Breckler, 1994). These indirect measures have been used in a large number of social–psychological studies on the causes and consequences of ambivalence (see, e.g. Sparks *et al.*, 2001; Conner *et al.*, 2003).

However, one disadvantage of an indirect measure is that it contrasts ambivalence with all other possible attitudinal positions. We propose that, as an alternative to constructing one single indirect ambivalence measure, the positive and negative dimensions could be combined to identify four distinct attitudinal groups (cf. Cacioppo and Berntson, 1994; Margolis, 1996; Poortinga, 2004). Here, we use the combined risk and benefits measures to explore the distribution across the four distinct attitudinal positions (see Fig. 2.1; cf. Cacioppo and Berntson, 1994; Margolis, 1996).

First, a combination of high perceived risks and low perceived benefits can be assumed to reflect a *negative* attitudinal position towards a particular issue. Likewise, people who think that GM food is beneficial and at the same time think that it is not risky most probably have a generally *positive* attitude towards GM food. For both of these groups the positive and negative attitudes are consistent with one another (a pattern also implied by studies of the influence of affective beliefs on risk and benefit perceptions; see Alhakami and Slovic, 1994).

For the *ambivalent* group, however, the positive and negative attitudes are in tension. That is, people belonging to this group feel that GM food and crops may be beneficial but at the same time also feel that they may be risky. As noted above, ambivalence seems to be an important under-researched topic, particularly because the public

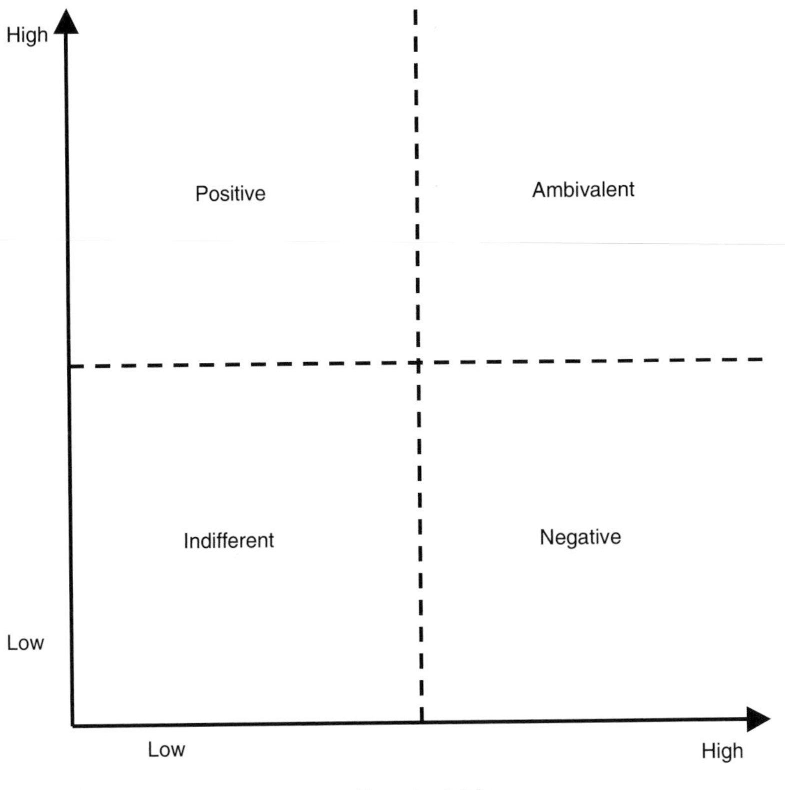


**Fig. 2.1.** A proposed typology of attitudes towards GM foods.

often indeed expresses *ambivalent* feelings about agricultural applications of biotechnology (see also Marris *et al.*, 2001).

Finally, people who think that GM food is neither beneficial nor risky may be said to be *indifferent*. That is, the absence of positive as well as negative attitudes most probably indicates that someone either has no (clear) opinion or is simply not interested about this particular issue.

Figure 2.2 illustrates the frequency distribution for our sample of individual responses in terms of their joint risk and benefit scores, as averaged across the two sets of items grouped by the factor analysis. These results suggest that, rather than simply being for or against, many individuals in the British population do indeed hold essentially ambivalent attitudes towards GM food and crops (at least in relation to the risk and benefit measures used here).

Inspecting Fig. 2.2, it is immediately apparent that there are very few individuals who are positive by our definition. The main distribution is essentially bimodal, with a major proportion (about 50%) clustering in the top right-hand *ambivalence* quadrant of the figure. A further 30% or so hold a clear *negative* pattern of attitudes, i.e. those who see significant risks but very few benefits. The factor

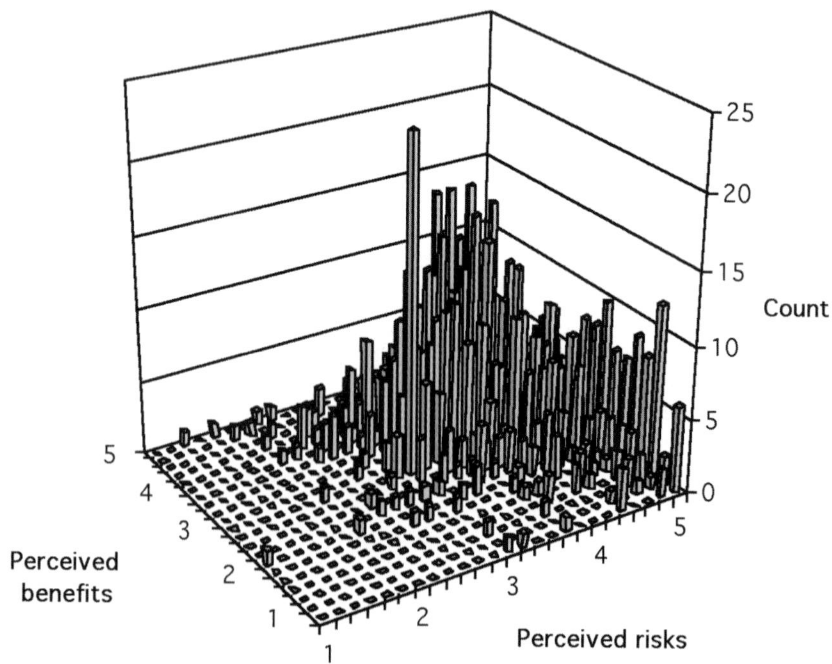


**Fig. 2.2.** Distribution of perceived risks and benefits of GM foods.

analysis results are also consistent with the earlier findings that most people are undecided about the more specific issue of GM food. In addition, Table 2.7 shows responses to an item that was specifically designed to measure these four distinct attitudinal positions on GM food.

People were asked to indicate which of the available options most closely describes their opinion about GM food, each representing one of the four attitudinal positions. We found that about one in ten people thought that GM food should be promoted, and that about one in three thought that GM food should be opposed. Just over one-half of the sample were not sure whether GM food should be promoted or opposed, while about one in ten people said they did not care whether GM food should be promoted or opposed (see Table 2.7). Taken as a whole, the results of this study suggest that public opinion about agricultural biotechnology is not a unitary whole, but fragmented with considerable ambivalence coexisting alongside outright opposition.

The suggestion that around onehalf of the population is ambivalent about GM food could further be confirmed with two direct measures of ambivalence. Table 2.8 shows that 56% of the respondents agreed with the statement: 'I have mixed feelings about GM food.' Even more people agreed with the statement 'I need more information to form a clear opinion about GM food' (84%). These latter findings strongly suggest that GM food is an issue about which many people have not yet made up their minds.

## Trust in the regulation of GM food

As described above, one contextual reason for unease in the UK about the safety of food products is the BSE crisis of the early 1980s. As argued before, people may well have become suspicious about the ability of the government to regulate a new technology such as GM food through their experiences with this previous crisis. From that perspective, the current concerns about the risks of GM food could be considered an issue of

**Table 2.7.** Four distinct attitudinal positions towards GM food (from UEA/MORI GM Food Survey, 2003; weighted dataset: *n*, 1363).

| Attitudinal position | Agreement (%) |
|---|---|
| GM food should be promoted (positive) | 9 |
| GM food should be opposed (negative) | 29 |
| I am not sure whether GM food should be promoted or opposed (ambivalent) | 53 |
| I don't care whether GM food should be promoted or opposed (indifferent) | 8 |

**Table 2.8.** Ambivalence, attitudinal certainty and need for information (from UEA/MORI GM Food Survey, 2003; weighted dataset: *n*, 1363).

| Ambivalence *v.* certainty | Strongly disagree (%) | Tend to disagree (%) | Neither/ nor (%) | Tend to agree (%) | Strongly agree (%) | Don't know (%) |
|---|---|---|---|---|---|---|
| I have mixed feelings about GM food | 10 | 12 | 18 | 40 | 16 | 3 |
| I need more information to form a clear opinion about GM food | 3 | 5 | 7 | 33 | 51 | 1 |
| There are so many arguments for and against GM food; I could be persuaded by any of them | 16 | 24 | 24 | 26 | 5 | 4 |

trust. Longstanding research in the risk perceptions domain suggests that distrust is indeed associated with low acceptability of some hazardous technologies, although the precise determinants of trust, and the direction of association between trust and risk perceptions, are less clear (see Johnson, 1999; Poortinga and Pidgeon, 2005).

Losing trust, as befell the UK government over the BSE crisis, may have far-reaching consequences. Once lost, trust may be difficult to rebuild (see, e.g. Slovic, 1993). Accordingly, trust has become an explicit objective for several UK government departments, most particularly in relation to food safety (ILGRA, 1998; House of Lords, 2000; Phillips *et al.*, 2000; HM Government, 2001; Cabinet Office Strategy Unit, 2002). The need to (re)build trust is recognized because trust is not only seen as an essential ingredient in obtaining acceptance of risk management decisions but because one can also argue that mistrust lies at the root of conflicts about the validity of risk assessments (ILGRA, 1998).

In line with previous studies, we found that the general levels of trust in the regulation of GM food to be very low (see Table 2.9). Only one in five (19 and 21%, respectively) agreed with the statements 'I feel confident that the British government adequately regulates GM food' and 'I am confident that the development of GM crops is being carefully regulated'. In contrast, about one-half of the respondents disagreed with these statements (55 and 45%, respectively).

The low levels of trust in risk regulation by government – as well as by the biotechnology industry – were reflected by the fact that a large majority would like to see the establishment of regulatory organizations separate from government and industry (see Table 2.9).

The responses to the two statements were practically similar. Whereas four out of five (79 and 80%, respectively) agreed, only 6 and 4%, respectively, disagreed that organizations independent from government and industry were required.

What also becomes clear from Table 2.9 is that labelling and liability arrangements are critically important to people. An overwhelming majority (94%) agreed that all food containing GM material should be labelled (only 1% did not think it is necessary to label GM products!). Labelling is probably such a popular measure because, in theory at least, it enables people to decide for themselves whether they want to buy GM products or not.

In addition, Table 2.9 also shows that almost four out of five (79%) of the respondents agreed that biotechnology companies should be made liable for any damage caused by GM products, whilst only one in twenty (5%) disagreed. As later confirmed by follow-up qualitative work, people feel that it is necessary to keep the biotechnology industry accountable for their activities in order to ensure that 'taxpayers don't end up picking up the pieces of some big company's mess' (see Poortinga, 2004).

## Evaluation of government GM food policy

In addition to the general expressions of trust in the regulation of GM food, we tried to identify the factors influencing these judgements. A range of factors appear to influence trust in risk-managing institutions, which can be summarized under the general concepts of *competence*, *care* and *consensual values* (Johnson, 1999). In this study, respondents were asked to evaluate government policy on GM food. The items used were designed to measure *competence*, *credibility*, *reliability*, *integrity* (vested interests), *care*, *fairness*, and *openness* (see Table 2.10). The statements were selected from previous research on risk perception and trust (e.g. Renn and Levine, 1991; Frewer *et al.*, 1996; Peters *et al.*, 1997; Johnson, 1999; Metlay, 1999).

In addition, two questions were included aimed at measuring the extent to which the government was seen as having the same values as respondents in the context of GM food (based upon the *Value Similarity* model proposed by Earle and Cvetkovich, 1995). Furthermore, two questions were used to assess whether people feel that the Government is biased towards a particular position regarding GM food.

The low levels of trust in the regulation of GM food are reflected in the negative evaluations given of the Government. Table 2.10 shows that people are fairly critical of the Government and its GM policies. In

**Table 2.9.** Regulation of GM food (from UEA/MORI GM Food Survey, 2003; weighted dataset: *n*, 1363).

| | Strongly disagree (%) | Tend to disagree (%) | Neither/ nor (%) | Tend to agree (%) | Strongly agree (%) | Don't know (%) |
|---|---|---|---|---|---|---|
| Trust in risk regulation | | | | | | |
| I feel confident that the British government adequately regulates GM food | 28 | 27 | 19 | 16 | 3 | 6 |
| I am confident that the development of GM crops is being carefully regulated | 18 | 27 | 27 | 19 | 3 | 5 |
| Independent regulatory organizations | 2 | 4 | 10 | 39 | 40 | 5 |
| Organizations separate from government are needed to regulate GM Food | | | | | | |
| Organizations separate from industry are needed to regulate GM Food | 1 | 3 | 10 | 38 | 42 | 5 |
| Labelling | | | | | | |
| All food containing GM material should be labelled | 0 | 1 | 3 | 29 | 65 | 1 |
| Liability | | | | | | |
| Biotechnology companies should be made liable for any damage caused by GM products | 1 | 4 | 13 | 31 | 48 | 3 |

**Table 2.10.** Evaluation of government regarding GM foods (from UEA/MORI GM Food Survey, 2003; weighted dataset: *n*, 1363).[a]

| | Strongly disagree (%) | Tend to disagree (%) | Neither/ nor (%) | Tend to agree (%) | Strongly agree (%) | Don't know (%) |
|---|---|---|---|---|---|---|
| Competence | | | | | | |
| The government is doing a good job with regard to GM food | 24 | 25 | 28 | 11 | 1 | 10 |
| The government is competent enough to deal with GM food | 27 | 27 | 19 | 19 | 2 | 5 |
| Credibility | | | | | | |
| The government distorts facts in its favour regarding GM food | 3 | 9 | 23 | 34 | 22 | 8 |
| Reliability | | | | | | |
| The government changes policies regarding GM food without good reasons | 3 | 9 | 30 | 30 | 17 | 11 |
| Integrity (vested interests) | | | | | | |
| The government is too influenced by the biotechnology industry regarding GM food | 2 | 7 | 25 | 35 | 20 | 10 |
| Care | | | | | | |
| The government listens to concerns about GM food raised by the public | 19 | 33 | 20 | 20 | 2 | 5 |
| The government listens to what ordinary people think about GM food | 31 | 35 | 15 | 11 | 2 | 6 |
| Fairness | | | | | | |
| I feel that the way the government makes decisions about GM food is fair | 23 | 25 | 31 | 11 | 1 | 9 |
| Openness | | | | | | |
| The government provides all relevant information about GM food to the public | 35 | 33 | 17 | 7 | 2 | 6 |
| Value similarity | | | | | | |
| The government has the same opinion as I have about GM food | 24 | 27 | 26 | 7 | 1 | 13 |
| The government has the same ideas as I have about GM food | 25 | 30 | 24 | 8 | 2 | 12 |
| Bias of government | | | | | | |
| The government wants to promote GM food | 2 | 6 | 23 | 39 | 21 | 8 |
| The government is not in favour of GM food | 24 | 33 | 26 | 4 | 2 | 9 |

[a] The scale ranged from 1, 'strongly disagree' to 5, 'strongly agree'.

general, about one-half of the respondents agreed with the negatively formulated statements or disagreed with the positively formulated statements. Table 2.10 also shows that a majority think that the government has different values than themselves with regard to GM food, and that the government is strongly biased in favour of GM food.

In order to examine the relationship between various (trust) concepts that are mentioned above, and which might be thought of as comprising distinctive 'dimensions' of trust, we conducted a PCA with Varimax rotation. This analysis enables us to see whether people's perception of government risk policies can be described by a limited number of underlying dimensions. Table 2.11 shows that the nine statements (excluding the items on 'value similarity' and 'bias of government') can be described by two main factors. These two factors accounted for about 68% of the variance of the original variables.

Most items loaded high on the first factor, which accounted for 43% of the variance. This factor was concerned with the items aimed at measuring competence, care, fairness and openness, and can be interpreted as a *general trust* factor. That is, it represents a general evaluation of government policy on GM food.

The second factor accounted for about 25% of the original variance and was concerned with the items: (i) 'the government distorts facts in its favour regarding GM food'; (ii) 'the government changes policies regarding GM food without good reasons'; and (iii) 'the government is too influenced by industry regarding GM food'. This factor reflects a sceptical view of how government GM policies are brought about and can be labelled as *scepticism*. These results are comparable to similar analyses conducted across five risk cases in an earlier national survey (see Poortinga and Pidgeon, 2003).

Reflecting on the results with regard to trust in the government and its GM policies, it can be argued here that the importance of full 'unconditional' trust tends to be exaggerated (Barber, 1983; O'Neill, 2002). As the public becomes more competent and knowledgeable it might deploy 'effective distrust', expressed through a level of underlying scepticism about government policies. 'Distrust' in this sense is not destructive, but can be seen as an essential component of political accountability in democratic societies.

As a result of qualitative research on public perceptions of government health and safety regulators, Walls *et al.* (2004) posit that what is frequently called trust or distrust exists along a continuum, ranging from uncritical emotional acceptance to (downright) rejection. Somewhere between these extremes a healthy type of distrust can be found that they call *critical trust*. Critical trust can be conceptualized as a practical form of reliance on a person or institution combined with some healthy scepticism (see also Poortinga and Pidgeon, 2003).

**Table 2.11.** Factor loadings after Varimax rotation.

| | Factor[a] | |
|---|---|---|
| | 1 | 2 |
| The government is doing a good job | **0.77** | –0.33 |
| The government is competent enough | **0.80** | –0.23 |
| The government distorts facts in its favour | –0.27 | **0.79** |
| The government changes policies without good reasons | –0.26 | **0.81** |
| The government is too influenced by industry | –0.23 | **0.78** |
| The government listens to concerns raised by the public | **0.74** | –0.23 |
| The government listens to what ordinary people think | **0.80** | –0.20 |
| I feel that the way the government makes decisions is fair | **0.82** | –0.28 |
| The government provides all relevant information to the public | **0.76** | –0.25 |
| Eigenvalue | 3.86 | 2.29 |
| Explained variance | 42.9 | 25.4 |
| Average agreement | 2.28 | 3.65 |
| Cronbach's α | 0.90 | 0.78 |

[a] The scales ranged from 1, 'totally disagree' to 5, 'totally agree'. Factor loadings higher than 0.50 are in bold. Factor interpretations: 1, general trust; 2, scepticism.

The current study provides further quantitative support for this conceptualization of trust. The two trust components that were found in this study suggest that different degrees of general trust might coexist with different degrees of scepticism. Based on these two independent components, a conceptual typology of trust can be proposed that ranges from full trust to a deep type of distrust or cynicism (see Fig. 2.3).

This typology raises some interesting questions both for risk policy and the direction of future trust research. For example, in policy terms decision-makers may sometimes be confusing expressions of 'critical trust' with outright 'distrust' or rejection. Of course these are not the same thing; nor do they necessarily demand similar policy responses. For example, instead of focusing on how to increase trust in risk management organizations, it could be more fruitful to pay attention to the interaction between institutional structures, agency behaviour and perceptions of trust. That is, what kind of relationship between people and risk management institutions is achievable and desirable? For a functioning society it could well be more suitable to have critical but involved citizens in many situations.

## Trust in information sources

As discussed above, a large proportion of the British public are not simply for or against, but have more complex views on agricultural

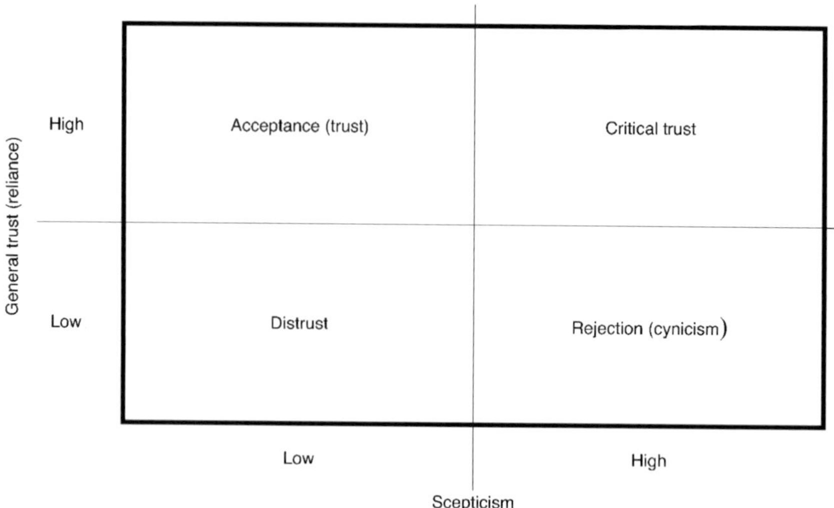


**Fig. 2.3.** A typology of trust in government regarding GM foods.

biotechnology; and many people feel that they need more information to form a clear opinion on this issue. With regard to a polarized issue such as GM food, trust in information from different sources may be very important for the dynamics of public attitudes. Several studies have shown that people differentially trust different 'types' of institutions. Whereas doctors, environmental and/or consumer organizations and 'quality' media are generally highly trusted, government ministers and departments, as well as 'industry and commerce', are typically distrusted (see, e.g. Frewer *et al.*, 1996; Poortinga and Dekker, 2002; Poortinga *et al.*, 2004). We found a similar pattern in the current study.

Table 2.12 shows that doctors, consumer organizations, environmental organizations and scientists working for universities were the most trusted information sources. More than three-quarters of the sample indicated that they trusted these information sources (a little or a lot) to tell the truth about GM food. More than one-half said that they trusted scientists working for environmental groups, the Food Standards Agency (FSA – the independent UK food safety regulator set up following the BSE crisis), friends and family, the Department for Environment, Food and Rural Affairs (DEFRA), 'people from your local community' and farmers to tell the truth about GM food.

The least trusted sources were government scientists, scientists working for the biotechnology industry, local authorities, the biotechnology industry, food manufacturers, the EU and the national government. All these sources were trusted by fewer than half of the sample.

**Table 2.12.** Trust in information sources regarding GM foods (from UEA/MORI GM Food Survey, 2003).

| Source | Distrust a lot (%) | Distrust a little (%) | Neither/ nor (%) | Trust a little (%) | Trust a lot (%) | Don't know (%) |
|---|---|---|---|---|---|---|
| Doctors | 1 | 2 | 14 | 42 | 39 | 1 |
| Consumer organizations | 2 | 5 | 13 | 43 | 33 | 3 |
| Environmental organizations | 2 | 6 | 14 | 45 | 31 | 1 |
| University scientists | 2 | 4 | 17 | 46 | 29 | 1 |
| Scientists working for environmental groups | 2 | 7 | 15 | 48 | 25 | 2 |
| FSA[a] | 3 | 6 | 14 | 45 | 26 | 3 |
| Friends and family | 1 | 3 | 23 | 30 | 40 | 2 |
| DEFRA[b] | 5 | 7 | 18 | 44 | 20 | 3 |
| Local community | 2 | 5 | 33 | 42 | 14 | 3 |
| Farmers | 5 | 10 | 26 | 36 | 19 | 2 |
| Government scientists | 15 | 21 | 19 | 34 | 8 | 2 |
| Scientists working for the biotech. industry | 14 | 21 | 23 | 28 | 9 | 4 |
| Local authorities | 10 | 17 | 33 | 31 | 5 | 2 |
| Biotechnology industry | 15 | 20 | 25 | 27 | 8 | 5 |
| Food manufacturers | 17 | 26 | 21 | 29 | 5 | 2 |
| The European Union (EU) | 20 | 19 | 25 | 25 | 7 | 3 |
| The national government | 23 | 25 | 19 | 25 | 5 | 2 |

[a] The Food Standards Agency.
[b] The Department of Environment, Food and Rural Affairs.

A PCA with Varimax rotation confirmed the previous finding of Poortinga *et al.* (2004) that people differentiate between government organizations, the industry and independent non-government organizations (which Poortinga *et al.* termed 'watchdogs'). As expected, watchdogs were the most trusted information sources, while both government organizations and the biotechnology industry were slightly distrusted to tell the truth about GM food (see Table 2.13). A fourth factor also differentiated personal sources (friends and family, people from your community), who were also highly trusted.

The results of this study reflect a typical pattern of trust in institutions. The remarkably stable pattern found across time, place and (risk) contexts seems to support the *functional view of trust* that trust in institutions depends largely upon the function someone attributes to a certain body in a particular situation (see, e.g. Cronkhite and Liska, 1976; McComas and Trumbo, 2001). Interestingly, the two government regulators (the FSA and DEFRA) both loaded highly, alongside farmers, on the factor representing trust in industry, suggesting that these

**Table 2.13.** Factor loadings after Varimax rotation.

| | Factor[a] | | | |
|---|---|---|---|---|
| | 1 | 2 | 3 | 4 |
| Doctors | –0.04 | 0.26 | 0.43 | 0.45 |
| Consumer organizations | 0.09 | 0.06 | **0.61** | 0.27 |
| Environmental organizations | 0.09 | 0.02 | **0.65** | 0.40 |
| University scientists | 0.15 | 0.17 | **0.79** | –0.05 |
| Scientists working for environmental groups | 0.12 | 0.08 | **0.82** | 0.06 |
| FSA[b] | 0.15 | 0.58 | 0.47 | 0.08 |
| Friends and family | –0.00 | 0.09 | 0.05 | **0.81** |
| DEFRA[c] | 0.25 | **0.56** | 0.39 | 0.07 |
| Local community | 0.16 | 0.02 | 0.25 | **0.74** |
| Farmers | –0.02 | **0.59** | 0.15 | 0.38 |
| Government scientists | **0.75** | 0.35 | 0.12 | –0.04 |
| Scientists working for the biotech. industry | 0.41 | **0.76** | 0.06 | –0.06 |
| Local authorities | **0.69** | 0.26 | 0.14 | 0.30 |
| Biotechnology industry | 0.40 | **0.76** | 0.04 | –0.02 |
| Food manufacturers | 0.41 | **0.56** | –0.09 | 0.23 |
| The European Union (EU) | **0.79** | 0.09 | 0.22 | –0.00 |
| The national government | **0.85** | 0.25 | 0.08 | 0.02 |
| Eigenvalue | 3.05 | 2.86 | 2.84 | 1.95 |
| Explained variance | 17.9 | 16.8 | 16.7 | 11.5 |
| Average agreement | 2.84 | 3.28 | 3.95 | 3.87 |
| Cronbach's α | 0.85 | 0.82 | 0.77 | 0.64 |

[a] The scales ranged from 1, 'totally disagree' to 5, 'totally agree'. Factor loadings higher than 0.50 are in bold. Factor interpretations: 1, trust in government institutions; 2, trust in industry; 3, trust in watchdogs; 4, trust in personal sources.

[b] The Food Standards Agency.

[c] The Department of Environment, Food and Rural Affairs.

organizations or groups of people may be perceived as being part of a wider 'biotechnology complex', including the biotechnology industry, government departments and the producers of food.

High trust ratings do not, of course, necessarily mean that people consider environmental and consumer organizations as more reliable or in other regards more 'trustworthy' than other information sources. Indeed, additional qualitative work (reported in Poortinga, 2004) showed that people are also critical of information received from environmental organizations. People are aware that NGOs, just like the government or the biotechnology industry, have their own stake in any controversy, and that they may at times also be selective in the presentation of information about GM food.

These results strengthen the idea posited by the House of Lords (2000) that expressed trust is not necessarily the same thing as the actual credibility of the communication itself. Expressions of trust may also stand for approval of an altogether different kind. For instance, many surveys show relatively high levels of trust in pressure group science and scientists (see Worcester, 2001). This may not necessarily mean that most people truly believe pressure group science to be more reliable or 'independent' than science sponsored by government or industry: most people know that pressure groups are as dependent on subscriptions and donations as are companies on orders (and governments on votes). Yet these expressions of trust must mean something; specifically, they may well signify approval that the pressure group plays a counterbalancing role against government and industry (House of Lords, 2000, para. 2.30).

## Involvement in decision-making

After people had given their judgements of trust in the various information sources, they were asked to what extent they agreed that the same people or organizations *should* be involved in making decisions about GM food (see Table 2.14). It appeared that a large majority (more than 80%) agreed that environmental organizations, the FSA, the general public, scientists working for environmental groups, consumer organizations and doctors should be involved in making decisions about GM Food.

Slightly fewer people (but still more than 60%) felt that DEFRA, scientists working for universities, farmers, local communities, food manufacturers, government scientists and the national government should be involved in making decisions about GM Food.

The lowest levels of agreement were found for local authorities, the biotechnology industry, scientists working for the biotechnology industry and the European Union (EU). However, even here (with these relatively distrusted institutions), more than one-half of the sample felt that these groups should be involved in making decisions about GM Food. This clearly illustrates that people think that all views should be heard and (thus) that a wide range of stakeholders should be involved in making decisions about GM Food.

Figure 2.4 plots the percentages of respondents who agree or strongly agree that the group or agency is trusted to tell the truth alongside (for each institution or group, respectively) the judgements about the extent to which they should be involved in decision-making. The pattern of judgements here is clear – with the least trusted organizations having much higher involvement ratings compared to their trust ratings. So much so that, with the relatively distrusted institutions, more than one-half of the sample felt that these groups *should* be involved in making decisions about GM food.

**Table 2.14.** Involvement in decision-making (from UEA/MORI GM Food Survey, 2003; weighted dataset: *n*, 1363).

| | Strongly disagree (%) | Tend to disagree (%) | Neither/ nor (%) | Tend to agree (%) | Strongly agree (%) | Don't know (%) |
|---|---|---|---|---|---|---|
| Environmental organizations | 1 | 3 | 8 | 43 | 41 | 3 |
| FSA[a] | 1 | 2 | 7 | 42 | 42 | 4 |
| The general public | 1 | 4 | 11 | 41 | 40 | 2 |
| Scientists working for environmental groups | 1 | 3 | 10 | 45 | 36 | 3 |
| Consumer organizations | 1 | 4 | 10 | 43 | 37 | 4 |
| Doctors | 1 | 4 | 11 | 43 | 37 | 2 |
| DEFRA[b] | 2 | 3 | 9 | 42 | 37 | 4 |
| University scientists | 1 | 3 | 13 | 46 | 31 | 3 |
| Farmers | 4 | 7 | 13 | 38 | 35 | 3 |
| Local community | 2 | 8 | 17 | 42 | 26 | 3 |
| Food manufacturers | 8 | 15 | 11 | 40 | 22 | 3 |
| Government scientists | 7 | 11 | 15 | 40 | 22 | 3 |
| The national government | 9 | 10 | 15 | 41 | 21 | 3 |
| Local authorities | 6 | 14 | 20 | 40 | 16 | 4 |
| Biotechnology industry | 8 | 12 | 18 | 39 | 16 | 6 |
| Scientists working for the biotechnology industry | 8 | 13 | 18 | 37 | 18 | 6 |
| The European Union (EU) | 13 | 12 | 16 | 34 | 19 | 4 |

[a] The Food Standards Agency.
[b] The Department of Environment, Food and Rural Affairs.

These data demonstrate that, despite overt statements of distrust, amongst the general public a wide variety of input to decision-making is desired. This pattern can again be interpreted within the framework provided by the idea of critical trust, in that scepticism goes hand in hand with a pragmatic reliance. Even if people are sceptical of the information that they receive from (in particular) government and industry sources, they still feel that those organizations have an important role to play in making decisions about GM food.

## Conclusion

This chapter has given an overview of current public perceptions of agricultural biotechnology in the UK, together with related facets such as trust in government regulation and in communication sources. At first sight it seems that, as is often assumed, people have very negative

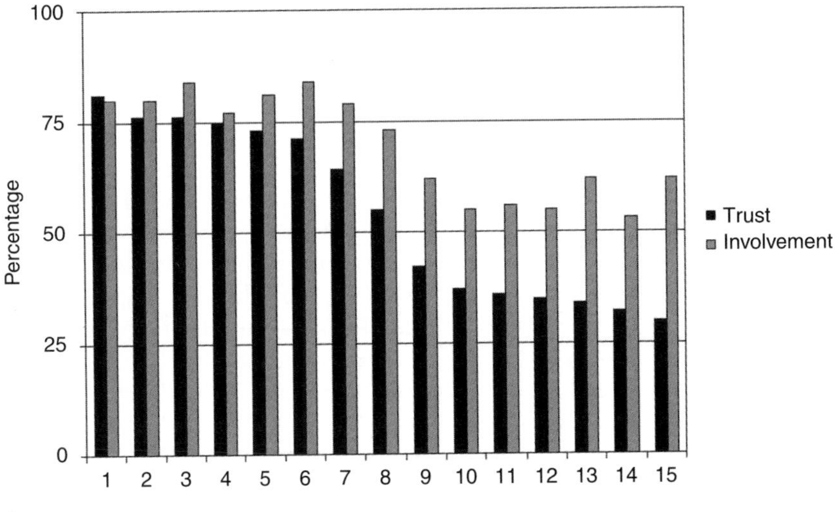


Key
1. Doctors
2. Consumer organizations
3. Environmental organizations
4. University scientists
5. Scientists working for environmental groups
6. Food Standards Agency (FSA)
7. Department of Environment Food and Rural Affairs (DEFRA)
8. Farmers
9. Government scientists
10. Scientists working for the biotech. industry
11. Local authorities
12. Biotechnology industry
13. Food manufacturers
14. The European Union (EU)
15. The national government

**Fig. 2.4.** Trust in various information sources and agreement about involvement in decision-making regarding GM foods.

views of GM food and crops. Using a number of general evaluative measures, this study shows that public opinion is indeed skewed towards the negative. More people oppose than support GM food, and many people are concerned about developments in this application of modern biotechnology. Moreover, a large number of respondents think that GM food has unknown consequences and poses risks to future generations, and they do not feel able to control any risks to themselves associated with GM Food.

However, this does not necessarily mean that there is widespread opposition to GM food. The results of this study suggest that, rather than simply for or against, attitudes appear nuanced and complex. When people were asked which of four statements best described their own opinion, more than half of the sample indicated that they were not sure whether GM food should be promoted or opposed. Also, when asked directly, a large majority agreed that they had mixed feelings about GM food and that they need more information to form a clear opinion about this subject. Indirect evidence for widespread ambivalent feelings was found by exploring the distribution across *perceived risk* and *perceived benefit* dimensions for GM food and crops, as a majority appreciated the (potential future) benefits at the same time as expressing concerns about possible risks of GM.

Overall, the results of this study suggest that public opinion about GM food is not a unitary whole, but fragmented with considerable ambivalence co-existing alongside both outright opposition and weak support. Importantly, available trend data also seems to suggest *increasing ambivalence* and *attitudinal uncertainty* in the UK regarding support or otherwise for GM food.

As was expected from research on public perceptions of other controversial risk issues, trust does seem to play an important, if complex, role in people's perceptions of agricultural biotechnology. In line with previous research, this study found that trust in the regulation of GM food was at (very) low levels. These low levels of trust are reflected in the evaluation of the government and its GM policies on a wide range of measures. In general, a majority disagreed with 'positive' and agreed with the 'negative' trust-relevant items, indicating an overall lack of endorsement of government and its policies with regard to GM food.

The evaluation of government could be summarized by two underlying and independent dimensions: namely, a *general trust* and a *scepticism* dimension, respectively. The two trust components that were found in this study show that different degrees of general trust can coexist with different degrees of scepticism. This suggests that even where expressed trust in an institution appears high, critical sentiments almost always coexist regarding such things as organizational motives or available resources, an idea which has been labelled critical trust (see Walls *et al.*, 2004).

Based on these findings, Poortinga and Pidgeon (2003) proposed a typology of trust that ranges from full trust to a deep form of distrust. We have also argued that, where distrust is an expression of scepticism about rather than downright rejection of government policy, such a situation should be seen as an important component of political accountability of new technologies and their regulation in democratic societies.

This study has also shown that, on average, doctors, consumer organizations, environmental organizations, scientists working for universities, scientists working for environmental groups – as well as the FSA – are the most trusted sources to tell the truth about GM food. The least trusted sources were government institutions (such as the national government, the EU, local authorities and government scientists), as well as industry sources (such as food manufacturers, the biotechnology industry and scientists working for the biotechnology industry).

In line with the findings of Poortinga *et al.* (2004), people differentiated between government organizations, the biotechnology industry, independent non-government organizations (which Poortinga *et al.* termed 'watchdogs') and personal sources. We argue that the remarkably stable trust judgements across contexts and studies supports the *functional view of trust* that trust in institutions largely depends upon the function someone attributes to a certain body in a particular situation. This also implies that high trust ratings do not necessarily mean that people consider environmental and consumer organizations as more reliable or trustworthy than other information sources. It may also be that the relatively high levels of trust in non-governmental organizations signify a form of approval of their counterbalancing role against government and industry.

An interesting result of this study is that the major differences in trust in the various information sources largely disappeared when people were asked how much they thought the sources should be involved in making decisions about GM food. In fact, more than one-half of the respondents agreed that *all* included organizations *should* be involved in decision-making. This clearly shows that people feel that that all views should be heard to create a successful decision-making process about this still-controversial technology.

It also provides some additional support for the idea of *critical trust.* While people may be critical of (in particular government and industry) sources with regard to the reliability of information about GM food, they still feel that these bodies have an important role to play in the decision-making process.

Overall, this chapter has shown that public opinion towards the controversial issue of agricultural biotechnology is more complex than is often assumed. Negative evaluations of GM food do not necessarily mean than people are unequivocally opposed to any development in this area, and if someone is critical of a particular information source it does not automatically imply that this person completely rejects all information. Our study suggests that ambivalence in attitudes – as well as complex trust judgements – are widespread with regard to the complex issue of agricultural biotechnology.

Of course, our results are specific to the UK in the early part of the 21st Century, so should not be assumed to readily transfer in the same detail to other contexts (and particularly other countries and populations). Conceptually, however, the idea of critical trust may well hold explanatory power in other contexts and eras. As a final commentary on the results, there is currently much debate as to the usefulness of adopting participatory processes for addressing stakeholder conflicts over risk issues in general (Stern and Fineberg, 1996) and biotechnology in particular (Pidgeon *et al.*, 2005; also Part II, this volume).

Our results suggest that people do endorse the broad principle of inclusiveness in decision-making about biotechnology. However, the challenges of doing this, with biotechnology or any other controversial new technology, should not be underestimated. Designing analytic-deliberative processes that can accommodate the plurality of views found here, from outright opposition and support through to more ambivalent and nuanced, sets a major challenge for the science and technology studies as well as the public policy communities.

## Acknowledgements

The research reported in this article was supported by the Leverhulme Trust under the *Understanding Risk* programme, as well as the UK Economic and Social Research Council (including its Science in Society programme). We wish to thank in particular Michele Corrado and Claire O'Dell of MORI for detailed assistance with aspects of the survey design and its detailed administration.

## References

Alhakami, A.S. and Slovic, P. (1994) A psychological study of the inverse relationship between perceived risk and perceived benefit. *Risk Analysis* 14 (6), 1085–1096.

Barber, B. (1983) *The Logic and Limits of Trust.* New Rutgers University Press, Brunswick, New Jersey.

Bassili, J.N. (1996) Meta-judgemental *versus* operative indexes of attitudinal attributes: the case of measures of attitude strength. *Journal of Personality and Social Psychology* 71 (4), 637–653.

Breckler, S.J. (1994) A comparison of numerical indices for measuring attitude ambivalence. *Educational and Psychological Measurement* 52 (2), 350–365.

Cabinet Office Strategy Unit (2002) *Risk: Improving Government's Capability to Handle Risk.* HMSO (Cabinet Office), London.

Cacioppo, J.T. and Berntson, G.G. (1994) Relationship between attitudes and evaluative space: a critical review, with emphasis on the separability of positive and negative substrates. *Psychological Bulletin* 115 (3), 401–423.

Cacioppo, J.T., Gardner, W.T. and Berntson, G.G. (1997) Beyond bipolar conceptualisations and measures. The case of attitudes and evaluative space. *Personality and Social Psychology Review* 1 (1), 3–25.

Conner, M. and Sparks, P. (2002) Attitudes and ambivalence. *European Review of Social Psychology* 12, 37–70.

Conner, M., Povey, R., Sparks, P. and James, R. (2003) Moderating role of attitudinal ambivalence within the theory of planned behaviour. *British Journal of Social Psychology* 42, 75–94.

Cronkhite, G. and Liska, J. (1976) A critique of factor analytic approaches to the study of credibility. *Communication Monographs* 43 (2), 92–107.

Durant, J., Bauer, W. and Gaskell, G. (eds) (1998) *Biotechnology in the Public Sphere: a European Source Book.* Science Museums Publications, London.

Earle, T.C. and Cvetkovich, G.T. (1995) *Social Trust. Towards a Cosmopolitan Society.* Praeger, London.

Einsiedel, E., Allansdottir, A., Allum, N., Bauer, M.W., Berthomier, A., Chatjouli, A., Cheveigne, S.D., Downey, R., Gutteling, J.M., Kohring, M., Leonarz, M., Manzoli, F., Olofson, A., Przestalski, A., Rusanen, T., Seifert, F., Stathopoulou, A. and Wagner, W. (2002) Brave new sheep – the clone named Dolly. In: Bauer, M.W. and Gaskell, G. (eds) *Biotechnology: the Making of a Global Controversy.* Cambridge University Press, Cambridge, UK.

Ewen, S.W.B. and Pusztai, A. (1999) Effect of diets containing genetically modified potatoes expressing *Galanthus nivalis* lectin on rat small intestine. *The Lancet* 354, 1353–1354.

Finucane, M.L., Alhakami, A.S., Slovic, P. and Johnson, S.M. (2000) The affect heuristic in the judgement of risks and benefits. *Journal of Behavioral Decision Making* 13 (1), 1–17.

Frewer, L.J., Howard, C., Hedderley, D. and Shepherd, R. (1996) What determines trust in information about food-related risks? Underlying psychological constructs. *Risk Analysis* 16 (4), 473–485.

Gaskell, G. (2000) Agricultural biotechnology and public attitudes in the European Union. *Agbioforum* 3 (23), 87–96.

Gaskell, G., Bauer, M.W., Allum, N., Lindsey, N., Durant, J. and Lueginger, J. (2001) United Kingdom: spilling the beans on genes. In: Gaskell, G. and Bauer, M.W. (eds) *Biotechnology 1996–2000. The Years of Controversy.* Science Museum, London.

Grabner, P., Hampel, J., Lindsey, N. and Torgersen, H. (2001) Biopolitical diversity: the challenge of multilevel policy-making. In: Gaskell, G. and Bauer, M.W. (eds) *Biotechnology 1996–2000. The Years of Controversy.* Science Museum, London.

Grove-White, R., Macnaghten, P., Mayer, S. and Wynne, B. (1997) *Uncertain World: Genetically Modified Organisms, Food and Public Opinion in Britain.* University of Lancaster Centre for the Study of Environmental Change, Lancaster, UK.

Harvey, M. (1999) *Genetic Modification as a Bio-socio-economic Process* (CRIC discussion paper No. 31). CRIC, University of Manchester, Manchester, UK.

HM Government (2001) *Response to the Report of the BSE Inquiry.* HM Government, London.

Horlick-Jones, T., Walls, J., Rowe, G., Pidgeon, N.F., Poortinga, W. and O'Riordan, T. (2004) *A Deliberative Future? an Independent Evaluation of the GM Nation? Public Debate about the Possible Commercialisation of Transgenic Crops in Britain, 2003* (Understanding Risk Working Paper 04-02). Centre for Environmental Risk, Norwich, UK.

House of Lords (2000) *Science and Society, 3rd report.* HMSO, London.

ILGRA (1998) *Risk Communication: a Guide to Regulatory Practice* (Inter-departmental Liaison Group on Risk Assessment). HSE Books, London.

Jasanoff, S. (1997) Civilization and madness: the great BSE scare of 1996. *Public Understanding of Science* 6 (3), 221–232.

Johnson, B.B. (1999) Exploring dimensionality in the origins of hazard-related trust. *Journal of Risk Research* 2 (4), 325–354.

Kasperson, R.E., Renn, O., Slovic, P., Brown, H.S., Emel, J., Goble, R., Kasperson, J.X. and Ratick, S. (1988) The social amplification of risk: a conceptual framework. *Risk Analysis* 8, 177–187.

Langford, I.H. (2002) An existential approach to risk perception. *Risk Analysis* 22 (1), 101–120.

Lassen, J., Allansdottir, A., Liakopoulos, M., Mortensen, A.T. and Olofson, A. (2002) Testing times – the reception of Roundup Ready soya in Europe. In: Bauer, M.W. and Gaskell, G. (eds) *Biotechnology: the Making of a Global Controversy*. Cambridge University Press, Cambridge, UK.

Levidow, L., Carr, S. and Wield, D. (1999) Regulating biotechnological risk, straining Britain's consultative style. *Journal of Risk Research* 2 (4), 307–324.

Loewenstein, G.F., Weber, E.U., Hsee, C.K. and Welch, N. (2001) Risk as feelings. *Psychological Bulletin* 127 (2), 267–286.

Margolis, H. (1996) *Dealing with Risk: Why the Public and the Experts Disagree on Environmental Issues*. University of Chicago Press, London.

Marris, C., Wynne, B., Simmons, P. and Weldon, S. (2001) *Public Perceptions of Agricultural Biotechnologies in Europe* (FAIR CT98-3844). University of Lancaster Centre for the Study of Environmental Change, Lancaster, UK.

McComas, K.A. and Trumbo, C.W. (2001) Source credibility in environmental health-risk controversies: application of Meyer's credibility index. *Risk Analysis* 21 (3), 467–480.

Metlay, D. (1999) Institutional trust and confidence: a journey into a conceptual quagmire. In: Cvetkovich, G.T. and Löfstedt, R.E. (eds) *Social Trust and the Management of Risk*. Earthscan, London.

O'Neill, O. (2002) *A Question of Trust*. Cambridge University Press, Cambridge, UK.

PDSB (2003) *GM Nation? The Findings of the Public Debate* (Report by the Public Debate Steering Board). Department of Trade and Industry, London: (available at http://www.gmnation.org.uk).

Peters, R.G., Covello, V.T. and McCallum, D.B. (1997) The determinants of trust and credibility in environmental risk communication: an empirical study. *Risk Analysis* 17 (1), 43–54.

Phillips, L., Bridgeman, J. and Ferguson-Smith, M. (2000) *The Report of the Inquiry into BSE and Variant CJD in the UK*. The Stationary Office, London.

Pidgeon, N.F., Kasperson, R.K. and Slovic, P. (eds) (2003) *The Social Amplification of Risk*. Cambridge University Press, Cambridge, UK.

Pidgeon, N.F., Poortinga, W., Rowe, G., Horlick-Jones, T., Walls, J. and O'Riordan, T. (2005) Using surveys in public participation processes for risk decision-making: the case of the 2003 British GM nation? Public Debate. *Risk Analysis* 25 (2), 467–479.

Pidgeon, N.F., Hood, C., Jones, D., Turner, B.A. and Gibson, R. (1992) Risk perception. In: *Risk – Analysis, Perception and Management: Report of a Royal Society Study Group*. The Royal Society, London, pp. 89–134.

Poortinga, W. (2004) Public perceptions and trust in the regulation of genetically modified food. PhD thesis, School of Environmental Sciences, University of East Anglia, Norwich, UK.

Poortinga, W. and Dekker, P. (2002) Voedselveiligheid: een kwestie van vertrouwen? [Food safety: a matter of trust?]. In: Bronner, A.E., Dekker, P., Hoekstra, J.C., Leeuw, E.D., Poiesz, T., Ruyter, K.D. and Smidts, A. (eds) *Recente Ontwikkelingen in het Marktonderzoek*. De Vriescheborg, Haarlem, Netherlands.

Poortinga, W. and Pidgeon, N.F. (2003) Exploring the dimensionality of trust in risk regulation. *Risk Analysis* 23 (5), 961–972.

Poortinga, W. and Pidgeon, N.F. (2004a) *Public Perceptions of Genetically Modified Food and Crops, and the GM Nation? Public Debate on the Commercialisation of Agricultural Biotechnology in the UK* (Understanding Risk Working Paper 04-01). Centre for Environmental Risk, Norwich, UK.

Poortinga, W. and Pidgeon, N.F. (2004b) Trust, the asymmetry principle and the role of prior beliefs. *Risk Analysis* 24 (6), 1475–1486.

Poortinga, W. and Pidgeon, N.F. (2005) Trust in risk regulation: cause or consequence of the acceptability of genetically modified food. *Risk Analysis* 25 (1), 199–209.

Poortinga, W., Bickerstaff, K., Langford, I.H., Niewöhner, J. and Pidgeon, N.F. (2004) The British 2001 foot and mouth crisis: a comparative study of public risk perceptions, trust and beliefs about government policy in two communities. *Journal of Risk Research* 7 (1), 73–90.

Renn, O. and Levine, D. (1991) Credibility and trust in risk communication. In: Kasperson, R.E. and Stallen, P.J.M. (eds) *Communicating Risks to the Public*. Kluwer, The Hague, Netherlands.

Robinson, G. (1997) Genetically modified foods and consumer choice. *Trends in Food Science and Technology* 8 (3), 84–88.

Simmons, P. and Weldon, S. (2000) The GM controversy in Britain: actors, arenas and institutional change. *Politeia* 16 (6), 53–67.

Slovic, P. (1993) Perceived risk, trust and democracy. *Risk Analysis* 13 (6), 675–682.

Slovic, P. (2000) *The Perception of Risk*. Earthscan, London.

Slovic, P., Finucane, M.L., Peters, E. and MacGregor, D. (2004) Risk as analysis and risk as feelings: Some thoughts about affect, reason, risk, and rationality. *Risk Analysis* 24 (2), 311–322.

Sparks, P., Conner, M., James, R., Shepherd, R. and Povey, R. (2001) Ambivalence about health-related behaviours: an exploration in the domain of food choice. *British Journal of Health Psychology* 6, 53–68.

Stern, P.C. and Fineberg, H.V. (1996) *Understanding Risk: Informing Decisions in a Democratic Society*. US National Academy of Sciences Press, Washington, DC.

Walls, J., Pidgeon, N.F., Weyman, A. and Horlick-Jones, T. (2004) Critical trust: understanding lay perceptions of health and safety risk regulation. *Health, Risk and Society* 6 (2), 133–151.

Worcester, R.M. (2001) Science and society: what scientists and the public can learn from each other. *Proceedings of the Royal Institution* 71, 97–160.

Zwick, M. (2005) Risk as perceived by the German public: pervasive risks and 'switching' risks. *Journal of Risk Research* 8 (6), 481–498.

## Appendix A

Data for this study were collected between 19 July and 12 September 2003. A quantitative survey was administered in the UK (England, Scotland and Wales) by the market research company MORI. A national quota sample of 1363 people aged 15 years and older was interviewed face-to-face in their own homes. The interviews were carried out using fully trained and supervised market research interviewers and took, on average, about 30 min to complete.

The overall sample was made up of a core British sample of 1017 interviews, a booster survey in Scotland of 151 interviews and a booster survey in Wales of 195 interviews (fieldwork for main survey: 19 July–26 August; fieldwork for Scottish booster: 11–26 August; fieldwork for Welsh booster: 11 August–12 September). For the total sample of 1363, the 95% confidence interval on a 50% finding is +/–3% around the obtained percentage, reducing to +/–2% for a 30 or 70% finding.

The survey sample was run in Enumeration Districts (EDs) or constituencies that were randomly selected with a probability proportional to the size of the population. Interviewers approached selected addresses within these EDs until they reached the quotas for gender, age and work status. The quotas reflected the actual profiles of the EDs. A maximum of one interview per address was conducted. The booster surveys were conducted in order to be able to compare public perceptions of GM food in England, Scotland and Wales. All frequency data have been weighted back to the profile of the British population in terms of age, sex, social class and region.

3

# German Reactions to Genetic Engineering in Food Production

H.P. Peters[*] and M. Sawicka[**]

*Research Centre Jülich, Programme Group Humans, Environment, Technology, 52425 Jülich, Germany;*
*e-mail: [*]h.p.peters@fz-juelich.de; [**]m.sawicka@fz-juelich.de*

## Introduction

No other issue, except for nuclear power, has led to such a heated controversy in Germany during the past few decades as biotechnology. Unlike the debate about nuclear power, however, the social debate about biotechnology resulted in a differentiated outcome: most applications of genetic engineering in the medical field are well accepted (exceptions being applications that affect human reproduction), while genetic engineering in food production is rejected by the majority of the population.

Although legally possible, no genetically modified food is presently sold in German shops. Not only are the majority of consumers very critical of GM food but critics of biotechnology, especially Greenpeace, are also threatening the first shops to put GM food on their shelves with protests and boycotts.

In this chapter, we attempt to explain public attitudes to food biotechnology and to describe the context of the controversy. Our perspective is guided by the conviction that there is no simple, single-factor explanation of these attitudes but that many factors have contributed to the evolution of current public attitudes to food biotechnology. Furthermore, we will argue that these attitudes cannot be understood as the result of a cost–benefit analysis from a consumer perspective, nor as a credibility crisis, but rather as the outcome of ascribing symbolic meanings which are part of German popular culture.

We will show that the GM food issue is related to a broad and very diverse range of cultural elements, shared experiences and symbols ranging from concepts of nature and the nuclear power controversy to the so-called 'Dutch tomato' and EU agricultural policy – and even the perceived economic imperialism of the USA.

We will also look at Germany in the European context – and in many respects Germany is a fairly typical European country as far as food biotechnology is concerned. The European perspective is useful, even for a specifically German case study, because the European Union (EU) is an important regulator of agriculture in general and food biotechnology in particular for its member states. Furthermore, one of the main data sources about German public opinion is provided by the Eurobarometer surveys of public attitudes towards biotechnology. Where possible, we will also make references to US attitudes towards food biotechnology.

The outline of the chapter is as follows. We will first give a brief historical account of the development of the current legal framework for food biotechnology in Germany (and the EU) and an overview of the protest movement against it. Then we will describe what we label the 'attitudinal syndrome' – a network of semantically related congruent attitudes and beliefs regarding (food) biotechnology. A core part of this chapter is the description of fields and issues in popular culture constituting the grounds on which attitudes towards food biotechnology are formed. This analysis will also emphasize the cultural relativity of attitudes towards food biotechnology. Finally, we will address German media coverage of biotechnology and discuss its possible influence on public opinion.

## Background

The following section will give a short overview of the most important developments in the political and societal dealing with agricultural biotechnology from its beginnings up to the present. The survey is based on several publications in this area, including Hampel *et al.* (1998), Aretz (2000), Hampel *et al.* (2001), Torgersen *et al.* (2002) and Hampel (2005). References to these authors will not be explicitly cited.

### Regulation of food biotechnology

The regulatory development accompanying scientific and commercial progress made by agricultural biotechnology in Germany can be traced back to the 1970s. Starting in 1970, the German government's first reactions to genetic engineering were aimed at promoting biotechnology and providing incentives. Regulation was regarded as a matter of the scientific system itself rather than the political and administrative system. It was up to the scientific community to establish a regulatory framework, which at this time mainly focused on laboratory safety. The approach changed slightly after the first *in vitro* nucleic acid was successfully recombined in 1973, which is regarded as the first milestone in agricultural biotechnology. The speed of technological developments raised concerns among the international scientific community regarding possible safety hazards of genetic engineering. In 1975 (Asilomar Conference II) an agreement about preventive safety measures for the laboratory work was agreed upon by the scientific community.

As a result of the scientific discourse in Germany, as in other industrialized countries, an expert committee was appointed with the aim of advising the government and defining safety guidelines for biotechnological research. Their recommendations were included in the guideline on protection from dangers resulting from genetic engineering (*Richtlinien zum Schutz vor Gefahren durch in-vitro neukombinierte Nukleinsäuren*) that was passed in 1978. These first regulations in this area were only binding for publicly funded research.

The self-regulation policy changed after the start of the commercial use of genetic engineering and public resistance ushered in by the birth of the first test-tube baby in Germany in the early 1980s. As a consequence, two commissions were established in 1984 in order to discuss legal and moral issues regarding *in vitro* fertilization (Benda Commission), as well as to evaluate the opportunities and risks of genetic engineering (Enquete Commission on genetic engineering). This has led to a widening of the focus of the discussion from laboratory safety and embryo protection to a broader context. The recommendations of both commissions were incorporated into the first German law on genetic engineering that was introduced in May 1990.

Germany was the second country, after Denmark, to introduce a genetic engineering law of this kind (Meyer, 2003, p. 74). The aim of the law was, in the first place, to protect the life and health of humans, animals, plants and the environment in general from possible risks of genetic engineering and its applications, as well as to establish a legal framework for research, development, utilization and promotion of genetic engineering. Among other aspects, the law was intended to regulate the release of genetically modified organisms, including procedural rules and liability provisions. Interestingly enough, unlike then, the law was supported by industry, who were afraid of disadvantages for German industries' future after the licensing of a plant producing human insulin using genetically modified bacteria failed because of the missing regulatory framework.

The year 1990 also marked the beginning of the harmonization of the regulation of genetic engineering within the EU. The European Commission passed a set of directives on biotechnology (Contained Use Directive and Deliberate Release Directive). The new guidelines gave occasion to amend the German law. Thereby, concessions were made to the German biotech industry that criticized the former law as being too restrictive and thus endangering the industry's economic competitiveness in 1993. These concessions concerned the simplification and curtailment of the permission procedure and limitation of public involvement.

In 1997, the Novel Food Regulation (No. 285/97) came into force in the EU countries. This regulates the registration and labelling of new types of food, including genetically modified food. As a consequence laboratories for detecting genetically modified food emerged. Detected products containing traces of genetically modified ingredients were removed from the market by the producers.

In the following years, several food scandals in Europe – in particular mad cow disease (BSE) – deeply shocked consumers' general confidence in food safety. The increasing public awareness and resistance to genetically modified food reinforced pressure on national governments. Subsequently, the environmental ministers of five countries (France, Greece, Italy, Denmark and Luxembourg) achieved a *de facto* moratorium on registrations of genetically modified organism by their blocking minority, starting in 1998. They demanded the introduction of further regulations regarding labelling, traceability and liability.

The moratorium lasted until 2003 when – due to pressure especially from the US, Canadian and Argentinian governments, who regarded the moratorium as an unjustified trade restriction – the EU environmental ministers agreed upon two new regulations on the labelling and traceability of genetically modified food and cattle fodder (Regulations No. 1929/2003 and No. 1930/2003), which became effective on 17 April, 2004.

According to the new laws, all food and food ingredients that contain more than a defined limit of 0.9% of genetically modified organisms have to be identified as such by a label on the product. This much stricter regulation replaced the Novel Food Regulation in the field of genetically modified organisms. Besides these regulations regarding the approval of genetically modified food and cattle fodder, an EU directive on deliberate release of genetically modified organisms into the environment (No. 2001/18) had already been passed in 2001.

The new German law on genetic engineering is in force since February 2005. It is based on several EU directives and guidelines and is intended to regulate the approval, labelling, traceability, coexistence, responsibilities and laboratory work on genetically modified organisms. While some EU regulations – for instance, those regarding labelling and approval – were valid immediately in all member countries, some directives solely give guidelines that need to be translated into national law, which is subject to quite different national interpretations. The latter category includes the regulation of the coexistence of conventional and genetically modified seed farming. Germany has implemented quite strict regulations in comparison to other countries, permitting stakeholders to negotiate their own rules of co-existence. A harmonization of this regulation by the EU is planned.

## Protests against biotechnology and critical events

Any consideration of the German controversy about agricultural biotechnology should take into account prior events that partly determined the context of the discussion in advance. The birth of the first test-tube baby in Germany in 1982, which has led to a heated discussion on genetic engineering, is regarded as one of the first and most influential events (Gill, 1991). Although not directly related to the new technology, the event gave rise to a debate about the inviolability of embryos, moral permission to intervene in the origins of human life and paramount claims of scientific progress.

A movement against reproductive and genetic engineering resulted, headed by feminist and left-wing groups (Gill, 1991). In 1984 the attempt to build a new plant for the production of human insulin using genetic engineering methods by the chemical company Hoechst led to a long-lasting public and legal confrontation. Opposition against this plant, however, was motivated by the more general concern of using genetic engineering without a social consensus (Thielemann, 1995).

Likewise, in the mid-1980s, the controversy about nuclear power reached its height. Opposition was mainly directed against the new type of hazard resulting from scientific progress, which was regarded as being guided by scientific and economic interests disregarding the social and environmental consequences (Aretz, 1999).

Criticism of biomedical biotechnology and nuclear power was transferred to the new issue of food biotechnology. On the one hand, general opposition to scientific progress was immediately valid for the new technology. On the other hand, there was a clear tendency among the protesting groups towards a dedifferentiation of the conflicting object and association with nuclear energy and reproduction techniques (Aretz, 1999). An indicator for this transfer is the similarity of symbols of the anti-nuclear and anti-GM food movements. By taking old, well-known symbols of the anti-nuclear movement and changing them slightly the protesters managed to transfer the old conflict pattern to the new issue (see Fig. 3.1). This illustrates the fact that opposition to food biotechnology is based not only on the perception of specific risks but on very fundamental and general objections.

The actual controversies regarding genetic engineering with plants first arose in 1990 about Germany's first field experiments with transgenic petunias in Cologne. The protesters were mainly concerned with the influences of the genetically modified plants on the environment and – due to the unexpected result of the test – the reliability of scientific prognoses. The test was accompanied by demonstrations and even some trespassing on the plants themselves. The protest was subsequently organized by several groups and initiatives including the 'Gen-ethisches Netzwerk', the 'Bund für Umwelt und Naturschutz (BUND)' and several civil action groups and anti-release initiatives like 'Arche GENoah'. The range of activities of these groups encompassed information campaigns, conferences, petitions against GMOs, field occupations and even crop destruction, cynically called 'harvest help'.

Greenpeace entered the debate in 1996. This was the phase when public controversy over agricultural biotechnology reached its height, triggered by the introduction of the first genetically modified

**Fig. 3.1.** Icons of the anti-nuclear and anti-GM food movement in Germany, showing much similarity. The slogans read: 'Atomic power – no thank you' (left) and 'GM food – no thank you' (right).

agricultural product, the Monsanto soybean, into the European – and thus German – market and the first cloned sheep, Dolly, which was born in 1997. The first ships transporting genetically modified soybeans to the port of Hamburg were escorted and decorated with banners by Greenpeace. Since this time the initially broad-based resistance in Germany has become dominated by Greenpeace as the main actor (Hampel *et al.*, 2001).

The organization initiated shoppers' boycotts of genetically engineered products, stuck warning signs on suspected products and published a shopping guide 'Food Without Genetic Engineering'. Greenpeace's strategy of pillorying food companies and retailers turned out to be very effective. It is due to their initiatives that, even now, no retailer dares to be the first to stock genetically modified food on their supermarket shelves. To date there are no known foodstuffs containing genetically engineered ingredients in German supermarkets.

The great influence of Greenpeace in Germany may be explained by their popularity, credibility, professionalism and financial resources. Greenpeace is one of the best known NGOs and the most important environmental NGO in Germany in terms of donations (€41.5 million in 2004).[1]

In comparison to many other institutions, NGOs enjoy high credibility in Germany (see below). Additionally, their campaigns and strategies are very professional. The reluctance of producers and retailers to stock GM foodstuffs is due to Greenpeace's potential to damage the image of companies and cause severe economic losses.

Germany is regarded as having had one of the most heated and controversial debates on agricultural biotechnology and genetic engineering in comparison to other European countries and the USA (e.g. van den Daele, 1991; European Federation of Biotechnology, 1994; Hampel *et al.*, 2000a). At least in the early phase, the debate in Germany was very polarized. Aretz (2000) explains this by pointing to the structural background of the debate in Germany in comparison to that in the USA. He identifies a strong division between, on the one hand, influential industrial scientific and political associations that shared a favourable view of mainstream science and, on the other, minor critical groups without direct access to the political decision-making process. This separation was caused by system-inherent accentuation that supported strong polarization in Germany, while in the US the system was open to particular interest groups from the beginning.

As a consequence of the exclusion of critical groups from direct

1 Annual Report, 2004 (Greenpeace, 2005). For the ranking see an older evaluation (Gill, 2003).

political influence they were forced to take the approach of mobilizing public opinion. A good example of this is the intervention of the Green Party in the controversy. After entering the German Parliament in 1984, the Green Party tried to influence the new regulation by calling the Enquete Commission on opportunities and risks of genetic engineering into being. When their attempts to use this instrument to ban the new technology failed, the Green Party started to mobilize the public (Torgersen *et al.*, 2002).

In the following years, Germany experienced heated debates that calmed down when the Green Party became part of the coalition government. The Federal Minister for Food, Agriculture and Consumer Protection, Renate Künast, initiated a discourse on agricultural biotechnology (*Diskurs Grüne Gentechnik*) in 2001/2002, for which she invited participation from a broad spectrum of protagonists and stakeholders. Due to the incompatibility of the participants involved, the discourse ended without the expected consensus.

Since the EU countries agreed upon formal regulation and labelling of GMO food, the debate on agricultural biotechnology is more focused on concrete, consumer-oriented issues, such as the presence of genetically modified ingredients in certain foodstuffs. More general concerns are still present, but expressed in the debate on more practical questions of how to deal with food biotechnology.

## Public Opinion Regarding Food Biotechnology

### Predominantly negative attitudes

The following description of the German public attitude to food biotechnology is mainly based on three representative public opinion surveys: (i) the survey by the Stuttgart Academy of Technology Assessment in 1997 (Hampel and Pfenning, 1999; Hampel *et al.*, 2000b); (ii) the Eurobarometer 58.0 survey in 2002 (Gaskell *et al.*, 2003);[2] and (iii) the German counterpart of the US survey on food

[2] The Eurobarometer 58.0 data for Germany have been re-analysed by the authors using the publicly available data file provided by the Central Archive for Empirical Research, University of Cologne, and using weights (W14) to compensate for biases in sociodemographic sample characteristics and for the disproportionate sampling scheme. About 1000 interviews were conducted in West Germany as well as in East Germany, while the proportion of the East German population of the total population is only about 25%. If no specific source is mentioned for results from the Eurobarometer survey, these results are based on the authors' own analyses.

biotechnology by the Food Policy Institute, Rutgers University, New Jersey, USA, in 2004 (Hallman *et al.*, 2004).[3]

Across several surveys most positive statements about food biotechnology tend to be rejected and critical statements tend to be accepted by a majority of survey respondents in Germany (see, for example, Fig. 3.2 and Table 3.1). From a pragmatic point of view, and using the semantics of survey questions as a criterion, it is thus quite safe to say that the public attitude to food biotechnology is predominantly negative

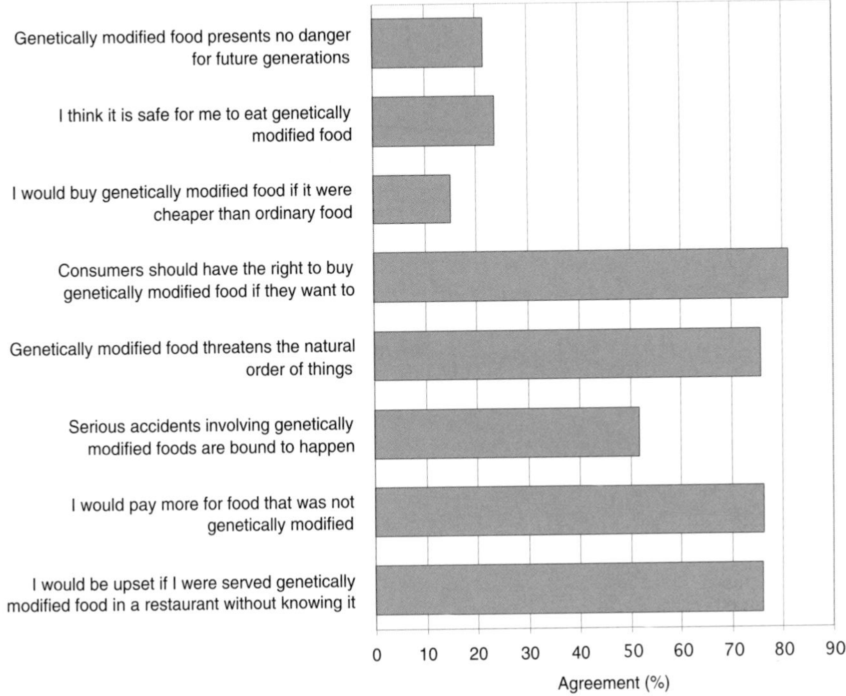


**Fig. 3.2.** Opinions of the German population on GM food (from representative survey conducted in 2004 by the Food Policy Institute, Rutgers University, New Jersey, USA, in collaboration with the authors).

3 The US survey and its German counterpart are part of the project 'Evaluating Consumer Acceptance of Food Biotechnology in the USA', with William K. Hallman as principal investigator (grant #2001-52100-11203 from the US Department of Agriculture). In this chapter we make (almost) exclusive use of the German data of the comparative US–German telephone survey, which was designed to test two specific hypotheses about intercultural differences between Germany and the USA. Results of the comparison will be published elsewhere (see also Lang *et al.*, 2005). For an analysis of the US part of the survey see Hallman *et al.* (2004).

**Table 3.1.** Agreement to evaluative items about genetically modified food in Germany and in the European Community (from Eurobarometer 58.0, 2002).[a]

| | Germany[b] (*n*, 1044) | | | | European Union[c] (*n*, 7962) | | | |
|---|---|---|---|---|---|---|---|---|
| | Tend to agree (%) | Tend to disagree (%) | Don't know (%) | Total (%) | Tend to agree (%) | Tend to disagree (%) | Don't know (%) | Total (%) |
| Genetically modified food will be useful for me and other consumers | 23.2 | 60.6 | 16.3 | 100.0 | 32.0 | 51.9 | 16.1 | 100.0 |
| Genetically modified food will be useful in the fight against world hunger | 47.2 | 31.8 | 21.0 | 100.0 | 46.6 | 34.5 | 18.9 | 100.0 |
| Genetically modified food and crops will only be good for industry and not for the consumer | 47.9 | 32.3 | 19.8 | 100.0 | 43.5 | 35.7 | 20.8 | 100.0 |
| In the long run, a successful [German etc.] genetically modified food industry will be good for the economy | 38.4 | 37.2 | 24.4 | 100.0 | 38.8 | 33.7 | 27.4 | 100.0 |
| Genetically modified food poses no threat to future generations | 20.6 | 52.0 | 27.4 | 100.0 | 19.4 | 52.4 | 28.2 | 100.0 |
| Eating genetically modified food will be harmful to my health and my family's health | 47.4 | 23.6 | 29.0 | 100.0 | 46.3 | 25.6 | 28.0 | 100.0 |
| Genetically modified food threatens the natural order of things | 62.7 | 22.9 | 14.4 | 100.0 | 64.6 | 19.8 | 15.6 | 100.0 |
| I think it is safe for me to eat genetically modified food | 23.3 | 58.1 | 18.7 | 100.0 | 28.1 | 51.4 | 20.5 | 100.0 |
| I will be able to choose whether I eat genetically modified food or not | 59.8 | 23.0 | 17.2 | 100.0 | 55.5 | 23.6 | 20.9 | 100.0 |
| Whatever the dangers of genetically modified food, future research will deal with them successfully | 33.9 | 35.8 | 30.3 | 100.0 | 39.9 | 30.0 | 30.1 | 100.0 |
| Current regulations are sufficient to protect people from any risks linked to genetically modified food | 22.8 | 50.0 | 27.2 | 100.0 | 25.4 | 45.7 | 28.8 | 100.0 |
| Growing genetically modified crops will be harmful to the environment | 43.8 | 28.5 | 27.7 | 100.0 | 44.6 | 27.5 | 27.9 | 100.0 |

[a] This question was only asked in a split-half subsample.
[b] Using weights (W14) to compensate disproportional stratified East/West German subsamples. Results are representative for the German population.
[c] EU in 2002 (15 member states), including Germany, using weights (W11) to compensate disproportional stratified country subsamples. Results are representative for the EU population.

in Germany. Only 23% of the German respondents of the 2004 survey 'strongly' or 'somewhat' agreed with the statement 'I think it is safe for me to eat genetically modified food', for example. Consequently, distributions of attitude scores based on Likert scales including equal numbers of positively and negatively framed items tend to have their maximum and arithmetic mean in the 'negative' half of the scale.

Gaskell *et al.* (2003, p. 14) put Germany in the category of *weak opposition* with respect to food: 'Use modern biotechnology in the production of foods, for example, to make them higher in protein, keep longer or improve the taste'; and *weak support* with respect to crops: 'Taking genes from plant species and transferring them into crop plants, to make them more resistant to insect pests.' With respect to attitudes towards food biotechnology, Germany is a rather typical European country, although slightly more negative than the European population on average in most items (see Table 3.1). In particular, Germans are less likely than Europeans in general to agree that genetically modified food will be useful for them (23 *versus* 32%) and also somewhat less convinced that it is safe for them to eat GM food (23 *versus* 28%).

The recent Eurobarometer 2005 survey 'Europeans, Science and Technology' places Germany at about the EU average regarding responses to the item: 'Food made from genetically modified organisms is dangerous' (European Commission, 2005, p. 63). Fifty-one per cent of the German respondents agreed with this item, 14% disagreed. The rest said 'neither agree nor disagree' or 'don't know'. France is even more critical towards genetically modified food, while the UK is much less negative. The Eurobarometer on biotechnology of the same year sees support for GM food in Germany at about the same level as in France, lower than support in the United Kingdom and below the EU average (Gaskell *et al.*, 2006).

Compared to the USA, the public attitude to food biotechnology is significantly more critical in Germany (as in most European countries) (e.g. Gaskell *et al.*, 1999; Lang *et al.*, 2005; Sawicka, 2005). In our 2004 survey, only 23% of the German, but 53% of the US, respondents agreed that it was safe to eat GM food, and 76% of the German respondents said that they would pay more for non-GM food, as compared to 61% in the USA. In seven of the eight items shown in Fig. 3.2, the US respondents on average were significantly less negative than the Germans. (The only exception was the item 'Serious accidents involving genetically modified foods are bound to happen', which finds significantly higher approval in the USA than in Germany, probably a consequence of the US 'September 11' trauma.)

The 1997 Stuttgart survey and the Eurobarometer survey clearly show very distinct public attitudes towards different applications of genetic engineering (Hampel *et al.*, 2000b; Gaskell *et al.*, 2003). These differences reveal a pattern that has been labelled the 'red–green distinction' (Bauer, 2005a): Medical (red) applications are generally much more readily accepted than (green) applications in agriculture and food production. The latter applications are more critically assessed if the genetic modifications concern animals than if they concern plants. Bauer (2005a) showed that the red–green distinction in the evaluation of genetic engineering increased somewhat in Germany (as in most European countries) between 1996 and 1999.

The available time-series data reveal that attitudes towards food biotechnology in Germany became somewhat more negative between 1991 and 2002 (Hampel and Pfennig, 1999; Gaskell *et al.*, 2003). The recent EU regulation on GM food, which addresses consumer concerns with respect to labelling, may have produced a shift in public opinion towards a more positive attitude. The public debate about food biotechnology seems to have calmed down since then. If this calming down resulted in a less negative public opinion, however, it is not yet reflected in our 2004 survey, which took place in May, shortly after the EU regulation became effective in April 2004. On the contrary, the most recent Eurobarometer on biotechnology shows a sharp decrease in support of GM food in Germany between 2002 and 2005 (Gaskell *et al.*, 2006).

## Sociodemographic correlates as indicators of cultural factors

Analysing inter-individual differences in attitudes towards genetically modified food is not the primary goal of this chapter. Rather, the analysis of inter-individual differences serves as a means to generate hypotheses about general factors (i.e. cultural references) shaping public opinion. According to our analytical perspective, attitude differences between different sociodemographic groups give rise to the question of which possible semantic links to a cultural element might be stronger or weaker in those groups, thus identifying that particular element as a general factor of influence.

In general, the statistical associations of sociodemographic variables with attitudes towards food biotechnology are quite weak in terms of explained variance of attitudes. Some stable patterns have consistently been found in several surveys.

### *Gender*

As in Europe in general (Gaskell *et al.*, 2003) and in the USA (Hallman *et al.*, 2002), there is a gender difference in German attitudes towards GM food. Women on average tend to be more critical of GM food than men (see Table 3.2). While one might be tempted to explain this relationship by traditional gender roles, according to which males are primarily responsible for economic welfare, females for the health care and nutritional welfare of the family (leading to concerns regarding GM food), it should be noted that, in Germany, the gender difference is not less among the younger and better-educated respondents, where one might expect traditional gender roles to be less pronounced.

The critical attitudes towards GM food among women have, at least partly, a cultural foundation outside the traditional gender role context. Values established by the Women's Movement among the well-educated female population might play a role here – such as scepticism about technology and a special focus on self-determination, with a particularly critical opinion of dependence on the abstract (male-dominated) expert systems (Giddens, 1991) involved in the development, use and monitoring of food biotechnology.

### *Age*

Furthermore, both the Eurobarometer 58.0 and our 2004 survey consistently show a considerable age effect: while opposition to

**Table 3.2.** Variation of attitudes towards genetically modified food in Germany with socio-demographic variables.

| | | EB 58.0[a] (2002) *n*, 932 | | FPI[b] (2004) *n*, 963 | |
|---|---|---|---|---|---|
| | | mean | standard error | mean | standard error |
| Gender | Male | −1.1 | 0.3 | −2.6 | 0.2 |
| | Female | −2.7 | 0.3 | −4.3 | 0.2 |
| Age | <30 | −1.0 | 0.5 | −1.6 | 0.3 |
| | 30–44 | −2.2 | 0.4 | −4.1 | 0.3 |
| | 45–59 | −2.4 | 0.5 | −4.1 | 0.3 |
| | >60 | −2.0 | 0.4 | −3.7 | 0.3 |
| Educational level[c] | Low | −3.2 | 0.4 | −3.9 | 0.2 |
| | Medium | −1.8 | 0.3 | −3.6 | 0.3 |
| | High | −1.7 | 0.5 | −3.1 | 0.6 |
| | Still in education | −0.2 | 0.7 | −1.7 | 0.5 |
| Region | Former West Germany | −2.2 | 0.2 | −3.8 | 0.2 |
| | Former East Germany | −1.0 | 0.5 | −2.5 | 0.4 |

[a] Split half version A of Eurobarometer 58.0, only German data, weights W14 used. The attitude scale was constructed (after adequate recoding) as sum scale from question 13, items 1–12. Value range: −12 to +12, mean = −1.9, standard deviation = 6.4.

[b] German data of the US–German survey in 2004 conducted by the Food Policy Institute, Rutgers University, NJ, in collaboration with the authors. The attitude scale is a Likert-type scale based on four positively and four negatively framed items. Value range: −12 to +12, mean = −3.5, standard deviation = 4.8.

[c] Educational level was operationalized differently in the two surveys and thus is not exactly comparable.

innovative technologies was particularly strong among the younger generation for quite some time in Germany under the influence of the anti-nuclear and environmental movement, this picture has completely changed in the meantime. There is a strong difference between the 14–29-year age group and the 30–59-year age group, the younger respondents on average having much less negative attitudes towards GM food (see Table 3.2). The oldest age group (60 and older) is also somewhat less negative than the middle age group.

The particularly negative attitudes of the 30–59-year age group are probably caused by the cultural impact of the anti-nuclear and environmental movement. The active and passive members of that movement are now getting older, however. Prominent members of the anti-nuclear and environmental movement in the meantime are, or have been, members of federal and state governments, examples being the former federal ministers Joschka Fischer (foreign ministry) and Jürgen Trittin (environment).

In a way, their critical attitudes towards certain technologies in general and toward food biotechnology in particular are mainstream sentiments – but less so among the new generation. Among young voters, support for the Green Party is declining, for example. Inglehart's socialization thesis (Inglehart, 1977) may be an explanation of why the younger age cohorts, who were socialized during a time of economic crisis and high unemployment, are more concerned with the state of the economy and with the attempt to make the most of the available resources than with – in the case of food biotechnology – rather abstract environmental and health concerns.

This age effect, namely that the 14–30-year-old age cohorts are significantly less negative towards food biotechnology than the older respondents, is not found in the 1997 Stuttgart survey (see Hampel *et al.*, 1997). Perhaps, as an age-related attitude gap, it is caused by a recent development. The younger generation may respond more readily by value or attitude changes to the economic crisis and are perhaps more susceptible to a changing innovation and risk culture fostered by the recently initiated political discussion about the 'modernization' of the German economy and job market.

### *Educational level*

The Eurobarometer 58.0 shows a clear relationship between attitudes towards food biotechnology and educational level. Higher educational levels are associated with more positive attitudes (see Table 3.2). The FPI survey of 2004, however, indicates barely any such association. We do not have a clear explanation for this difference, especially given the similarity of the surveys regarding the other sociodemographic variables.

Perhaps the difference is an artefact produced by different operationalizations of 'level of education'. Educational level in the Eurobarometer survey was simply operationalized by the age at which school/university education ended.

The duration of the educational period, however, can be influenced by many factors that have nothing to do with the educational level achieved, such as military service for men or the repetition of classes in high school because of low grades (possible in Germany), which may suggest a higher educational level. In the FPI survey, educational level was measured by a categorization of educational certificates obtained at different kinds of schools and programmes.

Level of education is an ambiguous sociodemographic variable. It can indicate degrees of general knowledge or cognitive abilities. It can, furthermore, indicate different socialization environments in school and in peer groups – especially in Germany, where pupils are separated into different schools at the age of 10 years, based on their performance in class. And, finally, level of education can stand for the social status of the respondents but also – especially in Germany with its embarrassingly high correlation between success at school and family status (PISA-Konsortium Deutschland, 2005) – for the status of the respondents' parents and thus for the kind of socialization environment in the family.

### *East versus West Germany*

Specifically for Germany, another aspect of political culture has to be mentioned as relevant for attitudes towards GM food. It is well known that attitudes towards science and technology in general are more positive in the eastern parts of Germany (former East Germany) than in the west of Germany (former West Germany), and the same is true of attitudes towards GM food (see Table 3.2). Although the reunification of East and West Germany took place more than 15 years ago, their different histories of political culture still have an effect on the attitudes and opinions of the population even today.

There are several possible reasons for the difference between the two parts of Germany:

**1.** East Germany remained largely unaffected by the public controversies about the negative side effects of modernization such as environmental destruction and technological risks of unprecedented magnitude, which took place in West Germany.
**2.** Although nobody had to go hungry in East Germany, the population experienced a frequent scarcity of many everyday goods. Agricultural productivity in East Germany thus remained an important goal and was never de-legitimized as in West Germany (see discussion of EU

agricultural policy below). Agricultural overproduction is a cultural experience that is not shared by the East German population.

**3.** The organization of agriculture in East Germany was based on large farms and a high level of mechanization, while in West Germany relatively small farms were, and are still, more common. The image of highly productive, industrialized agriculture is thus a negative image for West Germans, but probably less so for the East German population.

## Awareness and knowledge

Attitudes depend on subjective knowledge: there is no doubt about that. However, it is not clear what kind of knowledge is relevant for attitudes towards biotechnology. According to a widespread belief, discrepancies between the assessment of technologies by experts and attitudes on the part of the population are caused by the knowledge gap between experts and laypeople. This has always been a bold assumption, because experts and laypeople differ in more respects than just knowledge level – in the amount of control over a technology and in the benefits they draw from their involvement in technological development, for example.

Some studies, however, find a relationship between issue-specific expert knowledge (e.g. measured by the Eurobarometer knowledge items) and attitudes, better informed people being more positive towards food biotechnology than the less informed (see, for example, Hallman *et al.*, 2002). The German data reveal a different pattern, however. Awareness and (issue-specific expert) knowledge are positively associated with the *extremeness* of attitudes. Those with very positive as well as those with very negative attitudes towards food biotechnology show a higher issue awareness and greater knowledge than those with moderate attitudes (see Figs 3.3 and 3.4). This pattern has also been observed in other scientific/technical controversies (Peters, 2000a).

The often implied causal relationship between knowledge and attitudes is not as obvious as it seems. It is possible that a higher degree of (issue-specific expert) knowledge changes beliefs and subsequently modifies attitudes, but it is also possible that both are dependent on a third variable (e.g. awareness) without a causal relationship between them, and it is possible that certain attitudes modify people's information behaviour and thereby level of knowledge. The latter effect might be caused by a need to rationalize one's attitude, by motives of persuading others to adopt one's own opinion or by the wish to defend one's opinion against criticism from others. Especially the two last-mentioned motives might be particularly relevant for people with

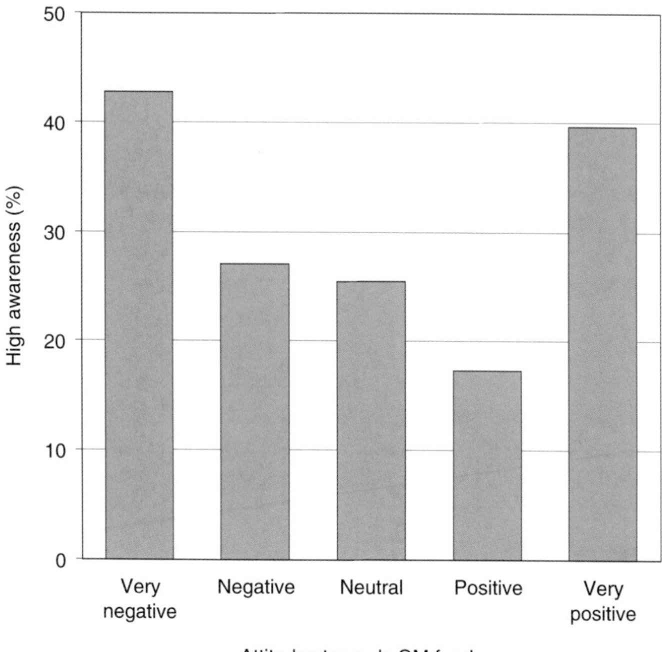


**Fig. 3.3.** Awareness of the GM food issue for different attitudinal groups in Germany (from representative survey conducted in 2004 by the Food Policy Institute, Rutgers University, New Jersey, USA, in collaboration with the authors).

extreme attitudes and thus might help to explain the u-shaped relationship between knowledge and attitudes (Peters, 2000a).

Only an experimental approach can ultimately confirm a causal relationship. Such a 'field experiment' took place in Germany. In 1997 the German state of Bavaria started an information campaign on biotechnology with a mobile genetic engineering laboratory ('BioTech mobil'), enabling high schools to offer their students 'hands on' experience with genetic engineering. This information campaign was carefully evaluated by a team from the Institute for Communication and Media Research of the University of Munich (Schweiger and Brosius, 1999). In a three-wave panel design, the researchers surveyed the participating students 1 week before they participated in the 'hands-on' course, 1 week afterwards and again 4 months later. A control group of students who did not participate in the course was also surveyed in the first and third waves.

The analysis showed that the 'hands-on' course only slightly changed beliefs and attitudes. Furthermore, the only significant changes were mixed: a slight reduction in reported fear of the risks of genetic

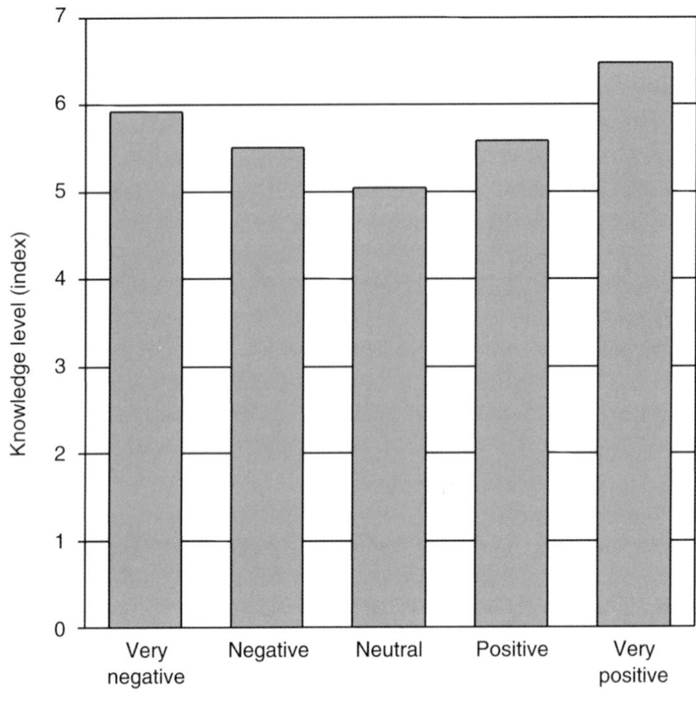


**Fig. 3.4.** Knowledge about modern biotechnology for different attitudinal groups in Germany (from Eurobarometer 58.0, 2002; split-half subsample).

engineering and a decreased desirability of medical applications of genetic engineering and genetically modified plants (Schweiger and Brosius, 1999).

These findings concerning the relationship of knowledge and attitudes are consistent with our assumption that it is not the more or less 'correct' perception of specific risks and benefits that determines attitudes towards food biotechnology, but the contingent semantic anchoring of these attitudes in the cultural context. Subjective beliefs about risks and benefits are then selected from the available information in order to justify and defend the attitude.

## Issue-specific Trust and Credibility as Part of an Attitude Syndrome

Several authors have argued that trust in institutions involved in the development, application and regulation of biotechnology – and the credibility of sources of information about this technology – are important factors in risk perception and attitude formation (e.g. Peters, 1999; Siegrist, 2000; Hornig Priest *et al.*, 2003).[4] Two plausible arguments can be raised in support of this hypothesis:

**1.** Classical and modern persuasion theories (e.g. Hovland *et al.*, 1953; Petty and Cacioppo, 1986) consider communicator credibility a critical variable in the persuasion process. Under certain conditions, credible communicators are more persuasive than non-credible communicators. The relative credibility of proponents and opponents of food biotechnology will thus influence whether pro-arguments or contra-arguments are more accepted as true by the public. The distribution of credibility among several possible information sources about 'modern biotechnology' in Germany shows that, in the first place, NGOs (consumer and environmental organizations) are trusted as information sources and, in the second place, the medical profession and universities (see Table 3.3).

**2.** Another argument related to 'trust' points to the role of institutions and experts in the development, use, monitoring and regulation of technologies. Since not even the proponents deny a potential risk of genetic engineering, the question is whether the risks can be controlled. This raises the question of the trustworthiness of the people who are responsible for the safety of a technology: scientists and industry, politicians who decide about regulations, even the media who contribute their watchdog approach to the societal monitoring of risks. Government, industry and the European Commission, important actors in the development and regulation of biotechnology, have the lowest trust ratings[5] of all institutions and groups. NGOs, the medical

[4] The conceptual distinction between 'trust' and 'credibility' is made in different ways in the literature on trust. Briefly, we define (general) trust as the default expectation of good performance in the actions or a positive outcome from a decision by an (individual or collective) actor without sufficient reasons to justify that expectation. We define credibility as a special case of trust(worthiness), where the expected good performance refers to communication: the quality of information and the honesty and reliability of information sources.

[5] The question in the Eurobarometer 58.0 survey was 'Do you think they [i.e. different people and groups involved in various applications of modern biotechnology and genetic engineering] are doing a good job for society or not doing a good job for society?'. This is not exactly a question about trust, since trust refers to the *expectation* of good performance in the future or in unobservable situations. However, the observation of previous performance is probably an important indicator for people in assigning trust. We may hence use the answers to the Eurobarometer question on 'performance' as a proxy for perceived trustworthiness.

**Table 3.3.** Credibility of sources of information about 'modern biotechnology' in Germany and correlations with attitudes towards genetically modified food (from Eurobarometer 58.0, 2002).

| | 'Trust to tell the truth'[a] (%) | $r_{pbis}$[b] |
|---|---|---|
| Consumer organizations | 57.6 | 0.05 |
| Environmental organizations | 51.9 | −0.14** |
| Animal welfare organizations | 26.5 | −0.12** |
| The medical profession | 49.9 | 0.15** |
| Farmers' organizations | 7.8 | 0.06 |
| Religious organizations | 8.0 | −0.14** |
| National government bodies | 10.2 | 0.14** |
| International institutions (not companies) | 19.9 | 0.15** |
| A particular industry | 4.5 | 0.10** |
| Universities | 35.6 | 0.14** |
| Political parties | 2.8 | −0.01 |
| Television and newspapers | 27.0 | 0.10** |

[a] Question: 'Now I would like to ask you which of the following sources of information, if any, you trust to tell you the truth about modern biotechnology' (selection from presented cards, multiple choices possible).
[b] Point biserial correlations between (dichotomously coded) credibility of information sources and the attitude towards GM food (correlations only calculated for a split-half subsample).
** Significant at the 1% level.

profession, academic science and the media are the institutions with the highest perceived trustworthiness (see Table 3.4).

The pattern of correlations between trust and attitudes towards food biotechnology in Tables 3.3 and 3.4 shows that people selectively trust organizations and groups which publicly represent stances that are congruent with the respondents' attitudes. People in favour of food biotechnology, for example, tend to trust industry, science and the medical profession more than do people opposed to biotechnology; people with a critical attitude trust environmental organizations more than people with a positive attitude towards biotechnology. This pattern of selectively ascribing trust and credibility also exists in Europe in general (Hornig Priest *et al.*, 2003) and was earlier found in German public opinion towards nuclear power (Peters, 1992). Actually, it is to be expected that this pattern is quite universal in controversies.

The authors of most studies discussing correlations between trust/credibility and attitudes explicitly or implicitly interpret these correlations as indicators of a (causal) impact of trust/credibility on attitudes. As discussed above, there are good theoretical reasons to

**Table 3.4.** Trust in social institutions and groups 'involved in various applications of modern biotechnology and genetic engineering' in Germany and correlations with attitudes towards genetically modified food (from Eurobarometer 58.0, 2002).

| | 'Doing a good job'[a] (%) | $r_{pbis}$[b] |
|---|---|---|
| Newspapers and magazines reporting on biotechnology | 64.4 | 0.11** |
| Industry developing new products with biotechnology | 41.3 | 0.42** |
| University scientists doing research in biotechnology | 66.4 | 0.29** |
| Consumer organizations checking products of biotechnology | 74.2 | 0.03 |
| Environmental groups campaigning against biotechnology | 62.4 | −0.26** |
| Our government making regulations on biotechnology | 43.5 | 0.16** |
| Shops making sure our food is safe | 54.8 | 0.14** |
| Farmers deciding which types of crops to grow | 57.4 | 0.13** |
| Scientists in industry doing research in biotechnology | 52.8 | 0.38** |
| Medical doctors keeping an eye on the health implications of biotechnology | 75.9 | 0.14** |
| Organizations of patients or their relatives looking after patients' interests | 70.8 | −0.01 |
| The European Commission making laws on biotechnology for all European countries | 44.0 | 0.21** |

[a] Question: 'Do you think they [the mentioned groups involved in various applications of modern biotechnology and genetic engineering] are doing a good job for society or not doing a good job for society?'
[b] Point biserial correlations between (dichotomously coded) trust in the mentioned groups and the attitude towards GM food (correlations only calculated for a split-half subsample).
** Significant at the 1% level.

expect such a direction of influence between trust and attitudes. However, the observed correlations do not prove that impact. The correlations could also be caused by a reverse causal relationship or a more complex linking of trust and attitudes.

In fact, two well-known psychological theories would predict an impact of pre-attitudes on credibility and trust: (i) the theory of cognitive dissonance (Festinger, 1957), i.e. the selective perception and acceptance of information that fits pre-existing attitudes; and (ii) the principle of congruity (Osgood and Tannenbaum, 1955), i.e. the tendency to evaluate semantically related concepts (i.e. attitudes and trust in actors related to the attitude object) in a congruent way and to modify one or more of these concepts if they are incongruent.

These effects can explain the intuitively plausible tendency of people to like/trust/believe people and institutions that have beliefs and attitudes similar to their own, and to dislike/distrust/disbelieve those people and institutions acting according to beliefs and attitudes with which they themselves disagree. A recent study by Frewer *et al.*

(2003) indeed concluded that existing attitudes towards biotechnology determine trust rather than trust determining attitudes.

We find it most plausible that the observed correlations between issue-specific trust and attitudes are produced by a combination of impacts in both directions. Früh and Schönbach (1982) called such bi-directional relationships 'transactional' rather than causal. Trust and attitudes – and relevant beliefs about advantages and disadvantages – form an *attitude syndrome*, i.e. a network of semantically related concepts that support each other and vary jointly. The attitude syndrome has to be considered as a meta-variable. The variables that are part of the syndrome are endogenous variables – linked by two-way relationships and feedback loops. In particular, this means that trust and credibility cannot explain attitudes.

According to this model, a trust rating of an institution or group involved in biotechnology is just another measurement of the attitude syndrome – the correlations between issue-specific trust and attitudes thus being trivial. In order to find true (i.e. non-trivial) predictors for attitudes one has to look for exogenous variables, i.e. variables that are truly independent and not themselves affected by attitudes towards biotechnology.[6]

## Attitude Formation as Cultural Anchoring

### Analytical perspective

Since the German population has no concrete experience with food biotechnology, we expect attitudes to be the result of symbolic attributions rather than the result of balancing concrete costs and benefits. A resource for symbolic attributions is culture. According to the toolbox model of culture (Swidler, 1986), the respective culture provides tools to make sense of new events, objects or problems.

We consider this cultural sense-making of food biotechnology as a specific form of 'framing', as conceptualized by Gamson and Modigliani (1989). It is based on contextualization (i.e. treating food

[6] The argument of a transactional relationship between trust and attitudes is only valid for issue-specific trust, i.e. for trust in institutions with a particular relevance for biotechnology (i.e. the biotech. industry) or when the survey question explicitly refers to biotechnology as the field for which the trust ratings are requested. Trust may also be considered and measured issue-independently as part of the political culture. If conceptualized in this latter way trust may well be an endogenous variable and a possible predictor of attitudes towards biotechnology (Lang *et al.*, 2005).

biotechnology as an issue of risk, economy, food quality, ethics, etc.) and – more specifically – on the instantiation of existing cultural schemes (i.e. making food biotechnology a specific case of a more general semantic pattern, such as the Frankenstein myth).

Figure 3.5 gives an example of cultural anchoring (framing) of food biotechnology by visual communication. This image accompanied a media story on the political controversy about food biotechnology and was aimed at summarizing some of the concerns regarding food biotechnology (which were not addressed in the article itself).

The picture refers to three cultural elements: (i) the kind of vegetable shown in the picture and its perfect shape recalls the proverbial 'Dutch tomato', which in German popular culture stands for fruit and vegetables that look very appetizing but taste like water, thus deceiving consumers about the low food quality; (ii) the monstrous size of the tomatoes frightens and furthermore reminds one of the excessive and senseless agricultural productivity, a major problem in the European Union in recent decades; and (iii) the farmers dressed in protective suits trigger an association with danger, implying that the genetically modified foods pose a health threat.

The term 'Frankenfood' for genetically modified food is another example of cultural anchoring. It relates food biotechnology to the Frankenstein myth, which serves as an interpretation folio. Implied is the following: with genetically modified food, science steps outside

**Fig. 3.5.** Cultural anchoring of GM food by an image (from *Die Zeit*, 9 October 2003).

natural and moral boundaries. It develops products with unanticipated characteristics, which then turn out to be unmanageable hazards.

The idea that the cultural anchoring of food biotechnology determines the resulting attitudes is only a framework. What links are actually made in the debate by the protagonists of both sides (and the media) and which of these links members of the public individually select as relevant are highly contingent and thus an empirical question.

In the following, we describe some cultural contexts relevant for the sense-making of food biotechnology in Germany – without any claim of exhaustiveness. The discussion of these contexts is partly based on a non-systematic observation of the issues in German media coverage of food biotechnology, a reception study in which the thoughts of recipients stimulated by media stories on biotechnology were recorded and analysed (Peters, 2000b), and the analysis of correlations between attitudes towards food biotechnology and other scales in the available public opinion surveys.

## Relevant Cultural Contexts

### Innovation, risk and controversies on technologies

Several authors have argued that, in an international comparison, the German population is particularly negative with respect to innovative technologies (e.g. Noelle-Neumann and Hansen, 1988); others, however, have presented evidence that Germany is not a special case in that respect (e.g. Kistler, 1991). The recent Eurobarometer 2005 survey on public attitudes towards science and technology suggests that, within the European context, Germany is not particularly negative towards science and technology in general. Regarding the item 'Science and technology make our lives healthier, easier and more comfortable', Germany is actually among the most optimistic European countries (European Commission, 2005).

The analysis of the Eurobarometer 58.0 data shows moderate but significant statistical associations between attitudes towards food biotechnology and attitudes towards such diverse technologies as nuclear energy, space exploration, nanotechnology, mobile phones, computers and information technologies and the Internet. These associations suggest that there is, indeed, a general dimension in the perception of innovative technologies and/or a spill-over effect from the evaluation of one technology to the evaluation of another technology, perhaps produced by the similarity of issue cultures and the generalized relevance of individual references to that culture.

Hazardous technologies have provoked major political controversies in Germany. The nuclear power controversy in Germany has lasted for three decades and – as described above – prepared the ground for subsequent controversies about other technologies such as genetic engineering. Criticism of modern technologies is mostly expressed in the semantic and social structures developed in the nuclear power controversy. These include a resourceful environmental movement, powerful NGOs such as Greenpeace, the establishment of research institutes producing 'counter-expertise' and, finally, the establishment of a new political party, the Green Party.

## Control, trust, power, self-confidence and self-determination

While Germany conforms to the European average regarding beliefs about and attitudes towards most subjects, it differs from most other European countries (and the USA) in one respect: national pride or patriotism. Compared to most other Western countries, Germany is quite low in several dimensions of national pride – probably a lasting effect of the retrospective horror engendered by the Nazi regime, the genocide of the Jews and responsibility for the outbreak of World War II (Almond and Verba, 1963).

Smith and Kim (2006), using data from the 2003/2004 National Identity Study of the International Social Survey Program (ISSP), placed West Germany and East Germany at rankings 17 and 19, respectively, of a group of 21 nations with respect to *general national pride* (the USA, Australia and Austria occupying rankings 1, 2 and 3, respectively).

This important feature of German political culture probably causes distrust in the ability of political and societal institutions to keep 'dangerous powers' under control. This cultural belief is obviously generalized to the societal management of hazardous technologies such as nuclear power, stem cell research and biotechnology. The 'Frankenfood' label summarizes these concerns which – as the widespread international use of this term indicates – are by no means limited to Germany, but are perhaps particularly strong there.

A recent study by the German Office of Technology Assessment (TAB) concluded that general scepticism regarding the controllability of technological developments and the opportunity for citizens to participate in the shaping of technology is widespread in Germany (Hennen, 2002). An analysis based on the data of the 1997 Stuttgart survey also showed that attitudes towards biotechnology were, indeed, strongly associated with perceptions of controllability and trust in experts and institutions who play a role in the control of biotechnology (Peters, 1999).

The controllability issue concerns science and industry in particular. These institutions are perceived as having most influence on the use of genetic engineering, while political institutions and agencies are considered even less influential than the media and NGOs (Peters, 1999, p. 230). While academic science is generally trusted to a certain extent, trust in industry is extremely low (see Table 3.3). The low trust ratings are only part of a more general negative image of industry in Germany, in particular of multinational companies. Only 12% of the German population agreed with the statement 'What is good for business is good for the citizens', but 77% agreed that 'Multinational companies are too powerful nowadays'.

The correlations between evaluative beliefs about industry and economy and attitudes towards food biotechnology indicate the relevance of cultural images of industry and economy for these attitudes (see Table 3.5). Critical attitudes towards food biotechnology are thus part of a more general pattern of criticism of industry and the role of the economy in society. Positive attitudes also reflect a more positive assessment of the relevance of the economy for individual and national well-being.

The perception of being at the mercy of economic forces with a generally bad reputation may also imply a critical attitude towards the role of powerful, US-based multinational companies (e.g. Monsanto) and the role of the US Government in putting pressure on Europe to allow the import of genetically modified crops. The German public perceives this as US imperialism in international trade and an attempt to undermine the self-determination of countries, farmers and consumers.

**Table 3.5.** Beliefs about business and the economy in Germany and their correlations with attitudes towards genetically modified food (from Eurobarometer 58.0, 2002).

| | 'Tend to agree' (%) | $r_{pbis}$ [a] |
|---|---|---|
| Economic growth brings better quality of life | 65.0 | 0.22** |
| What is good for business is good for the citizens | 12.1 | 0.13** |
| Multinational companies are too powerful nowadays | 77.2 | −0.10** |
| Private enterprise is the best way to solve Germany's problems | 41.8 | 0.19** |

[a] Point biserial correlations between (dichotomously coded) opinions on industry/business and the attitude towards GM food (correlations only calculated for a split-half subsample).
** Significant at the 1% level.

## Agriculture and nutrition

One of the 'promises' of food biotechnology is higher productivity, e.g. the cheaper production of more food per acre with less fertilizer and pesticides. In the German media this has led to ironic word creations, such as 'turbo carp' for genetically modified carp that grow particularly fast and become particularly large, and 'turbo cow' as a label for the use of the growth hormone rBST, produced by genetic engineering to increase milk production in cows (see also Fig. 3.5).

As already briefly mentioned in the discussion of differences between the eastern and western parts of Germany, the promise of greater productivity held out by agricultural biotechnology in Germany is not perceived as a benefit but rather as a problem.

One of the main and continuing public debates concerning the EU was, and is, the debate about agricultural policy. Influenced by the experience of food scarcity and hunger during and after World War II, one of the major goals of the then European Economic Community (EEC) was to secure sufficient food supply for the member states. Along the way, this goal somehow gained a momentum of its own and led to overproduction, overflowing storerooms and the actual destruction of food – a nasty image for people who, in their youth, had learnt that one doesn't play around with food but shows respect for it.

As a side effect, the implementation of the goal of secured food supply in the EU produced a system of regulation in the agricultural and food sector that led to problematic economic incentives for food producers. The EU is still struggling to deregulate that field. The 'promise' of high productivity by the use of agricultural biotechnology from the (West) German perspective conjures up exactly these past images of bulldozers burying mountains of perfect apples.

Furthermore, the industrialization of agriculture carries images of mass production, standardization of products, reduction of variety and regional specificity. This is in line with the EU regulation of 'food quality', which introduced norms for the size and shape of vegetables and fruit, for health and safety aspects (e.g. pesticide levels) but not for flavour. In Germany, the proverbial 'Dutch tomato' serves as the stereotype for the kind of fruit and vegetables resulting from this misguided EU agricultural policy. The references between genetically modified food and the 'Dutch tomato' indeed show that the negatively framed issue culture of industrialized food production is associated with food biotechnology and contributes to its negative evaluation. Prominent food critics (e.g. Wolfram Siebeck in the influential weekly newspaper *Die Zeit*) have argued against genetically modified food, pointing to the decrease in food quality to be expected from applying agricultural biotechnology.

Consumers in Germany thus do not perceive the benefits of food biotechnology in terms of more economical food production, nor do they expect an improvement in quality. Scholderer (2000) hence concludes, on the basis of a survey among 2000 consumers from Denmark, Italy and Germany, that the lack of perceived benefits of food biotechnology, from the consumers' point of view, is the main direct determinant for negative attitudes in each of these countries.

Of course, the German public is aware that in many parts of the world it is not the flavour of food but its availability and nutritional value that are the problem. While, from a technical point of view, agricultural biotechnology could contribute to a solution of the global hunger problem, this argument is frequently rejected by critics of biotechnology. Critics, in turn, argue that introducing agricultural biotechnology in Third World countries would cause a dependency of local farmers on multinational companies, thus increasing the hunger problem, and that seed producers will invest in the development of new breeds only when they see a profitable market – and that would be in the industrialized countries, not in the Third World countries.

Furthermore, critics see the world hunger problem as being caused by unfair international trade relations between developed and less developed countries, by political mismanagement in these countries, as a late consequence of colonialism or as a distribution problem of the world's food resources. They doubt that (bio)technical fixes for political problems would work.

Despite these arguments, the German public acknowledges the potential of agricultural biotechnology for Third World countries. 'Genetically modified food will be useful in the fight against world hunger' is the only item stating a benefit of food biotechnology that gains substantially more support than rejection (see item 2 in Table 3.1). Although the semantic links of food biotechnology to the world hunger issue are not strong at the moment, the world development issue provides a context in which a more positive evaluation of agricultural biotechnology – at least for Third-World countries – seems possible.

## Concepts of nature

Culturally embedded concepts of nature are of high relevance for the assessment of food biotechnology, i.e. inter-individual variations of these concepts explain part of the variance in attitudes towards biotechnology (see Table 3.6). Two studies found that a stronger appreciation of nature in Germany than in the USA explained a considerable part of the country difference in attitudes towards biotechnology between these two countries (Lang *et al.*, 2005; Sawicka, 2005).

The leading ideals and approaches that shape people's expectations of food today are naturalness, pureness and authenticity (Spelsberg, 2002). This preference is well recognized by the marketing departments of food producers. Their advertisements present rural and small-scale production, where the farmer's wife herself stirs the freshly picked strawberries into the yogurt. Milking machines or the mechanical addition of liquid eggs or artificial flavours on production lines is left out in favour of 'happy cows' and traditional, small-scale manufacturing processes.

Ironically, it is the food producers and their marketing strategies themselves that support the consumers' values and expectations that give rise to a critical assessment of food biotechnology. The marketing departments of food companies in Germany actually work for Greenpeace and cultivate values which critics of biotechnology can turn against food biotechnology. The demand for natural food aroused in this way and the turning of food technology into a taboo subject inevitably leads to disappointment when the harsh reality of industrialized food production – often in food scandals – becomes obvious for the general public (Spelsberg, 2002).

Apart from the fact that German consumers use the 'naturalness' of food (i.e. the absence of non-transparent technical processes in its production) as a genuine quality criterion for food, two other aspects of concepts of nature are potentially relevant for making sense of food biotechnology. Large sections of the German public view genetic engineering as an intervention in the essence of nature by fiddling

**Table 3.6.** Beliefs about nature in Germany and their correlations with attitudes towards genetically modified food (from Eurobarometer 58.0, 2002).

| | 'Tend to agree' (%) | $r_{pbis}$[a] |
|---|---|---|
| Modern technology has upset the balance of nature | 73.0 | −0.29** |
| Exploiting nature is unavoidable if humankind is to progress | 23.5 | 0.11** |
| Nature is fragile and easily damaged by human actions | 85.9 | −0.05 |
| Nature can withstand human actions | 17.5 | 0.16** |

[a] Point biserial correlations between (dichotomously coded) opinions on nature and the attitude towards GM food (correlations only calculated for a split-half subsample).
** Significant at the 1% level.

around with the blueprints of biological organisms – with or without a religious connotation (see item 7, Table 3.1).

Finally, agricultural biotechnology is seen as having potentially adverse environmental impacts thus negatively affecting 'natural' organisms or ecosystems. This places the issue of agricultural biotechnology at the centre of attention for the German environmental movement and the culturally deep-rooted concern about the environment. Scholderer (2000) presents evidence that consumer perceptions of the risks of food biotechnology are indeed strongly influenced by attitudes towards nature.

## Mass Media and their Impact on Public Opinion

### Media coverage

Several content analyses of the German media coverage of biotechnology give an overview of, among other aspects, the intensity and evaluative tone of this coverage, the sources used by the journalists and the 'frames' in which the information was packed (e.g. Kepplinger *et al.*, 1991; Ruhrmann *et al.*, 1992; Hampel *et al.*, 1998; Aretz, 1999; Kohring *et al.*, 1999; Bauer *et al.*, 2001; Gutteling *et al.*, 2002; Kohring and Matthes, 2002). A comprehensive description of the results of these studies is far beyond the scope of this chapter. In the following, we thus focus on only a few aspects, namely on the intensity and positive–negative valence of the coverage and on the actors mentioned and sources used by journalists.

#### *Intensity*

A longitudinal content analysis (1973–1999) of two German newspapers, the conservative elite newspaper *Frankfurter Allgemeine Zeitung* and the equally respected but more 'leftist' weekly newsmagazine *Der Spiegel*, conducted as part of a European cooperation to study media coverage of biotechnology in an international context, showed a strong increase in the intensity of coverage. The number of articles published on biotechnology increased slowly but steadily from 1973 to 1989, decreased temporarily around 1991, rose slowly again until 1996 and then showed a rapid increase up to the end of the analysis period in 1999 (Hampel *et al.*, 2001).

The relative proportion of articles dealing with agricultural and food biotechnology decreased from 1992–1996 to 1997–1999 (Kohring and Matthes, 2002). This is probably not the result of decreasing interest in food biotechnology but of an increasing media interest in biomedical applications. A comparative content analysis by Kohring and Görke (2000) showed that, in the period from 1991 to 1996, agricultural/environmental applications of biotechnology had a greater share of the total coverage of biotechnology in Germany than in the USA, France or the UK.

### *Valence*

Perhaps surprisingly for most observers, content analyses of media coverage measuring the evaluation of biotechnology in different ways concluded that the German media portrayed biotechnology in total as predominantly positive – or at least neutral, emphasizing benefits more than risks, with variations between different media and newspaper sections (Kepplinger *et al.*, 1991; Merten, 1999; Kohring and Görke, 2000; Gutteling *et al.*, 2002). According to Hampel *et al.* (2001), the relative frequency of risks mentioned in the coverage increased somewhat and the frequency of reported benefits decreased between 1992 and 1999, although benefits by far outweighed risks during the whole period.

Unfortunately, most publications on the media coverage of biotechnology do not clearly distinguish between the evaluation of different applications. As we have seen above, public attitudes differ strongly between different applications of biotechnology. We should expect that the media would show similar differences in the coverage of these applications. The content analysis of German TV reporting of biotechnology by Merten (1999), differentiating between applications of biotechnology, indeed reveals that the average evaluation of agricultural/food biotechnology is clearly more negative than the average evaluation of biomedical applications.

Bauer (2005a) argues that, in Europe, the press has increasingly differentiated between the evaluation of red (biomedical) and green (agricultural/food) applications of biotechnology (usually presenting the latter less positively). He found, however, that Germany (in addition to Austria and Sweden) was an exception to the European pattern. In Germany, the 'red/green press contrast' decreased between 1992–1996 and 1997–1999. In the latter period, he found about the same average evaluation of both red and green biotechnology in Germany. Bauer (2005a) explains these findings by the instigation of the debate on stem cell and embryo research, mostly reported critically in Germany and thus lowering the average evaluation of the otherwise more positively reportage on red biotechnology.

One methodological problem is the definition and measurement of media valence. Most content analyses refer to journalistic norms of 'balanced reporting' as indicators of neutrality or bias. Coverage of controversial issues is considered neutral, for example, if the journalists do not take a particular stance themselves, if the coverage contains the same proportion of positive and negative statements and if the same number of sources in favour of or opposed to the issue are quoted.

We think that this takes only the superficial meaning of media stories into account. A quasi-experimental reception study showed that media stories on biotechnology of a similar valence (according to the above-mentioned criteria) can evoke extremely different – even counter-intuitive – evaluative responses in the audience (Peters, 2000b). Our suspicion is that the most relevant evaluative meanings are not in the text itself but outside it, and that subtle references in the article to existing beliefs, values and experiences activate that meaning. We would thus expect that the negative framing of food biotechnology in German culture is also implicit in the media coverage.

The available content analyses for Germany sketch the picture of a positive to neutral coverage of biotechnology over a long time period, with agricultural/food applications being clearly portrayed more negatively than biomedical applications, except in the more recent phase during which research on the human genome, cloning and embryonic stem cells is also critically reported. Kohring and Matthes (2002) have found that, from 1992 to 1999, the media reporting of agricultural/food biotechnology changed from a discussion of 'pros and cons' to coverage of regulatory processes.

We would argue that even a matter-of-fact discussion of regulatory processes carries a negative slant: the aim of regulation is seen as protecting consumers (implying that there are risks), and the labelling of GM food is discussed in Germany as a means of allowing consumers to avoid genetically modified food (again implying that GM food is something negative) rather than as an issue of consumer choice.

### *Actors and sources*

The available content analyses agree that scientific actors and sources are most important in the coverage of biotechnology in general, more important than sources from politics, business and interest groups (e.g. Kohring and Görke, 2000; Hampel *et al.*, 2001). A survey of German journalists by Schenk and Sonje (2000) confirmed these findings and reveals that journalists tend to contact predominantly scientific sources 'directly involved in the development of genetic engineering methods and applications and who have a positive attitude towards this technology'.

Kepplinger *et al.* (1991) showed that scientists are not only quoted with respect to their factual knowledge but are also the main source of evaluations in the media coverage of biotechnology (apart from the journalists themselves). Again, the analyses do not adequately distinguish between different fields of biotechnology. Food biotechnology is a particularly controversial and politicized field and hence we would expect a much higher proportion of sources and actors mentioned from politics (national, EU and international), business and NGOs than in the biomedical fields.

### *Media effects*

Because the media are so deeply embedded in the structure of society and in the processes of cultural reproduction, it is quite safe to say that they are an important part of a system in which a public discourse among social actors evolves around issues and in which individuals are linked to the public sphere by two processes: the reception of media coverage by the audience and the anticipation of the audience's relevance structure by journalists. Several media effects are well accepted, such as agenda-setting, diffusion of knowledge and cultivation, i.e. the shaping of the recipients' subjective realities by characteristics of the media coverage.

While it is quite obvious that the media are an important means for average media users to generate opinions on and attitudes to subjects of which they have no direct experience, it is surprisingly difficult to actually prove these effects. Two reasons why it is difficult to conceptualize and measure the persuasive influences of media coverage outside laboratory settings are:

- the ubiquity of media in our media society, shaping a 'symbolic environment' and making it more and more pointless to look at the impact of individual media stories, papers or programmes as 'stimuli'; and
- the nonlinear and non-causal relationships in which media coverage, public relations and policy strategies of social actors, individual attitudes and popular culture are linked.

Researchers in Germany (and Europe) studying media impacts on public attitudes to technologies have been trying to demonstrate media effects. Two studies on the effects of German media coverage of biotechnology on the public have been published: (i) a large-scale comparison of survey data on attitudes towards biotechnology in 12 European countries (including Germany), with content analysis data on the media coverage in those same countries (for an overview see Bauer, 2005b); and (ii) a quasi-experimental reception study using media stories on different applications of biotechnology as stimuli (Peters, 2000b).

Members of the European media analysis team (coordinated by Martin Bauer) tested several media effect hypotheses, namely the intensity of coverage hypothesis of Allan Mazur (Gutteling, 2005), the knowledge gap hypothesis (Bonfadelli, 2005) and the cultivation theory (Bauer, 2005a). Peters used a constructivist paradigm, looking at inter-individual differences in sense-making of the same media stories and differential persuasive impacts, dependent on the sense-making process rather than story content.

Gutteling (2005), comparing readers of elite newspapers, readers of popular newspapers and non-readers and public attitudes between countries with different intensity of coverage, did not find any consistent relationship between exposure to media coverage and attitudes. This finding for Europe in general is also true of Germany in particular. Bonfadelli (2005) did not find the expected effect of increased media reportage on knowledge gaps. Contrary to the hypothesis, in almost all countries the knowledge gap, i.e. the difference in knowledge level between the highest and lowest educational groups, decreased between 1996 and 1999, while the amount of coverage increased. The only country showing no such decrease in the knowledge gap was Germany which is, however, one of the countries with the highest increase in coverage of biotechnology.

Cultivation effects concerning the 'red–green distinction', i.e. the difference in evaluation of 'red' and 'green' biotechnology, were identified by Bauer (2005a). Readers of elite newspapers followed changes in the elite newspaper coverage over time more closely than did non-readers. However, in other respects media coverage and the audience's perceptions appeared rather unrelated. Overall, the analysis by the Bauer team did not convincingly explain differences between European countries and the dynamic of public attitudes by media effects.

The reception study of Peters (2000b), who presented four articles on biotechnology to 338 test readers and three TV films to 51 test viewers recording the thoughts of these test recipients during the reception of the media stories, led to several results that did indeed question the assumption of a strong coupling of media content and audience attitudes. The cognitive responses to the media stories varied considerably between the test recipients; the main predictor of the valence of the responses was the test recipients' pre-attitudes.

This leads to the hypothesis that media influences are particularly strong in the first phases of an issue, when no firm attitudes exist or after critical events, when the pre-attitudes are severely shaken. In their cognitive responses, test recipients linked the media content to issues, personal experience and symbols of popular culture that were not mentioned in the media stories. Cognitive responses were sometimes incongruent with the arguments presented in the media story – the article categorized by a group of reviewers as most positive actually evoked the most negative responses.

Finally, Peters found a negativity effect, i.e. a tendency of people to generate more negative than positive thoughts (even when controlling for the slant of the media story and the pre-attitude of the recipient). Attitude change could be linked to the valence of responses evoked, i.e. the kind of sense-making modifies the persuasive impact of an article in a systematic way.

Hardly anybody would deny the relevance of the media to the formation of public attitudes towards biotechnology. However, the relationship between media content and effects seems to be contingent – a kind of weak coupling. The media provide material for individual sense-making that is selectively used and differently processed. Furthermore, we obviously don't adequately understand the persuasive cues in media stories evoking positive or negative cognitive responses leading to attitude change. There are certainly examples of very direct media effects under specific circumstances. The biotechnology issue, with its rich semantic links to culture and other issues, is obviously not an example of direct media effects.

## Outlook

It is hard to predict the further development of public attitudes to food biotechnology in Germany. At the moment, attitudes towards GM food are pretty negative and they are deeply rooted in German culture; it is thus not very likely that they will change quickly towards the positive. A change of public attitudes, according to our 'cultural anchoring' model, would require the strengthening or weakening of links between food biotechnology and cultural elements, the evolution of new links or cultural change.

Currently, a change in political culture is taking place in Germany that may have consequences for attitudes towards food biotechnology. The issue of economic competitiveness and the necessity of taking risks is becoming more pertinent. In a provocative public speech in 1997, the former German President, Roman Herzog, pointed to the necessity of German society being more optimistic and prepared to take risks.

The government of former Chancellor Schröder started a policy of deregulation of the job market and the social security system in Germany. This policy is currently being maintained by the present government under Chancellor Angela Merkel. While many citizens are not very happy about this reduction of the welfare state, the necessity for such a change is also widely recognized. Such a changing political culture may have consequences for the public assessment of technologies. Even the abandonment of nuclear power is currently being questioned by some politicians. This initiative may not be successful, but it shows that some politicians assume that this is now the right time to make such a move. With respect to GM food, however, the most recent Eurobarometer on biotechnology shows a sharp decline in support between 2002 and 2005 for Germany (Gaskell *et al.*, 2006).

In our discussion of the agricultural context of attitudes towards GM food, we briefly referred to the global hunger problem. We mentioned indications that the possible contribution of biotechnology to the solution of this problem is a plausible argument for many people in Germany. The critics of food biotechnology at the moment prevent a public debate on that argument and have developed a counter argumentation. However, a credible actor might be able to put that issue on the public agenda and gain public support for applications of food biotechnology that could plausibly support Third World countries (e.g. Golden Rice).

Currently, the attitudes of the population regarding food biotechnology as measured by surveys are political attitudes, i.e. they represent judgements of the respondents *as citizens* and not judgements *as consumers*. This political attitude prevents the market entry of GM food to date. It remains valid only, however, as long as GM products do not have significant advantages for consumers. It is hardly possible to infer from the critical public attitudes in Germany that GM food would be rejected in shops if it had advantages for consumers with respect to price, appearance, flavour, shelf life or nutritional value. If GM products with a considerable benefit for consumers were available in the future, increasing consumer acceptance would have consequences for political attitudes. If that situation were to occur, it would also lead to a delegitimization of organized protests against food biotechnology.

## References

Almond, G.A. and Verba, S. (1963) *The Civic Culture: Political Attitudes and Democracy in Five Nations.* Princeton University Press, Princeton, New Jersey.

Aretz, H.-J. (1999) *Kommunikation ohne Verständigung: Das Scheitern des öffentlichen Diskurses über die Gentechnik und die Krise des Technokorporatismus in der Bundesrepublik Deutschland.* Peter Lang, Frankfurt M., Germany.

Aretz, H.-J. (2000) Institutionelle Kontexte technologischer Innovationen: Die Gentechnikdebatte in Deutschland und den USA. *Soziale Welt* 51, 401–416.

Bauer, M.W. (2005a) Distinguishing red and green biotechnology: cultivation effects of the elite press. *International Journal of Public Opinion Research* 17, 63–89.

Bauer, M.W. (2005b) Public perceptions and mass media in the biotechnology controversy. *International Journal of Public Opinion Research* 17, 5–22.

Bauer, M.W., Kohring, M., Allansdottir, A. and Gutteling, J. (2001) The dramatisation of biotechnology in elite mass media. In: Gaskell, G. and Bauer, M.W. (eds) *Biotechnology 1996–2000: the years of controversy*. Science Museum, London, pp. 35–52.

Bonfadelli, H. (2005) Mass media and biotechnology: knowledge gaps within and between European countries. *International Journal of Public Opinion Research* 17, 42–62.

European Commission (2005) *Europeans, Science and Technology. Eurobarometer 2005*. Office for Official Publications of the European Communities, Luxembourg.

European Federation of Biotechnology (1994) *Biotechnology in Foods and Drinks*. Briefing paper 2, EFB Task Group on Public Perceptions of Biotechnology, Barcelona, Spain.

Festinger, L. (1957) *A Theory of Cognitive Dissonance*. Stanford University Press, Stanford, California.

Frewer, L.J., Scholderer, J. and Bredahl, L. (2003) Communicating about the risks and benefits of genetically modified foods: the mediating role of trust. *Risk Analysis* 23, 1117–1133.

Früh, W. and Schönbach, K. (1982) Der dynamisch-transaktionale Ansatz: Ein neues Paradigma der Medienwirkungen. *Publizistik* 27, 74–88.

Gamson, W.A. and Modigliani, A. (1989) Media discourse and public opinion on nuclear power: a constructionist approach. *American Journal of Sociology* 95, 1–37.

Gaskell, G., Allansdottir, A., Allum, N., Corchero, C., Fischler, C., Hampel, J., Jackson, J., Kronberger, N., Mejlgaard, N., Revuelta, G., Schreiner, C., Stares, S., Torgersen, H. and Wagner, W. (2006) *Europeans and Biotechnology in 2005. Patterns and Trends. Final Report on Eurobarometer 64.3*. A report to the EC Directorate-General for Research, Brussels.

Gaskell, G., Bauer, M.W., Durant, J. and Allum, N.C. (1999) World apart? The reception of genetically modified foods in Europe and the US. *Science* 285, 384–387.

Gaskell, G., Allum, N. and Stares, S. (2003) *Europeans and Biotechnology in 2002. Eurobarometer 58.0*. A report to the EC Directorate General for Research, Brussels.

Giddens, A. (1991) *The Consequences of Modernity*. Polity Press, Cambridge, UK.

Gill, B. (1991) *Gentechnik ohne Politik. Wie die Brisanz der Synthetischen Biologie von wissenschaftlichen Institutionen, Ethik- und anderen Kommissionen systematisch verdrängt wird*. Campus, Frankfurt M., Germany.

Gill, B. (2003) *Streitfall Natur. Weltbilder in Technik- und Umweltkonflikten*. Westdeutscher Verlag, Wiesbaden.

Greenpeace (2005) *Jahresrückblick 2004. Kampagnen + Strukturen + Bilanz + Erträge und Aufwendungen* (http://www.greenpeace.de/fileadmin/gpd/user_upload/allg_inhalte/JRB2004.pdf, accessed 19 January 2006).

Gutteling, J.M. (2005) Mazur's hypothesis on technology controversy and media. *International Journal of Public Opinion Research* 17, 23–41.

Gutteling, J.M., Olofsson, A., Fjaestad, B., Kohring, M., Görke, A., Bauer, M.W., Rusanen, T. and Rusanen, M. (2002) Media coverage 1973–1996: trends and dynamics. In: Bauer, M.W. and Gaskell, G. (eds) *Biotechnology – the Making of a Global Controversy*. Cambridge University Press, Cambridge, UK, pp. 95–128.

Hallman, W.K., Adelaja, A.O., Schilling, B.J. and Lang, J.T. (2002) *Public Perceptions of Genetically Modified Foods. Americans Know not what they Eat.* Food Policy Institute, Rutgers University, New Brunswick, New Jersey.

Hallman, W.K., Hebden, W.C., Cuite, C.L., Aquino, H.L. and Lang, J.T. (2004) *Americans and GM Food: Knowledge, Opinion and Interest in 2004.* Food Policy Institute, Rutgers University, New Brunswick, New Jersey.

Hampel, J. (2005) Technik-, Akzeptanz- und Risikodiskurse: Europäisierung der Debatten oder nationale Signaturen? *Technikfolgenabschätzung. Theorie und Praxis* 14, 78–85.

Hampel, J. and Pfennig, U. (1999) Einstellungen zur Gentechnik. In: Hampel, J. and Renn, O. (eds) *Gentechnik in der Öffentlichkeit: Wahrnehmung und Bewertung einer umstrittenen Technologie.* Campus, Frankfurt M., Germany, pp. 28–55.

Hampel, J., Keck, G., Peters, H.P., Pfenning, U., Renn, O., Ruhrmann, G., Schenk, M., Schütz, H., Sonje, D., Stegat, B., Urban, D., Wiedemann, P.M. and Zwick, M.M. (1997) *Einstellungen zur Gentechnik. Tabellenband zum Biotech-Survey des Forschungsverbunds 'Chancen und Risiken der Gentechnik aus der Sicht der Öffentlichkeit'.* Arbeitsbericht Nr. 87. Akademie für Technikfolgenabschätzung in Baden-Württemberg, Stuttgart, Germany.

Hampel, J., Ruhrmann, G., Kohring, M. and Görke, A. (1998) Germany. In: Durant, J., Bauer, M.W. and Gaskell, G. (eds) *Biotechnology in the Public Sphere. A European Sourcebook.* Science Museum, London, pp. 63–76.

Hampel, J., Klinke, A. and Renn, O. (2000a) Beyond 'red' hope and 'green' distrust. Public perception of genetic engineering in Germany. *Notizie di POLITEIA XVI* 60, 68–82.

Hampel, J., Pfenning, U. and Peters, H.P. (2000b) Attitudes towards genetic engineering. *New Genetics and Society* 19, 233–249.

Hampel, J., Pfennig, U., Kohring, M., Görke, A. and Ruhrmann, G. (2001) Biotechnology boom and market failure: two sides of the German coin. In: Gaskell, G. and Bauer, M.W. (eds) *Biotechnology 1996–2000: the Years of Controversy.* Science Museum, London, pp. 191–203.

Hennen, L. (2002) *Technikakzeptanz und Kontroversen über Technik. Positive Veränderung des Meinungsklimas – konstante Einstellungsmuster.* Working report No. 83, Office of Technology Assessment at the German Parliament (TAB), Berlin.

Hornig Priest, S., Bonfadelli, H. and Rusanen, M. (2003) The 'Trust Gap' hypothesis: predicting support for biotechnology across national cultures as a function of trust in actors. *Risk Analysis* 23, 751–766.

Hovland, C.I., Janis, I.L. and Kelley, H.H. (1953) *Communication and Persuasion. Psychological Studies of Opinion Change.* Yale University Press, New Haven, Connecticut.

Inglehart, R. (1977) *The Silent Revolution. Changing Values and Political Styles Among Western Publics.* Princeton University Press, Princeton, New Jersey.

Kepplinger, H.M., Ehmig, S.C. and Ahlheim, C. (1991) *Gentechnik im Widerstreit. Zum Verhältnis von Wissenschaft und Journalismus.* Campus, Frankfurt M., Germany.

Kistler, E. (1991) Eurosklerose, Germanosklerose? – Einstellungen zur Technik im internationalen Vergleich. In: Jaufmann, D. and Kistler, E. (eds) *Einstellungen zum technischen Fortschritt. Technikakzeptanz im nationalen und internationalen Vergleich.* Campus, Frankfurt M., Germany, pp. 53–70.

Kohring, M. and Görke, A. (2000) Genetic engineering in the international media: an analysis of opinion-leading magazines. *New Genetics and Society* 19, 345–363.

Kohring, M. and Matthes, J. (2002) The face(t)s of biotech in the nineties: how the German press framed modern biotechnology. *Public Understanding of Science* 11, 143–154.

Kohring, M., Görke, A. and Ruhrmann, G. (1999) Das Bild der Gentechnik in den internationalen Medien – eine Inhaltsanalyse meinungsführender Zeitschriften. In: Hampel, J. and Renn, O. (eds) *Gentechnik in der Öffentlichkeit: Wahrnehmung und Bewertung einer umstrittenen Technologie.* Campus, Frankfurt M., Germany, pp. 292–316.

Lang, J.T., Peters, H.P. and Sawicka, M. (2005) The impact of socio-cultural factors on attitudes toward genetically modified food: Comparing Germany and the USA. Paper presented at the Annual Meeting of the American Sociological Association, 15 August 2005, Philadelphia, Pennsylvania.

Merten, K. (1999) Die Berichterstattung über Gentechnik in Presse und Fernsehen – eine Inhaltsanalyse. In: Hampel, J. and Renn, O. (eds) *Gentechnik in der Öffentlichkeit: Wahrnehmung und Bewertung einer umstrittenen Technologie.* Campus, Frankfurt M., Germany, pp. 317–339.

Meyer, H. (2003) Vorsorge oder business as usual? Internationale rechtliche Rahmenbedingungen der Grünen Gentechnik. *Politische Ökologie* 81/82, 74–77.

Noelle-Neumann, E. and Hansen, J. (1988) Medienwirkungen und Technikakzeptanz. In: Scharioth, J. and Uhl, H. (eds) *Medien und Technikakzeptanz.* R. Oldenbourg Verlag, Munich, Germany, pp. 33–76.

Osgood, C.E. and Tannenbaum, P.H. (1955) The principle of congruity and the prediction of attitude change. *Psychological Review* 62, 42–55.

Peters, H.P. (1992) The credibility of information sources in West Germany after the Chernobyl disaster. *Public Understanding of Science* 1, 325–343.

Peters, H.P. (1999) Das Bedürfnis nach Kontrolle der Gentechnik und das Vertrauen in wissenschaftliche Experten. In: Hampel, J. and Renn, O. (eds) *Gentechnik in der Öffentlichkeit: Wahrnehmung und Bewertung einer umstrittenen Technologie.* Campus, Frankfurt M., Germany, pp. 225–245.

Peters, H.P. (2000a) From information to attitudes? Thoughts on the relationship between knowledge about science and technology and attitudes toward technologies. In: Dierkes, M. and von Grote, C. (eds) *Between Understanding and Trust. The Public, Science and Technology.* Harwood Academic Publishers, Amsterdam, pp. 265–286.

Peters, H.P. (2000b) The committed are hard to persuade. Recipients' thoughts during exposure to newspaper and TV stories on genetic engineering and their effect on attitudes. *New Genetics and Society* 19, 365–381.

Petty, R.E. and Cacioppo, J.T. (1986) *Communication and Persuasion. Central and Peripheral Routes to Attitude Change*. Springer, New York.

PISA-Konsortium Deutschland (ed.) (2005) *PISA 2003. Der zweite Vergleich der Länder in Deutschland – Was wissen und können Jugendliche?* Waxmann, Munster, Germany.

Ruhrmann, G., Stöckle, T., Krämer, F. and Peters, C. (1992) *Das Bild der 'Biotechnischen Sicherheit' und der 'Genomanalyse' in der deutschen Tagespresse (1988–1990)*. Discussion paper No. 2. Office of Technology Assessment at the German Parliament (TAB), Bonn.

Sawicka, M. (2005) Die Rolle von Naturbildern bei der Meinungsbildung über Grüne Gentechnik – eine deutsch-amerikanische vergleichsstudie. *Umweltpsychologie* 9, 126–145.

Schenk, M. and Sonje, D. (2000) Journalists and genetic engineering. *New Genetics and Society* 19, 331–343.

Scholderer, J. (2000) Kampagnen zur Gentechnik und ihre Wirkung auf den Verbraucher. In: von Schell, T. and Seltz, R. (eds) *Inszenierungen zur Gentechnik*. Westdeutscher Verlag, Opladen, Germany, pp. 214–222.

Schweiger, W. and Brosius, H.-B. (1999) *Von der 'Gentomate' zur Gentechnikakzeptanz. Eine Panelstudie zu Einstellungseffekten eines rollenden Gentechniklabors an Gymnasien*. GSF-Bericht 8/99. GSF-Forschungszentrum, Munich, Germany.

Siegrist, M. (2000) The influence of trust and perceptions of risks and benefits on the acceptance of gene technology. *Risk Analysis* 20, 195–203.

Smith, T.M. and Kim, S. (2006) National pride in comparative perspective: 1995/96 and 2003/04. *International Journal of Public Opinion Research* 18, 127–136.

Spelsberg, G. (2002) Konfliktfeld Grüne Gentechnik. *Akademie-Journal* 1, 32–34.

Swidler, A. (1986) Culture in action. Symbols and strategies. *American Sociological Review* 51, 273–286.

Thielemann, H. (1995) 'Horrorwesen' oder 'kleine Mitarbeiter'? Risikokommunikation am Beispiel der Insulinherstellung mit Hilfe gentechnisch veränderter Bakterien. MSc thesis, Westfälische Willhelms-Universität Münster, Germany.

Torgersen, H., Hampel, J., von Bergmann-Winberg, M.-L., Bridgman, E., Durant, J., Einsiedel, E., Fjaestad, B., Gaskell, G., Grabner, P., Hieber, P., Jelsoe, E., Lassen, J., Marouda-Chatjoulis, A., Nielsen, T.H., Rusanen, T., Sakellaris, G., Seifert, F., Smink, C., Twardowski, T. and Kamara, M.W. (2002) Promise, problems and proxies: twenty-five years of debate and regulation in Europe. In: Bauer, M.W. and Gaskell, G. (eds) *Biotechnology – the Making of a Global Controversy*. Cambridge University Press, Cambridge, UK, pp. 21–94.

van den Daele, W. (1991) Risiko-Kommunikation: Gentechnologie. In: Jungermann, H., Rohrmann, B. and Wiedemann, P.M. (eds) *Risikokontroversen: Konzepte, Konflikte, Kommunikation*. Springer, Berlin.

# 4

# Mass Media and Public Perceptions of Red and Green Biotechnology: a Case Study from Switzerland

H. Bonfadelli,* U. Dahinden** and M. Leonarz***

*Institute of Mass Communication and Media Research (IPMZ), University of Zurich, Switzerland;*
*e-mail: *h.bonfadelli@ipmz.unizh.ch; **u.dahinden@ipmz.unizh.ch; ***martina.leonarz@bluewin.ch*

## Introduction

The chapter describes and analyses the controversy about 'red' and 'green' biotechnology in Switzerland that was strongly influenced by an ongoing intensive public debate about biotechnology because of the specific Swiss political system with its mechanisms of direct democracy. These led, in 1992 and 1998, to two national referenda on biotechnology regulation. An upcoming follow-up initiative, calling for a 5-year biotech. moratorium in agriculture in Switzerland, was accepted by the Swiss public on 27 November 2005.

After a general introduction, the following section will briefly describe biotechnology as a key economic and political issue for Switzerland, giving information concerning the shifting regulation of biotechnology in that country. In the third section, the impact of the political debate on media coverage will be analysed on the basis of a longitudinal content analysis. The existing data will be complemented by a new study dealing especially with aspects of visualization in Swiss television.

Section four will discuss the public frames of red and green biotechnology differently. This analysis is based on qualitative group discussion data. The final part will deal with media effects on public opinion. This analysis is based on quantitative data from three Eurobarometer surveys, carried out in Switzerland in 1997, 2000 and 2002/2003. The data allow comparisons between different applications of biotechnology (red and green) concerning dimensions such as benefits and risks, moral concerns and support. Therefore, it is possible to track the attitudes concerning red and green biotechnology over time and link them to media coverage.

## The Four Phases of Swiss Policy on Biotechnology

The public debate on biotechnology in Switzerland has special features in comparison to other industrial countries: The debate started relatively early, at the end of the 1980s. It was, and is, still a controversial debate that culminated within the context of regulation efforts of the green biotechnology in the late 1990s. Since then, specific subtopics of biotechnology have been the subject of an ongoing debate, both in the media and in the public domain (Bonfadelli *et al.*, 1998, p. 146).

Switzerland, like other European countries, looks back on an intense political and public debate on the issue. Biotechnology not only seems to be a bone of contention, but also an increasingly important and enduring public topic. There are two main explanations why biotechnology is of such importance to the Swiss: (i) the biotechnology industry is an important economic factor in Switzerland; and (ii) the Swiss political system of direct democracy provides favourable conditions for NGOs and other political actors to put new issues on the official political agenda. Therefore, we have to provide a few explanatory remarks about the Swiss system of direct democracy before we continue with a detailed description of the biotechnology debate in Switzerland.

The Swiss system of direct democracy allows citizens to launch a campaign to collect signatures for a so-called 'popular initiative' at the federal level. Provided they manage to gather the necessary number of signatures (150,000 within a specified time), any group of citizens may submit an initiative. It then has to be debated by the national parliament and the government, who can support or reject the initiative or write a counterproposal with similar, but typically less radical, aims. A national referendum on the initiative has to take place within 5 years of its submission. If there is a counterproposal, both suggestions have to be voted upon.

To simplify, four phases can be distinguished in the public debate on biotechnology in Switzerland (Hieber, 1999; Graf, 2003): (i) latency; (ii) growing concern; (iii) conflict; and (iv) normalization. The first phase of latency took place during the 1970s and 1980s. In this phase, modern biotechnology had been discovered and was known to an expert audience, but it was not an issue of public discussion or even controversy. The federal government had no intention of creating a specific, superimposed law on biotechnology. The dominant regulatory approach was self-regulation by the biotechnology industry and the biotech researchers.

This phase of latency was followed by a second phase of growing public concern and public debates about biotechnology. The year of 1987 can be defined as the beginning of this new phase. In that year, the so-called *Beobachter-Initiative* [Observer Initiative] was submitted, seeking to restrict the abuse of reproductive medicine and gene technology in humans. This first piece of legislation was motivated by a growing awareness of the lack of regulation in biotechnology, particularly that applicable to humans. It was accepted in 1992 as Article 24novies of the national constitution, with 74% voting in favour.

The third phase, from 1992 to 1998, was characterized by conflict and controversy that was triggered by the submission of the *Gen-Schutz-Initiative* (GSI) [Gene-Protection Initiative]. Originally submitted in 1992, the GSI was intended to protect living organisms and the environment from genetic manipulation. By 25 October 1993, the necessary number of signatures had been gathered. The initiative called for the prohibition of: (i) the production and sale of GM animals; (ii) the release of GM plants and animals into the environment; and (iii) the issuing of patents on GM plants and animals.

The year 1997 was characterised by the Swiss Parliament's attempts to develop an indirect counterproposal to the GSI. Swiss industrialists and the government decided to follow the strategy of not directly opposing the GSI, but rather planning and running a propaganda campaign for additional legislation. The pro-genetics lobbyists in parliament got together to draft a law to address the problem of regulating those areas of gene technology not already covered by existing Swiss legislation. On 4 March 1997, parliament issued a first draft of the Gene Law with the additional *Gen Lex Package* – modifications to various other laws, mainly within environmental legislation, but also those concerning, for example, agriculture and animal breeding.

Public opinion polls showed that the GSI had a realistic chance of being supported by a majority of the population. Since the Gene Law was not developed as a direct counterproposal to the GSI, the Swiss were not able to vote on it. In December of that same year, the federal government produced a second draft of the Gene Law in record time. The speed with which the Government worked was interpreted as an expression of its willingness to take the concerns of the Swiss population seriously and to develop strict regulations for gene technology.

The federal government became particularly active shortly before the vote on the GSI in June 1998. The Federal Government set up the

Swiss Ethics Committee on Non-Human Gene Technology (ECNH) on 27 April 1998, which was to be an expert committee with the task of advising the authorities in the field of non-human biotechnology and gene technology. The Government was responsible for selecting the members and the chair of the committee. The size of the committee was limited to 12 members from outside the government, with half of them university-trained ethicists. This last-minute step to set up a committee (2 months before the referendum) was criticized as a tactical move to wrong-foot the critics of gene technology.

The long-awaited vote on the GSI took place after a very intensive press and advertising campaign. Because the biotechnology industry and researchers in Switzerland believed their very existence was threatened, a vigorous campaign was mounted. Their campaign focused strongly on the positive aspects of red biotechnology, the loss of jobs in the pharmaceutical industry and the fear that Swiss biotech. research would be placed in jeopardy. The anti-gene-technology movement also had considerable financial resources. To the surprise of most political observers, the initiative was rejected by a large majority of Swiss voters (66.6%) although, as usual, the turnout was low: only 40.6% of those entitled to vote actually did (Bonfadelli *et al.*, 2001, p. 284).

In the *fourth policy phase* (1998 onwards), after the rejection of the GSI, the policy process focused on the development of the so-called *Gene Lex Package*, dealing with non-human biotechnology. This last phase can be labelled as normalization: the conflict surrounding biotechnology had not been resolved in any kind of compromise, but the management of this conflict had been institutionalized in specific laws and procedures that were reducing and normalizing the intensity of that conflict.

The third popular initiative on biotechnology [Agriculture free of genetically modified organisms] was submitted in that period (November 2000) and can be presented as an example of a normalized controversy: Various environmental, consumers' and farmers' organizations took up the call for a 5-year moratorium on the application of biotechnology in agriculture. This development must be seen against the background of recent BSE scandals. BSE is perceived to be an example of the negative consequences of insufficient attention to risk prevention in food production.

In comparison to the GSI, this latest initiative can be considered as relatively moderate: though the production and sale of genetically modified organisms (GMO) like crops is legal in Switzerland, there are almost no GMO products available in the shops, because there is no demand from the consumers. Therefore, an initiative that is asking to stop the production of GMO in agriculture is by no means radical, but rather institutionalizes the *status quo* of current consumer preferences.

To sum up, the process of policy development can be described as a procedure of trial and error and as a search for the least common denominator. Government regulation was never *proactive*, but rather *reactive* to the pressure and political resistance from NGOs. The activities of the NGOs were not successful at the very beginning, but can be interpreted as a search for those regulations that had a chance of being supported by the majority. While the most radical initiative (GSI) failed in the popular vote, the two more moderate proposals fared better: The *Beobachter-Initiative* triggered a legislation procedure that successfully passed the referendum (1992) and, on 27 November 2005, the campaign for a 5-year moratorium in agriculture was successful and the initiative was accepted with 55.7% voting in the affirmative.

## On the Media Agenda: Shifting from Red to Green

As stated in different studies, mass media are the first and most important source of information when it comes to judging new and complex technologies. Therefore, the mass media play a crucial role in creating and shaping public opinion on biotechnology. The reason for this is that it is a relatively new public issue and few people have the opportunity to experience the technology directly (Hornig-Priest, 1995; Weingart, 1998). In other words, people who want to learn about biotechnology and which benefits or risks the new technology includes depend on media reporting.

In Switzerland, as in all democratic countries, citizens elect representatives to act on their behalf. Switzerland, however, gives its citizens the chance to take a direct part in decision-making as well through the popular initiative and the referendum, two instruments that make the Swiss political system probably the most extensive in the world. This fact increases the significance of an adequate and comprehensive media coverage, where different points of view of different actors on the issue become visible and make it possible for the public to get an idea and to vote on the issue.

The following part teases out how biotechnology has been portrayed in the Swiss media over a period of 30 years. Especially emphasized will be the coverage in recent years and how the reportage has changed respectively. Since TV reporting makes up a considerable proportion of the sample, stress will be also laid on the visualization of the topic. The framing concept serves as a theoretical approach, which opens up the possibility of presuming how recipients read, see and evaluate biotechnology.

## Theoretical approach: the potential of framing

Framing theory has an interdisciplinary origin that draws on insights from sociology, psychology and communication, just to mention a few key disciplines. Although it is not new as a theoretical concept, interest in framing has grown, particularly in social science in the past 10 years (Reese, 2001). These interdisciplinary roots are one reason for the terminological diversity – if not confusion – that is associated with this theoretical perspective. But, on the other hand, the broad possibilities of the framing approach have resulted in a great deal of empirical work which has been done so far, especially in the field of mass communication research. For this contribution, we will not elaborate on the terminological discussion but refer to a definition of framing that is especially useful for the analysis of media texts, as well as reception processes.

But what does framing mean? Along the simple question of 'what is going on here' (Goffman, 1977), Erving Goffman tried to explain how people organize everyday experiences. From his point of view, people structure social and personal experience according to certain principles. These principles contain elements which he called frames. Goffman's basic idea of how we deal with information can be adapted for media research.

Referring to Goffman's work, Todd Gitlin (1980) defined frames as 'persistent patterns of cognition, interpretation and presentation, of selection, emphasis and exclusion, by which symbol-handlers routinely organize discourse, whether verbal or visual'. Frames can be understood as cognitive structures that journalists use to cope with an issue and use to compose a news story. But framing can also mean a mental map for recipients to read and handle media stories. In the early 1990s, Iyengar (1991), Neuman *et al.* (1992), Iyengar and Simon (1993) and other communication researchers proved with their work that media frames have an impact on how people interpret media stories and talk about certain political issues.

Taken together, frames are patterns of interpretation that can be identified in all phases of mass communication, including public relations, journalism, media content and media effects. They have similar functions on all these levels: They structure information and reduce complexity. Frames are not issues or themes, but general patterns that can be applied to any issue. Frames usually consist of four elements: problem definition, causal interpretation, moral evaluation and/or treatment recommendation. Similar to the concept of bias, frames do evaluate options, but they are not a synonym to bias because frames are more differentiated by giving information on the evaluation criteria (Dahinden, 2005).

In the media context at hand, framing is regarded as being a structure within the media text that organizes central ideas of an issue and which has the function of constructing meaning, incorporating news events into its interpretative package. These media frames are the result of the news production process, including organizational and personal (journalistic) factors.

The news coverage on biotechnology – mostly in newspapers – has been analysed in many studies. Some of them focus on media frames and try to detect media patterns in media texts. The results show, depending on the methodological approach, different frames. Not only are the different modes of media coverage responsible for these differences (e.g. different countries with different biotech. policies, different kinds of newspapers), but also other factors like the sampling and, most of all, the way how frames have been defined and determined. The biggest differences become obvious in comparing studies that extract frames in a deductive way with studies that choose an inductive way to locate media frames.

In general, the results from different empirical work suggest that biotechnology is framed in some dominant ways. Two main factors that seem to influence the framing are the time of the discussion and, above all, the field of application. This can be shown exemplarily In the EU project, where the frame *progress* prevails in most of the opinion leading newspapers in different countries for a good part of the time of the investigation period. This framing describes biotechnology in positive terms and celebrates it as a new development with a potential for science and health – a technology that serves the society.

Especially in recent times, the frame *progress* refers mostly to the human field of application (red biotechnology), whereas green biotechnology has been more and more framed in ambivalent, even negative, frames. The frame *public accountability*, meaning 'call for public control, public involvement and regulatory mechanism' (Durant *et al.*, 1998), as well as the *ethical* frame, have both become vital. Differentiated results for the Swiss press will be depicted below.

## Basic data of the analysis

The following findings rest on two different data sets. One set contains articles from the opinion-leading newspaper in Switzerland, the *Neue Zürcher Zeitung*, which is currently read by 8.5% of Swiss. In the context of the EU project (see above), 711 articles between 1973 and 2002 have been analysed. The sample comprises random weeks over these years and gives an idea about the absolute number of articles published in the issues. The general increase since the mid-1980s expresses the mainstream trend of biotechnology, which is no longer a scientific matter but has also become an economic and a political issue. Due to the two initiatives in 1992 and 1998, biotechnology has been on the media agenda. The analysis of this sample is based on a codebook that has been applied in the European project. It comprises, among others, the variable *frame*. In other words, the different frames have been sought in the text as a whole and collated to the given categories (deductive).

The second set encompasses broadcast programmes from the Swiss Public Television in Switzerland. The sample counted 282 news items, from which 191 have been broadcast in the prime-time news. An additional 91 items stem from different telecasts. Televised once a week, they lay stress on specific subtopics of biotechnology and provide the viewers with broader and deeper information on medical application, gene therapy, stem cells, etc. (health magazine), on declaration, contamination and food safety (consumers' magazine) or on scientific breakthroughs, transgenic animals, xenotransplantation and such (science magazine).

In contrast to the first sample (print), the frames in the TV items have been established in a two-step procedure. First, possible frames have been analysed in an inductive way, taking into consideration different features of the text (including visual text) – actors, themes, visualization, metaphors, advantages, disadvantages). These preliminary frame arrangements have been further analysed in a qualitative heuristic manner. The aim was to take into consideration the frame-influencing information that lies in the interplay of spoken/written text and visual text. Due to technical problems, the sampling does not represent the quantity of the actual TV programme broadcast on biotechnology. Hence, the following findings will focus mainly on qualitative results.

## Print media: from progress to public accountability

Looking back to the 1970s and early 1980s, biotechnology appears in the *Neue Zürcher Zeitung* mostly in the dominant frame *progress*. In almost every second article, a forward-looking perspective prevails. The frame stands for a homogeneous, positive media coverage that encompasses a scientific discourse with dominant actors from universities and the research area, touching on topics like basic research, inheritance issues and transgenic organisms (Leonarz and Schanne, 1999). In the course of the increasing debate in the public and on the political agendas, the media coverage shifts towards a more ambivalent and heterogeneous frame that shows all facets of the political discourse.

The frame *public accountability* underlines the account of a broad discussion in public. It implies the call for regulation and control, public involvement and participation. The frame remains on a high level even now and mirrors somehow the Swiss political system, where citizens decide at the ballet box over such things as biotechnology: Political actors need media to inform the public – mass media become an important tool not only for informing the public, but also for convincing them to vote either for or against a certain issue.

The frame *public accountability* is therefore mostly a political frame with politicians, NGOs and the public voice as main actors (Leonarz, 2002). In the present case of Switzerland, the topics within this frame comprise the topics of the initiatives or the issue on the political agenda at that corresponding time: reproduction, childbearing *in vitro* (from the beginning of the 1990s), transgenic plants, GM food, environmental safety. GMO release can be linked to the second initiative in the late 1990s. Recently, in the context of governmental attempts to regulate green biotechnology (*Gen Lex*, 2003) as well as the new law on embryonic stem cell research (2004), the corresponding themes are dominant.

As shown in Table 4.1, there are a number of other frames that appear continuously over the whole period of time. They never rate high, but express alternative ways of framing biotechnology with other meanings. The frame *economic,* for example, puts biotechnology in a positive light and underlines its economic potential. Most frequent are actors from the industrial sector, mainly the pharmaceuticals. Main topics are economic prospects and opportunities, as well as pharmaceutical products and vaccines.

A negative framing of biotechnology can be expressed by *ethical* and *Pandora's box.* The ethical discussion has always been in the media. The frame *Pandora's box*, however, has gained in influence in recent years. Describing biotechnology in the context of 'a call for restraint in the face of the unknown risk, the "opening of flood gates" warning, unknown risks as anticipated threats, catastrophe warning' (Durant *et al.*, 1998), the frame illustrates the rising scepticism mostly against green biotechnology. It is not surprising that NGOs play an important role within this frame.

## Television: different framing becomes obvious

As mentioned before, the sample with the TV coverage on biotechnology was analysed with a different instrument. The inductive way of finding

**Table 4.1.** Dominant media frames over time, 1973–2002 (from *Neue Zürcher Zeitung*, an opinion-leading quality Swiss newspaper).

| Frame | 1973–1987 (%) (*n*, 57) | 1988–1996 (%) (*n*, 154) | 1997–1999 (%) (*n*, 188) | 2000–2002 (%) (*n*, 312) |
|---|---|---|---|---|
| Public accountability | 9 | 25 | 35 | 32 |
| Progress | 49 | 22 | 21 | 23 |
| Ethical | 7 | 10 | 10 | 7 |
| Economic | 5 | 10 | 12 | 13 |
| Pandora's box | 0 | 3 | 10 | 8 |
| Others | 30 | 30 | 12 | 17 |
| Total | 100 | 100 | 100 | 100 |

frames in the telecasts has generated a set of eight different frames. Although they do not match the frames used for the print analysis, there are some obvious analogies. Indeed, there is in fact more than one frame that shows parallels to the frame *public accountability*. Additionally, there is one frame that circumscribes a clear economic context and another that can be labelled as *progress*, showing partly the same traits as the *progress* frame mentioned above. The frames have been described in detail as follows:

### *Regulation of red biotechnology*

A rather heterogeneous frame with ambivalent valuation of biotechnology, it describes a restrained discourse in a political context. The topic is solely the human field of application, where biotechnology is depicted as an inevitable technology that needs to be regulated. Pros and cons appear and show that the topic can be valuated in different directions.

### *Ethical discourse on red biotechnology*

Also touching red biotechnology, this frame puts the new technology strictly in a negative corner and argues with ethical and moral reasons against it. This encompasses themes from *in vitro* fertilization (at the beginning of the discussion) to xenotransplantations, pre-implantation diagnostics and cloning. The frame appears in its clearest and strongest characteristics within the scope of human cloning. Metaphors are frequent and they all speak a plain language: 'Who has the right to play God and decide over the quality of life?'

### *Positive description of red biotechnology*

Showing the opposite view, this frame puts red biotechnology in a positive light and evokes the advantages for health. It is mostly a popular scientific discourse that stresses the practicability of the technology. There are hardly any moral aspects – critical voices are non-existent.

### *Regulation of green biotechnology*

This frame can be compared with the first frame, *regulation of red biotechnology*. However, it focuses on the extra-human field of biotechnology. It is mostly political news coverage that emphasizes decision-making politics in the long term and the need for regulation. The green field of biotechnology is considered to be something society can live with – there is no open opposition, however scarce the benefits. Actors formulate pros and cons. Reassuring arguments ('there will be no disadvantages') can be often found.

### *Opposition to and scepticism of green biotechnology*

This frame is clear and without ambiguity. According to the actors in this frame, green biotechnology has, not only, no benefits for society, but also shows obvious risks. There is no reason to tolerate GM food, GM crops or any research being done on plants. NGOs play an important role in this frame. They appear regularly. In particular, members from Greenpeace stage – with their well-known strategies – a scoop for TV coverage: Greenpeace members, for example, seize a cargo ship, they dress up as chickens to hand out flyers to consumers, unload manure on a testing field. The frame can be described as action-oriented, although the political structure remains obvious.

### *Economic*

This positive frame shows a clear reference to scientific factors. This is true for the dominant actors, themes and above-mentioned advantages. Profits and prospects of investments, take-overs and mergers are topics.

### *Progress*

In a scientific interrelation, this frame describes biotechnology in a homogeneous, positive way. Most of the reports are engaged in the red field of biotechnology, but not exclusively. Especially at the beginning of the debate, some of the telecasts refer to green biotechnology, and they do so without admonishing it. However, most of the items centre on red biotechnology. Scientists, mostly in a positively connotated role, explain complex scientific projects and underline the advantages of biotechnology for the purpose of research and for society.

*Regulation identification*

The rather new discourse about DNA fingerprinting can be collated in this frame. Along the lines of the other regulatory frames, this frame also underlines the political debate of the issue. Altogether, the tenor reads as follows: DNA fingerprinting has considerable advantages for society; regulation is necessary to prevent misuse.

## Green biotechnology: opposition and scepticism as dominant features

As stated before, the sampling did not provide reliable information about the intensity of broadcasting on biotechnology. But still, the data at hand show very clearly which frame dominates the TV coverage. The frame *opposition and scepticism against green biotechnology* appears most frequently. Its action-oriented characteristics seem to be telegenic and fit well into TV coverage. It is the most dominant frame not only in prime-time news but also in weekly broadcasts. The consumers' magazine, in particular, composes the frame in the most obvious way: Biotechnological application within the food sector shows benefits neither for the consumers nor for society as a whole. In comparison to the daily news, the consumers' magazine frames biotechnology negatively by journalistic quotes and formalistic editing.

In the daily news, however, the frame shows up in a more subtle way. From the journalistic side, only a few evaluating votes can be found. It is mostly the actors who set biotechnology in a sceptical frame – by quotes or simply by their appearance. Greenpeace, as mentioned before, plays the leading role. Other NGOs – most of them explicitly against biotechnology and only active in Switzerland – complement Greenpeace. The best-established heads from these NGOs have a considerable presence on TV – mostly because they not only act as members of an anti-biotechnology organization but also as political members of the federal parliament.

The whole TV coverage of green biotechnology shows few remarkable visualizations. Most of the film material backs up the spoken text. It shows people working in laboratories, different plants and cells underneath the microscope, crops in greenhouses, farmland, important buildings (like the Bundeshaus), food and shopping scenes.

The most frequently used shots are interview situations mostly in professional surroundings. Shots of demonstrations and manifestations and frequent elaborate displays by Greenpeace, leave a strong mark.

Nevertheless, among these images, some film material is remarkable. A typical shot, also used in the context of the human field of application, shows a fine cannula penetrating a cell under an optical microscope. This picture symbolizes a manipulative intrusion in nature. The still images that the daily news uses to introduce their reports also show typical features with sometimes negative connotations, like the soybeen in poisonous-green in a test tube or a picture of an evil-looking scientist.

### Possible implications: the increasing gap between red and green biotechnology

All frames, in the newspaper and especially on TV, show a dichotomy between the depiction of red and green biotechnology. Recapitulating, one can state that only red biotechnology appears in a positive frame, underlining its potential for health and society. Both subfields appear in an ambivalent frame with a strong affinity to regulation in a political context. Both subfields also appear in a frame with a clear negative connotation. The most pronounced frame, however, describes an oppositional and sceptical discourse of green biotechnology. As a dominant frame, the assumption is legitimate that it backs up the existing bipolar attitude in public: an increasing opposition toward green biotechnology while the acceptance of red biotechnology can be consolidated.

## Framing of Red and Green Biotechnology by the Public

How does the Swiss public perceive and evaluate biotechnology? These research questions will be answered in the following two sections of this chapter. First, we will present findings from focus groups that provide qualitative data (Dahinden, 2002a). Then, the discussion in the following section will be based on quantitative survey data.

Focus groups allow the investigation of public perceptions of new and complex issues about which most people do not yet have very strong and stable attitudes. Biotechnology, with its many fields of applications, is certainly such a new and complex issue. The openness of the discussion gives participants an opportunity to develop their thoughts and to express their uncertainty and ambivalence.

What are focus groups? A simple definition of focus groups is the following: focus groups are small discussion groups with six to twelve members. The debate within the group is led by an experienced moderator. The discussion is recorded on audio- or videotape, which is a prerequisite for the transcription and in-depth analysis (Krueger and King, 1998). Focus groups are very popular in applied research contexts (e.g. marketing, political campaigning, etc.). In recent years, focus groups have received increasing interest from academic researchers.

## Data collection and data analysis

Focus group participants were selected by random sampling from the telephone book. The four discussion groups, based on a total of 28 participants, had six to eight members and they were composed in order to maximize diversity within the group: for instance, to have women and men, young and old, poor and good educational levels, supporters and opponents of biotechnology. The discussions within the focus groups that took place in autumn and winter of 1999 were transcribed and analysed by means of software that supports the analysis of qualitative data (ATLAS.ti, 2005). This software facilitates the process of data analysis but, nevertheless, coding had to be done manually.

Focus groups produce a wealth of data (in our case, 130 pages of transcript). Therefore, coding is necessary in order to reduce this large quantity and to concentrate on the essence of discussion. Coding was done in a three-step process: (i) codes were identified in a process of induction from the empirical data. Coders read the transcripts and looked for topics and arguments that related to the theoretical research question; (ii) these inductive codes were systematized in a process of deduction by grouping them under more general categories (e.g. actors, applications of biotechnology, ethical arguments, etc.); and (iii) these patterns of interpretation were interpreted and classified as individual frames (Scheufele, 1999).

## Findings

There was a clear difference between attitudes towards medical *versus* agricultural applications of biotechnology. On the one hand, medical applications were evaluated in an ambivalent, if not positive, way. On the other, the agricultural and food applications were confronted with a predominant rejection. Since the focus of this book is on the use of biotechnology in agriculture, we will not expand on the perceptions of medical biotechnology, but it is important to note that the Swiss public does not consist wholly of opponents of this technology; indeed, attitudes tend to be much more diverse and ambivalent.

As appears above, we have applied framing theory as a perspective for the data analysis. Within framing research, there is a diversity of frame typologies (Dahinden, 2002b). Since this project used multiple methods (policy analysis, media content analysis, surveys and focus groups), we were applying the frame typology for the focus groups that had already been used in the content analysis. This deductive frame typology used eight categories: *progress, conflict, economic, moral, run-away, Pandora's box, nature-nurture, globalization* (Durant *et al.*, 1998).

All of these frames could also be found in the focus group discussions but, due to space restriction, we will focus in this chapter on those three frames that were especially relevant for discussion about agricultural biotechnology (an extended presentation and international comparison of focus group results can be found in Kronberger *et al.* (2001) and Wagner *et al.* (2001)).

- *Pandora's box*: warning before the catastrophe, call for restrictions against unknown risks and threats.
- *Moral*: evaluation of the issue on the background of explicit moral values (e.g. environmental values).
- *Conflict, public responsibility*: call for public regulation and democratic control of private organizations and interest.

The rejection of agricultural biotechnology was rarely justified by reference to specific biological or technical risks, but rather with a general fear of unexpected consequences and loss of control over this new technology. This attitude resembles the conclusion of the Greek mythological story of Pandora's box, which should not be opened in order to avoid harm and destruction to humankind. The following quote from a focus group participant (the Christian names of the participants have been altered to English names) gives an illustration of the wording of this perspective: 'For me it means a big ethical incision into God's creation. I'm not catholic but I think this way. And one could say, with Goethe: I can't get rid of the ghosts that I called upon' (Margret).

A second frame that was used to justify the objection to green biotechnology was the moral frame relating to moral or ethical concerns. These ethical arguments were not restricted to religious values, but were about nature and respect for the natural world. The following quote is typical for that frame: 'Nature did everything right, and no human being is more perfect than what comes out of nature. And that's why I'm against it. I can't imagine that we can make things better like that' (Mia).

In addition to these principal arguments, some participants also took a more pragmatic stand, saying that GM food currently offered no advantages with regard to either quality or quantity. The hope that global famine could be solved by means of biotechnology was mentioned several times, but not judged to be very realistic because participants thought that farmers would become increasingly dependent on seed-producing companies.

The role of these corporate actors is a key element within the third frame (*conflict*) that was found as an argumentative background for the rejection of agricultural biotechnology. Within this frame, the public debate on biotechnology was seen as a conflict between the powerful economic interests of agricultural companies and the relatively weakly organized interests of consumers. Against that background, the objection to biotechnology is justified by a political argument to restrict the influence of 'big business' and to force it to produce according to consumer preferences.

The following quotes are typical examples of this conflict frame: 'This company from the States, they just want to sell this maize also here in Europe – they produce more and more but nobody wants this maize, but they go on producing it. The power of these big players is so big – an individual has no say' (Martha). 'I don't think that we can build up a system of regulation in Europe that will work. There have been so many other examples that show that it will not work. I'm quite pessimistic. I don't believe a single word of what a politician is saying' (Victor).

These statements show a high degree of mistrust and criticism, but they also illustrate, at the same time, a feeling of powerlessness.

In sum, the negative attitudes towards agricultural biotechnology were not based on judgements of the specific risks and impacts of these applications, but on more general arguments from ethical or political reasoning. Further analysis of the attitudes of the Swiss toward green biotechnology will be given in the following section on survey results.

## Media and Public Opinion: Agenda-Setting, Knowledge Gaps, Selective Exposure and Priming Effects

Quantitative media effects are the focus of this last part of our contribution to the situation of biotechnology in Switzerland. The empirical evidence presented here illustrates, in particular, three newer *paradigms of media effects research*, namely: (i) the agenda-setting function of mass media; (ii) the knowledge gap hypothesis; and (iii) selective exposure to and priming effects of media coverage (Bonfadelli *et al.*, 2002).

## Theoretical considerations

The *agenda-setting theory*, formulated by Maxwell McCombs and Donald L. Shaw in the 1970s (McCombs and Shaw, 1972/1973), became one of the most influential perspectives of media effects research (Dearing *et al.*, 1996). It rests on the assumption that media do not tell people what to think, but what to think about. This represents a general shift from *attitudes* to *cognitions* as media effects phenomena. The agenda-setting function of mass communication means that mass media cover news events selectively on the basis of media gatekeeping processes. As a result, a media agenda is constructed according to set news values that can be interpreted as journalistic routines.

The agenda-setting thesis postulates, then, that the intensity of media coverage on a topic, measured by content analysis, correlates with the public agenda, measured as the perceived salience of the topic by the public. Thus, media are able to influence the perceived public agenda by the varying amount of media coverage. And, over time, agenda-setting theory claims that the issue salience perceived by people will increase as a function of intensifying media coverage or, respectively, will decrease if media coverage reduces. This general agenda-setting hypothesis was further refined and differentiated by empirical research (McCombs and Bell, 1996; McCombs and Reynolds, 2002).

Focussing on the *development of effects over time*, Kepplinger and his colleagues (1989) demonstrated that there are further types of agenda-effect phenomena besides the basic *linear* relationship between media agenda and public agenda. For example, a constellation they called '*echo model*' means that even if media coverage decreases, the salience of a topic can still persist at a high level on the public agenda. This effect course is typical after intensive media coverage of important technological accidents or catastrophes.

Another differentiation relates to the *type of topic*. World events or topics of media coverage usually vary in relation to their visibility or obtrusiveness, the degree of complexity, the occurrence of interpersonal communication or the possibility of direct personal experience. It is hypothesized here that agenda-setting effects will be especially strong namely for unobtrusive and ego-distant topics like genetic engineering. As a consequence, as media coverage of biotechnology increases – e.g. because public political discussion is intensifying or is triggered by a political initiative of NGOs like Greenpeace – the topic of biotechnology moves up in the hierarchy of the public agenda as well.

And, finally, the public agenda can be operationalized or measured in various ways by survey research. Usually, people are asked by the interviewer to discuss topics they are most concerned about or to rate topics according their personal relevance. Besides this *intra-personal agenda*, the *inter-personal agenda* can be measured by asking people how often they discuss a topic like biotechnology with other people.

Whereas agenda-setting research is concerned with the function of media to focus the attention of the public on specific issues of society, thus creating and homogenizing a sphere of public opinion, the *knowledge gap perspective*, formulated by Tichenor *et al.* (1970), concentrates additionally on the *diffusion of information in society* and relies on a more *dysfunctional view of mass media* (Bonfadelli, 1994).

If media coverage of biotechnology is increasing, this intensified diffusion of media information will stimulate information exposure and learning processes by media users. As a result, the knowledge gap hypothesis postulates not an equal information distribution but a *widening gap in knowledge* between the segments of good educational level and/or high socio-economic status and those of poor educational and/or social status.

This heterogeneous flow of information is explained by various influencing and mediating factors working together. Generally, well-educated people: (i) use the information-rich print media on a more regular basis; (ii) learn more efficiently and on a more encompassing level; (iii) are more politically interested; and (iv) are more integrated into inter-personal social networks.

There are levelling factors as well, however, namely when mass media present a *topic as controversial* and there is a high level of social and political conflict around it. In Switzerland this seems to be the case, especially, in the field of biotechnology (Bauer and Bonfadelli, 2002). On the other hand, biotechnology is a rather complex scientific topic that is certainly *distant* to the everyday world of people, a situation that will result normally in stronger knowledge gaps (Gaziano and Gaziano, 1996; Viswanath and Finnegan, 1996).

As a consequence of this complex situation in the field of biotechnology, an equal distribution of the so-called 'knowledge about' – namely in form of agenda-setting effects – could be expected on the one hand but, on the other, increasing knowledge gaps based on background 'knowledge of' biotechnology could be expected as well.

Besides cognitive media-effect phenomena like agenda-setting and knowledge gaps, mass communication influences attitudes of people as well. Based on *consonance models* of attitude research, like the dissonance theory formulated by Leon Festinger in the 1960s (Cotton, 1985; Aronson, 1997), it is commonly assumed in traditional media effects research that the main function of modern mass media is not to change people's opinion on a certain topic like biotechnology. As a consequence of selective media exposure and selective interpretation of media content, influenced by existing attitudes that function as predispositions, consonance media effects dominate. Therefore, media coverage normally will confirm and reinforce rather than change existing attitudes.

The situation is more complicated in the field of biotechnology, however, since this is a new and complex topic and most people have not yet established opinions and attitudes that could function as stabilizing predispositions. So, it is assumed that people will be influenced by political propaganda and mass media publicity, especially concerning new, complex and controversial topics like biotechnology, a topic with only limited 'knowledge about' and no well-established predispositions. Furthermore, people's capacity to process persuasive political messages is limited, and not all media users are motivated enough and/or have the mental ability to process media information on complex and rather abstract topics like biotechnology.

Thus, the so called *elaboration likelihood model* (Petty *et al.*, 1997, 2002) postulates that persuasive media effects are still possible by the so-called peripheral route, but these are based not on deeper information-processing of the presented arguments, but largely on peripheral cues such as prominent expert sources, catchy slogans or simple – but seemingly convincing – visual images.

Moreover, there could also be a *priming effect* (Iyengar and Simon, 1993) at work. This phenomenon was postulated in the context of agenda-setting research. Sometimes it is also called a *second-level agenda-setting process* (McCombs, 1998). If media coverage concentrates exclusively on a certain topic like positively valued red biotechnology (for instance, highlighting future health-related benefits of genetic engineering instead of negatively regarded green biotechnology), these messages will activate positive rather than negative images in the mind and will be used by people as an anchor for the evaluation of biotechnology. This is why public opinion for or against a biotechnology initiative or referendum can change as a consequence of topic-centred media coverage, even if there is selective media exposure.

## Hypotheses, methods and data

To answer the question of mass media influence in the field of biotechnology in Switzerland, we formulated the following leading hypotheses, based on agenda-setting, knowledge gap, selective exposure and attitude research:

1. Attitudes to biotechnology are not homogeneous or undifferentiated, but vary depending on the concrete applications of biotechnology. *Red biotechnology* is favoured because of positively valuated goals and usages, whereas most people are against *green biotechnology* because of perceived risks and a lack of positively valued applications.
2. Attitudes for or against biotechnology have changed over time and are more intensive in comparison to those in other European countries, as a result of media coverage and the biased political campaign triggered by the second public 'gene protection' initiative (GSI) in June 1998. Furthermore, attitudes have changed in a positive direction in the context of the intensive political campaign in favour of biotechnology, which is to say against the initiative.
3. Biotechnology is increasing over time as an important personal topic, as a result of the intensifying media coverage. Even if media coverage is decreasing, however, the personal relevance of biotechnology as an *echo effect* will stay high. At a certain point in time there is a correlation between perceived salience of biotechnology and intensity of media use.
4. In general, there will only ever be a modest mean level of knowledge about biotechnology. Knowledge will be distributed unevenly: people higher in educational and social status will be better informed. The level of information will be higher in Switzerland in comparison with other European countries because of the higher levels of conflict and more intensive political discussions of biotechnology in Switzerland.
5. Exposure to and interpretation of biotechnology will be selective according to the underlying attitude structure concerning biotechnology.

Our data are based on three representative Eurobarometer surveys, with comparable questions in all three language regions of Switzerland (French, German and Italian). Each wave consisted of more than 1000 personal interviews. The first survey was carried out in 1997 (*n*, 1033) between the end of May and the beginning of June, almost a year before the very controversial so-called *Gene-Protection Initiative* (GSI) was voted on, and about 3 months after the big media event known as 'Dolly' occurred. In Switzerland, this survey revealed a mostly controversial – but clearly negative – bias. The second survey took place in the summer of 2000 (*n*, 1010) about 2 years after the Swiss population rejected the heavily debated Gene-Protection Initiative by a 66.6% margin. And the third survey (*n*, 1026) was carried out between the beginning of December 2002 and the end of March 2003.

Consequently, the first survey documented a situation where public discussion of biotechnology was just starting to become increasingly intense and controversial in Switzerland, whereas the second and third surveys took place in very different situations. In 2000, there was no longer much public debate concerning biotechnology and there was no referendum on the political agenda. Thus there was much less media coverage and decreased involvement by the Swiss public. And, 2 years later, in the winter of 2002/2003, there was even less public debate and biotechnology was no longer a prominent topic of the media agenda.

## Results

### Differentiated attitudes towards biotechnology

In the survey of 2002/2003, people's attitudes to biotechnology were measured not only on a general level, but also regarding different applications of genetic engineering such as: (i) using genetic testing to detect diseases we might have inherited from our parents, such as cystic fibrosis; (ii) taking genes from plant species and transferring them into crop plants, to render them more resistant to insect pests; or (iii) using modern biotechnology in the production of foods – for example, to increase the protein content, to increase their shelf life or to alter their taste. And each application had to be rated according to the following four different criteria: (i) usefulness to society; (ii) risk to society; (iii) moral acceptance; and (iv) encouragement.

Table 4.2 shows that people in Switzerland do not judge biotechnology in a uniform way, but according to its specific applications. In general, red biotechnology is evaluated far more positively than green biotechnology. So, genetic testing as an application of red biotechnology is considered as useful and morally acceptable; its risks seem to be at a medium level and, as a consequence, this application should be encouraged. Contrary to this positive evaluation, applications of green biotechnology like modified crops – especially GM food – are evaluated in a negative way. In the public opinion, these applications are not useful, tend to be risky and are not morally acceptable. As a result, about half of the people asked in the survey of 2002/2003 expressed the opinion that these applications should not be encouraged.

Table 4.3 is based on the acceptance or rejection of a set of opinions dealing with benefits and strengths, but also with disadvantages or even threats connected with GM food. It is largely consistent with the data in

**Table 4.2.** Ambivalent attitudes towards red and green biotechnology (from Eurobarometer survey, 2002/2003, *n*, 1026).

| Statement | Application | Definitely agree (%) | Tend to agree (%) | Tend to disagree (%) | Definitely disagree (%) | Means[a] |
|---|---|---|---|---|---|---|
| Is useful for society | Genetic testing | 24 | 48 | 17 | 4 | +0.75 |
| | Modified crops | 10 | 42 | 29 | 9 | +0.15 |
| | Genfood | 10 | 25 | 38 | 19 | –0.31 |
| Is a risk for society | Genetic testing | 11 | 43 | 25 | 11 | +0.17 |
| | Modified crops | 17 | 46 | 24 | 3 | +0.51 |
| | Genfood | 20 | 40 | 23 | 6 | +0.44 |
| Is morally acceptable | Genetic testing | 22 | 42 | 17 | 7 | +0.54 |
| | Modified crops | 13 | 42 | 32 | 13 | +0.08 |
| | Genfood | 8 | 31 | 33 | 19 | –0.23 |
| Should be encouraged | Genetic testing | 21 | 43 | 17 | 8 | +0.51 |
| | Modified crops | 7 | 33 | 31 | 17 | –0.18 |
| | Genfood | 7 | 24 | 32 | 26 | –0.46 |

[a] Means are calculated with the following values: definitely agree, +2; tend to agree, +1; tend to disagree, –1; definitely disagree, –2; don't know, 0.

Table 4.2, but adds to the general picture concerning negative opinions towards GM food. Interestingly, the majority of people in Switzerland would not buy GM food even if it: (i) were *cheaper* than conventional food (87%); (ii) contained *less fat* (82%); (iii) *tasted better* (70%); (iv) contained *fewer pesticides* (57%); or (v) were produced *in an environmentally better way*. So, there seem to be almost no convincing, product-related arguments for GM food. In addition, GM food seems to be risky according to about two-thirds of the population and people do not seem to trust existing safety regulations.

Finally, the pros and cons seem to be more balanced on the societal level. But, even here, half of the population thinks that genfood or gencrops are useful for the industry but not for consumers, and less than half trust the industry's argument that genfood or gencrops will be useful to combat hunger in the Third World. To sum up, green biotechnology is still regarded as risky and without practical benefits and, morally, it is against the natural order of things. And this negative evaluation of green biotechnology exists, not only in Switzerland but in most European countries, although on a weaker basis.

## The changing climate towards biotechnology in Switzerland

In addition to the detailed evaluation of different applications discussed above, there was a question dealing with the perception of biotechnology in a more general way: 'Do you think biotechnology will improve our way

**Table 4.3.** Negative opinions towards GM food, 2002/2003.

| Statement | Acceptance (%) | |
|---|---|---|
| | Yes | No |
| Product-related arguments for GM *versus* non-GM food | | |
| I would buy GM food if it were cheaper | 11 | 87 |
| I would buy GM food if it contained less fat | 15 | 82 |
| I would buy GM fruits if they tasted better | 25 | 70 |
| I would buy GM food if it contained fewer pesticides | 35 | 57 |
| I would buy GM food if it were produced in environmentally better ways | 37 | 55 |
| Safety arguments for GM *versus* non-GM food | | |
| I think that the existing regulations guarantee safety | 17 | 65 |
| I think that eating GM food is safe for me | 20 | 63 |
| GM food will not be a danger for future generations | 16 | 58 |
| Societal arguments for GM *versus* non-GM food | | |
| GM food will be useful for me and other people | 24 | 64 |
| Genfood/Gencrops are useful for industry, but not for consumers | 51 | 34 |
| GM food will be useful to combat hunger in the Third World | 47 | 42 |
| In the long term, the GM food industry will be important for the economy of a country | 38 | 39 |
| Moral argument for GM *versus* non-GM food | | |
| GM food threatens the natural order of things | 79 | 11 |

of life in the next 20 years, will it have no effect or will it make things worse?' (see Fig. 4.1). In 1997 and 2000, about 40% gave a positive opinion of biotechnology in Europe and about 20% were more sceptical.

Contrary to these more or less stable attitudes in Europe, the climate of opinion was much less stable in Switzerland. The first measurement shows slightly more sceptical attitudes in 1997, whereas the climate towards biotechnology had become much more positive by 2000, after the intensive pro-campaign in Switzerland. Four years after the referendum was refuted, the positive public climate had diminished by about 15%, nearer the level of European values.

Whereas there was no difference concerning positive attitudes towards biotechnology in 1997 between those groups who had heard of biotechnology in the media and those without media contact, in 2000 the level of favourable attitudes towards biotechnology in the segment with media contact was significantly higher than in the segment without media contact: 62 *versus* 53%.

## Agenda-setting processes

The media agenda was measured on the one hand by the number of articles in newspapers in the previous 4 months before the survey was carried out, and on the other by the number of people who said in the

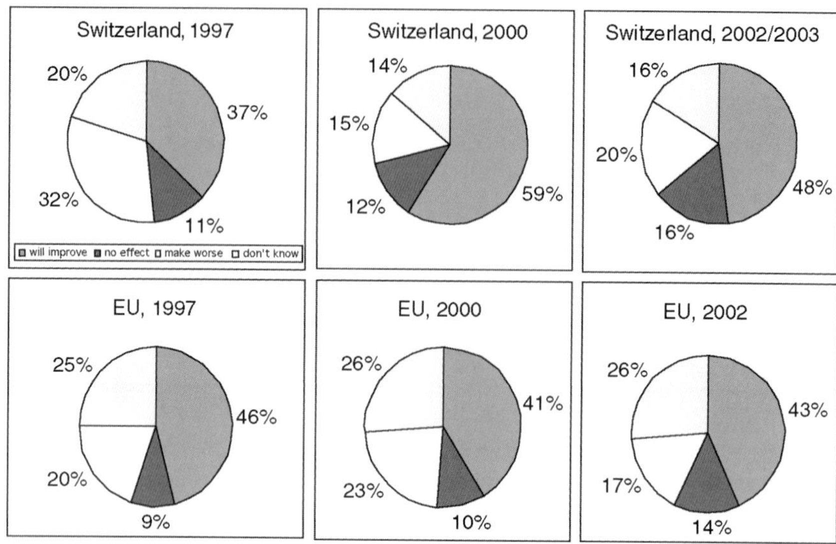


**Fig. 4.1.** Development of public opinion in Switzerland and the EU, 1997–2003.

survey that they had heard or read about biotechnology in the media in the previous 3 months. The public agenda was operationalized on one the hand by a question about the personal importance of biotechnology and, on the other, by the frequency of discussions about biotechnology with other people.

Table 4.4 presents mixed results concerning the agenda-setting function of media. The data show a marked quantitative increase in media agendas 1 year before the referendum for or against the *Gen-Schutz-Initiative*, and coverage was very controversial. But media coverage remained at a high level and increased even more after this political decision in 2000. Contrary to this increasing tendency of media coverage, media attention paid by people decreased slightly between 1997 and 2000, but increased again afterwards. The personal importance of biotechnology developed in a parallel way: It was highest in 1997 with a mean of 6.5, decreased to a mean of 5.0 in 2000 and increased again to 6.1 in 2002/2003. Contrary to these developments, discussions about biotechnology as an indicator of inter-personal agenda-setting did increase after the referendum in 1998, but decreased after 2000, possibly as a result of media coverage becoming less and less controversial.

## Knowledge gaps

Knowledge about biotechnology was measured on the basis of nine comparable schoolbook-type questions such as: 'Ordinary tomatoes do

**Table 4.4.** Increasing agenda-setting function of media, 1997–2003.

| Dimensions of agenda-setting | 1997 | 2000 | 2002/2003 |
|---|---|---|---|
| Media agenda | | | |
| Mean number of articles per newspaper p.a.[a] | 196 | 313 | 334 |
| Total number of articles p.a. (%) | 23 | 37 | 40 |
| Interviewee heard of biotech in last 3 months | | | |
| All media (%) | 79 | 59 | 72 |
| Print media (%) | 56 | 26 | 42 |
| Public agenda | | | |
| Personal importance of biotech (scale of 1–10) | 6.5 | 5.0 | 6.1 |
| Discussions about biotech.: often/sometimes (%) | 60 | 65 | 52 |

[a] Based on the eight biggest-selling newspapers in Switzerland.

not contain genes, while genetically modified tomatoes do: true or false?' In 1997, the mean knowledge was 5.6, which means that only about half of the population gave correct answers to the textbook-type knowledge questions; but this ratio was still significantly higher than in most other European countries (Bonfadelli, 2005).

The data in Table 4.5 show that knowledge decreased slightly between 1997 and 2000 from a mean of 5.6 to 5.1 correct answers and then increased again to 5.3. This tendency over time is consonant to the knowledge gap hypothesis that predicts increasing knowledge gaps when diffusion of information becomes more intensive, but closing gaps when intensity of conflict is high or media coverage is decreasing. In addition, there are knowledge gaps between the different educational segments according to knowledge gap theory. To sum up, as a result of the intensive media information flow in Switzerland, most people obviously acquired at least a minimal core understanding of this new phenomenon called biotechnology.

## Selective and attitude-consonant perceptions of media influences

Now, let's have a last look at media influences on attitudes. Forty-three per cent of the Swiss respondents in the second Eurobarometer

**Table 4.5.** Stable knowledge gaps between segments of education, 1997–2003.

| | | | Education | | | Knowledge gaps | |
|---|---|---|---|---|---|---|---|
| Knowledge measures | Year | Total | Low | Medium | High | Diff. | Corr. |
| Knowledge test: | 1997 | 5.6 | 4.9 | 5.3 | 6.6 | +1.7 | +0.31 |
| means (1–9 points) | 2000 | 5.1 | 4.4 | 5.1 | 5.7 | +1.3 | +0.20 |
| | 2003 | 5.3 | 4.5 | 5.2 | 6.1 | +1.6 | +0.31 |

survey in the year 2000 felt they had been influenced by the media coverage of biotechnology (see Table 4.6). Of those, 26% had become more critical of biotechnology and 17% had developed more positive attitudes towards it. At least, this was their personal perception. But this media impact seems to be mediated strongly by each individual's own original attitude towards biotechnology: 40% of those in favour of biotechnology reported that the media had influenced them to view biotechnology even more favourably, whereas 44% of those originally opposed to biotechnology shifted towards viewing it even more negatively. Thus, both groups seemed to view the media as having reinforced their attitudes.

These findings tie in with one conclusion from research on consonance models, postulating that different people use the same media content very selectively to support their existing attitudes – in this case toward biotechnology. Such predispositions function as frames that guide how people process new media content and the arguments presented for or against biotechnology.

## Conclusion

The case study of biotechnology through the 'watershed years' in Switzerland provided many differentiated insights into the complex interplay between intensified political debate and increasing media coverage, becoming more and more controversial on the one hand, while rising awareness, increasing public knowledge and shifting the climate of opinion on the other.

But our study raises many questions regarding political regulation and the active participation of the general electorate as well. Compared to other European countries, the public debate on biotechnology in Switzerland has been relatively intense. An interesting lesson to be

**Table 4.6.** Selective and attitude-consonant perception of influence, 2000.

| | Did media coverage influence attitude to biotechnology? (%) | | | | If yes, in what direction? (%) | |
|---|---|---|---|---|---|---|
| | Yes | No | Don't know | Total | In favour of | Against |
| All respondents (*n*, 1010) | 43 | 47 | 10 | 100 | 17 | 26 |
| Segments with different personal attitudes towards biotechnology | | | | | | |
| Positive | 52 | 41 | 7 | 100 | 40 | 12 |
| Ambivalent | 41 | 53 | 6 | 100 | 17 | 24 |
| Negative | 48 | 44 | 8 | 100 | 4 | 44 |

learned from this special democratic constellation is that the intense public controversy did not result in an anti-science shift but, instead, led to an increasing awareness and public understanding of biotechnology as a complex scientific topic, and to encouragement of an informed, democratic decision-making process. After the two referenda in 1992 and 1998, the debate has since slowed down somewhat. As a result, biotechnology seems to be covered by the media, after the advent of the new *Gen-Moratorium-Initiative* in November of 2005, on a much less emotionalized level than before.

## References


Aronson, E. (1997) A theory of cognitive dissonance. *American Journal of Psychology* 110, 127–137.

Bauer, M.A. and Bonfadelli, H. (2002) Controversy, media coverage, and public knowledge. In: Bauer, M.A. and Gaskell, G. (eds) *Biotechnology. The Making of a Global Controversy.* Cambridge University Press, Cambridge, UK, pp. 149–175.

Bonfadelli, H. (1994) *Die Wissenskluft-Perspektive. Massenmedien und gesellschaftliche Information.* Ölschlåger/UVK Verlag, Konstanz, Germany.

Bonfadelli, H. (2005) Mass media and biotechnology knowledge gaps within and between European countries. *International Journal of Public Opinion Research* 17 (1), 42–62.

Bonfadelli, H., Hieber, P., Leonarz, M., Meier, W., Schanne, M. and Wessels, H. (1998) Switzerland. In: Durant, J., Bauer, M. and Gaskell, G. (eds) *Biotechnology in the Public Sphere. A European Sourcebook.* Science Museum, London, pp. 144–161.

Bonfadelli, H., Dahinden, U., Leonarz, M., Schanne, M., Schneider, C. and Knickenberg, S. (2001) Biotechnology in Switzerland – from street demonstrations to regulations. In: Gaskell, G. and Bauer, M. (eds) *Biotechnology 1996–2000 – The years of the Controversy.* Science Museum, London, pp. 282–291.

Bonfadelli, H., Dahinden, U. and Leonarz, M. (2002) Biotechnology in Switzerland: high on the public agenda, but only moderate support. *Public Understanding of Science* 11 (2), 113–130.

Cotton, J.L. (1985) Cognitive dissonance in selective exposure. In: Zillmann, D. and Bryant, J. (eds) *Selective Exposure to Communication.* Hillsdale, New Jersey, pp. 11–33.

Dahinden, U. (2002a) Zwiespältige Beurteilung von Gentechnologie durch die Bevölkerung – eine Analyse von Argumentationsmustern mit Hilfe von Fokusgruppen. In: Bonfadelli, H. and Dahinden, U. (eds) *Gentechnologie in der Öffentlichen Kontroverse.* Seismo-Verlag, Zürich, Switzerland, pp. 97–112.

Dahinden, U. (2002b) Biotechnology in Switzerland – frames in a heated debate. *Science Communication* 24 (2), 184–197.

Dahinden, U. (2005) Framing: a decade of research experience. Paper presented at the *Annual Meeting of the International Communication Association*, 2005, New York.

Dearing, J.W. and Rogers, E.M. (1996) *Agenda-setting. Communication Concepts 6.* Sage, Thousand Oaks, California.

Durant, J., Bauer, M. and Gaskell, G. (eds) (1998) *Biotechnology in the Public Sphere.* Science Museum, London.

Gaziano, C. and Gaziano, E. (1996) Theories and methods in knowledge gap research since 1970. In: Salwen, M.B. and Stacks, D. (eds) *An Integrated Approach to Communication Theory and Research.* Erlbaum, Mahwah, New Jersey, pp. 127–143.

Gitlin, T. (1980) *The Whole World is Watching: Mass Media in the Making and Unmaking of the New Left.* University of California Press, Berkley, California.

Goffman, E. (1977) *Rahmen-Analyse. Ein Versuch, über die Organisation von Alltagserfahrungen,* 2nd edn. Suhrkamp-Verlag, Frankfurt, Germany.

Graf, N. (2003) Die Last von 30 Jahren Ökologiediskurs: Alte und neue Deutungsmuster in der Gentechnologiedebatte. In: Eisner, M., Graf, N. and Moser, P. (eds) *Risik Diskurse. Die Dynamik Öffentlicher Debatten über Umwelt- und Risikoprobleme in der Schweiz.* Seismo, Zürich, pp. 212–240.

Hieber, P. (1999) Gentechnologie in der Schweiz. In: Bonfadelli, H. (ed.) *Gentechnologie im Spannungsfeld von Politik, Medien und Öffentlichkeit.* Diskussionspunkt 37, IPMZ, Zürich, pp. 21–62.

Hornig-Priest, S. (1995) Information equity. Public understanding of science, and the biotechnology debate. *Journal of Communication* 45 (1), 39–54.

Iyengar, S. (1991) *Is Anyone Responsible? How Television Frames Political Issues.* University of Chicago Press, Chicago, Illinois.

Iyengar, S. and Simon, A. (1993) News coverage of the Gulf crisis and public opinion. A study of agenda-setting, priming, and framing. *Communication Research* 20 (3), 365–383.

Kepplinger, H., Gotto, K., Brosius, H. and Haak, D. (1989) *Der Einfluss der Fernsehnachrichten auf die Politische Meinungsbildung.* Freiburg, Münich, Germany.

Kronberger, N., Dahinden, U., Allansdottir, A., Seger, N., Pfenning, U., Gaskell, A., Rusanen, T., Montali, L., Wagner, W., Cheveigné, S., Diego, C. and Mortensen, A. (2001) The train departed without us – public perceptions of biotechnology in ten European countries. *Notizie di Politeia* 18 (63), 26–36.

Krueger, R. and King, J. (1998) *Involving Community Members in Focus Groups.* Sage, London.

Leonarz, M. (2002) Die Gentechnologie als kontroverses Medienthema. Eine Zeitungsanalyse von 1997 bis 1999. In: Bonfadelli, H. and Dahinden, U. (eds) *Gentechnologie in der Öffentlichen Kontroverse.* Zürich, pp. 25–43.

Leonarz, M. and Schanne, M. (1999) Gentechnologie als Medienthema. In: Bonfadelli, H. (ed.) *Gentechnologie im Spannungsfeld von Politik, Medien und Öffentlichkeit.* Seismo-Verlag, Zürich, Switzerland, pp. 63–98.

McCombs, M. (1998) Candidate images in Spanish elections: Second-level agenda-setting effects. *Journalism and Mass Communication Quarterly* 74 (4), 703–717.

McCombs, M. and Bell, T. (1996). The agenda-setting role of mass communication. In: Salwen, M. and Stacks, D. (eds) *An Integrated Approach to Communication Theory and Research.* Erlbaum, Mahwah, New Jersey, pp. 93–110.

McCombs, M. and Reynolds, A. (2002) News influence on our pictures of the world. In: Bryant, J. and Zillmann, D. (eds) *Media Effects. Advances in Theory and Research.* Hillsdale, New Jersey, pp. 1–18.

McCombs, M. and Shaw, D.L. (1972/1973) The agenda-setting function of mass media. *Public Opinion Quarterly* 36, 176–187.

Neuman, R., Just, M. and Crigler, A. (1992) *Common Knowledge. News and the Construction of Meaning.* University of Chicago Press, Chicago, Illinois.

Petty, R., Wegener, D. and Fabrigar, R. (1997) Attitude and attitude change. *Annual Review of Psychology* 48, 609–647.

Petty, R., Priester, J. and Briñol, P. (2002) Mass media attitude change: implications of the elaboration likelihood model of persuasion. In: Bryant, J. and Zillmann, D. (eds) *Media Effects. Advances in Theory and Research.* Erlbaum, Mahwah, New Jersey, pp. 155–198.

Reese, S. (2001) Prologue – framing public life a bridging model for media research. In: Reese, S., Gandy, O. and Grant, A. (eds) *Framing Public Life. Perspectives on Media and Understanding of the Social World.* Erlbaum, Mahwah, New Jersey, pp. 7–31.

Scheufele, D. (1999) Framing as a theory of media effects. *Journal of Communication* 49 (1), 103–122.

Tichenor, P., Donohue, G. and Olien, C. (1970) Mass media flow and differential growth in knowledge. *Public Opinion Quarterly* 34 (2), 159–170.

Viswanath, K. and Finnegan, J. (1996) The knowledge gap hypothesis: twenty-five years later. In: Burleson, B. and Kunkel, A. (eds) *Communication Yearbook 19.* Sage Publishing, Thousand Oaks, London/New Delhi, pp. 187–227.

Wagner, W., Kronberger, N., Gaskell, G., Allum, N., Allansdottir, A., Cheveigné, S., Dahinden, U., Diego, C.M., Lorenzo, M., Pfenning, U.R. and Seger, N. (2001) Nature in disorder – the troubled public of biotechnology. In: Gaskell, G. and Bauer, M. (eds) *Biotechnology 1996–2000 – The Years of the Controversy.* Science Museum, London, pp. 80–95.

Weingart, P. (1998) Science and the media. *Research Policy* 27, 869–879.

# 5

# Genetically Modified Foods: US Public Opinion Research Polls

W. Fink* and M. Rodemeyer**

*Science and Public Policy, The Pew Initiative on Food and Biotechnology (PIFB), Washington, DC, USA;*
*e-mail: *wfink@pewagbiotech.org; **michael@mrodemeyer.com*

## Introduction

Since 2001, the Pew Initiative on Food and Biotechnology (PIFB), a research and education project of the University of Richmond, Virginia, USA, has carried out four major national opinion polls to elicit and track US consumer views with regards to agricultural biotechnology. In addition to the polls, which are publicly available on its website, PIFB also commissioned in 2004 a series of focus groups to provide additional insight into US consumer opinion. As a result, PIFB's public opinion research is one of the richest publicly available sources of information about American attitudes toward genetically modified foods and other agricultural biotechnology issues since 2001. This chapter provides an overview of the findings of the PIFB polls and focus groups, combined with findings from other selected public opinion research.

PIFB contracted with the public opinion research firms, The Mellman Group and Public Opinion Strategies, to conduct four national telephone polls, in 2001, 2003, 2004 and 2005. Each survey polled about 1000 Americans selected through random digit dialling, with error rates of +/–3.1% at the 95% level of confidence. In each case, the surveys included some questions contained in the previous survey in order to track changes in opinion over time, although each survey also contained unique questions.

In order to gain additional insight into consumer perspectives, PIFB commissioned The Mellman Group and Public Opinion Strategies to conduct four focus groups in Philadelphia, Pennsylvania and Des Moines, Iowa in 2004, selecting a cross-section of the population who reported being responsible for grocery shopping. Combining the information from the focus groups with the results of the detailed surveys and other consumer survey information provides a rich picture of US consumer attitudes toward GM foods and genetic engineering.

## Summary of Findings: PIFB Poll Results, 2001–2005

Though this chapter will examine the results of our polls in detail, the results can be summarized as follows for those questions which have been tracked over the four survey years:

- Americans know relatively little about GM foods and biotechnology.
- Support for GM foods has been stable, but not strong; while overall opposition has been unstable, though strong opposition has remained steady.
- American opinion is malleable and will change given additional information about GM foods.
- Though they know little about GM foods and even less about how those foods are regulated, when asked about regulation most Americans would like to see more regulation of GM foods.
- Americans are more comfortable with the genetic modification of plants than of animals.
- Support for individual applications is sometimes much higher than the general categories of GM foods or GM animals.
- Americans are most supportive of those applications which they feel will be most beneficial to themselves, their family or to society as a whole; they are less supportive of applications that, on the surface, appear to help only businesses or individual industries.

## Informed Public Opinion?

In 2001, a little more than half of Americans reported having seen, read or heard not too much or nothing about GM foods (PIFB, 2001; Fig. 5.1). Two and three years later, that number had increased to more than three-fifths of those responding (PIFB, 2003, 2004). These numbers are reflected in other polls taken in the same timeframe (Hallman *et al.*, 2004; Harris Interactive, 2004). However, PIFB's most recent 2005 poll shows a slight decrease in those saying they had heard little or nothing about GM foods (PIFB, 2005a).

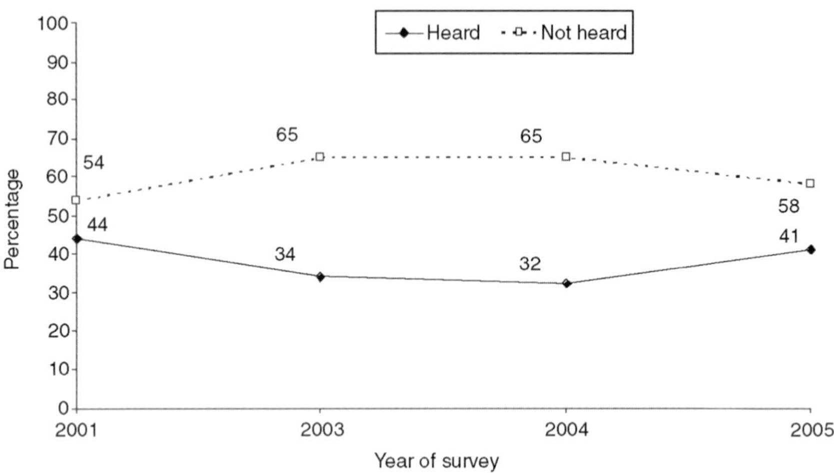


**Fig. 5.1.** Americans' awareness of GM food. Based on four PIFB surveys: 2001, 2003, 2004 and 2005 (*n*, 500 for 2001, 2003, 2004; *n*, 1000 for 2005). The question asked was: 'How much have you seen, read or heard about GM food that is sold in grocery stores?'

Conversely, the percentage of Americans who said that they had heard 'some' or a 'great deal' about GM foods dropped substantially from 2001 to 2003 and 2004, but then increased slightly in 2005 (see Fig. 5.1). In 2001, 44% of Americans reported having seen, read or heard recently about GM foods but, in 2003 and 2004, only 34 and 32%, respectively, said they had heard at least something about GM foods (PIFB, 2001, 2003, 2004). This drop in awareness has been attributed to the media's ever-changing attention to GM foods (PIFB, 2003).

In September of 2000, an unapproved variety of GM maize, StarLink™, was found to have entered the human food supply. Several food companies recalled their products containing yellow maize when it was discovered that they contained trace amounts of StarLink™ maize. These recalls and the subsequent furore over the loss of maize trade markets and who should bear the responsibility for the mistakes warranted a good deal of media attention. Additionally, because recalls involve urging consumers to return or discard items they have purchased, consumers had more of a reason to pay attention to the StarLink™ stories than subsequent news stories involving GM foods.

Since 2001, however, there has not been an equivalent headline-grabbing news story involving GM foods that would have engaged consumers as the StarLink™ story did. It is not surprising, then, that

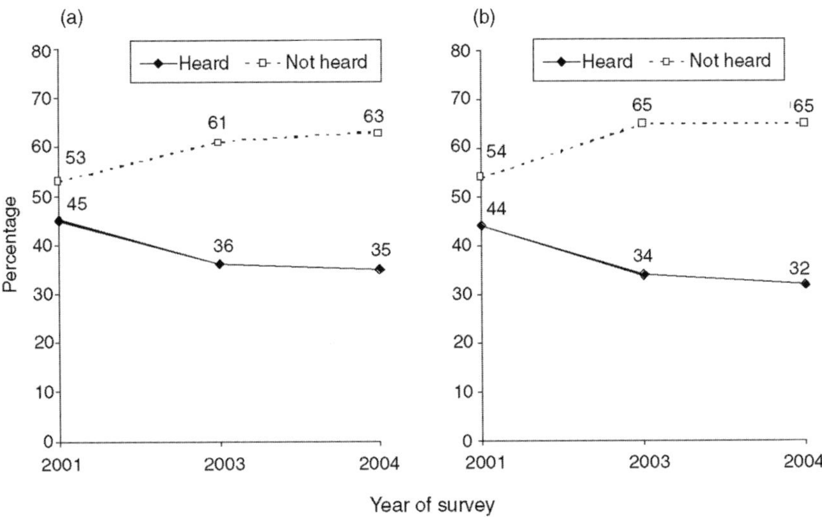


**Fig. 5.2.** Americans' awareness of (a) food derived from biotechnology and (b) GM food. Based on three PIFB surveys: 2001, 2003 and 2004 (*n*, 500). The question asked was: 'Have you seen, heard or read recently about biotechnology-derived/GM food?'

more people reported having seen, read or heard about biotech foods in 2001 than they did in 2003 or 2004.

PIFB's 2005 poll, however, showed an uptick in respondents reporting having heard, read or seen something about genetically modified foods in grocery stores (see Fig. 5.1). Unlike 2001, there does not seem to be a corresponding news story that would account for this rise in awareness. Something, however, has apparently caught Americans' attention, even if briefly. There was a fifteen percentage point decrease in those reporting that had heard 'nothing at all' and corresponding seven and eight percentage point increases in those reporting to have heard 'not to much' and 'some', respectively. Future polls or a media analysis could shed more light on what caused this increase in awareness.

Despite the lack of media stories about GM foods in recent years, consumers appear to retain knowledge from previously reported stories. Within two of the four focus groups held in 2004, the StarLink™ case was raised as an issue nearly 3 years after it had occurred. One Philadelphia man stated: 'I've been following the news. I can remember it was some stink storm when some genetically altered maize got loose in Minnesota in the general population. They're not approved across the board right now because they don't necessarily know how safe they are' (PIFB, 2004).

Awareness of GM food varies somewhat by demographic variables. Those who are completely or partially responsible for their household's food shopping are much more likely to know something about foods derived from biotechnology than are those who have no food shopping responsibility. Those with more education are also more likely to have seen, read or heard about GM foods. Men are slightly more likely than women to have seen, read or heard about GM foods. Of course, because these subgroups were sampled from less than 1000 people, the results are more prone to error (PIFB, 2001, 2003, 2004, 2005a).

There appears to be little difference in awareness in response to different terminology. The surveys used a split sample to test whether consumers responded differently to the terms 'genetically modified' and 'biotechnology.' The terminology used made, at most, a 4% difference in only one of the responses (PIFB, 2001, 2003, 2004) (see Fig. 5.2).

## Public Opinion about GM Foods

While a majority of Americans report having heard relatively little about GM foods over the last 5 years, their lack of awareness has not translated into an absence of opinions about GM foods.

One initial polling issue that must be addressed is the use of 'prompting' questions to elicit consumer opinions. The very act of 'prompting' may have an impact on the response. Indeed, this effect is demonstrated by consumer polling conducted by the International Food Information Council (IFIC) in June 2005, which used open-ended questions to elicit a list of food safety concerns from consumers. By using this method, the IFIC poll revealed that only one-half of one percent of respondents spontaneously identified food biotechnology with food safety issues (IFIC, 2005).

In contrast, other polls, including those of PIFB, prompt consumers to respond to specific questions about GM foods, with very different results. In the 2005 PIFB poll, for example, 30% of consumers said that GM foods were 'basically unsafe', 18% believing that 'strongly' (PIFB, 2005a).

As with all polling questions, one must be cautious about interpreting the results. The IFIC survey demonstrates that the safety of foods derived from biotechnology is not a 'top of mind' issue for most US consumers. Given the lack of awareness about GM food among US consumers, the finding is not that surprising. However, those findings do not necessarily mean that consumers believe that GM food is safe. The findings from prompted survey questions show consistent results over time and among different surveys, demonstrating real underlying concerns.

To get at this issue in another way, the PIFB 2001 poll listed a series of items that people often say they are concerned about when it comes to food safety (see Table 5.1). Respondents were then asked to state whether it was one of the things that worried them about food safety. By far, the biggest concerns for consumers are food poisoning, followed closely by food freshness and salmonellosis. Foods produced through biotechnology or GM foods registered at only half the rate of concern of food poisoning and food freshness. Even when combining the responses of 'one of the most' and 'great deal,' only one-third of respondents registered concern about these foods, compared to two-thirds for food poisoning and nearly three-quarters for food freshness (PIFB, 2001).

The 2001 PIFB survey also tested whether consumers had different prompted responses about safety depending on the use of either the term 'genetically modified foods' or 'biotechnology used in food production' (see Table 5.2). In the first question, Americans were asked whether they had a favourable or unfavourable opinion of irradiated foods, organic foods or, using a split sample, either GM foods or biotechnology used in food production. Americans had a largely favourable opinion of organic foods, with considerably lower ratings for the other three categories.[1]

The percentage of unfavourable responses increased for GM foods compared to 'biotechnology used in food production', but the 'biotechnology' term elicited more 'don't know' or 'never heard of' responses.

Moving from food as a whole to GM foods in particular, PIFB asked respondents to specify what they'd heard about GM foods (PIFB, 2003, 2004; Table 5.3). Analysts then coded the appropriate category for each response. Multiple responses were allowed, so the figures do not add up to 100. Excluding the categories of 'don't know' or 'other,' the idea that GM foods were unsafe topped the list in both years.

## Support and Opposition

Given that most Americans know very little about GM foods, it is not surprising that, when asked if they favoured or opposed the

1 Contrary to the previous discussion of the use of the terminology 'genetically modified' or 'biotechnology,' this question did reveal a difference in opinion between the two terms. This may be a result of both terms being used in the same question in comparison to the use of one term or the other in a split-sample question. In this same survey, however, there was little difference in response to how much people had heard, read or seen whether using the terms GM foods or biotechnology (PIFB, 2001).

**Table 5.1.** Public concerns about food safety (from PIFB survey, 2001).[a]

| Food safety concern | One of the most/great deal (net) | Not too much/not at all (net) | One of the most | Great deal | Some-what | Not too much | Not at all | Don't know |
|---|---|---|---|---|---|---|---|---|
| Food freshness | 71 | 10 | 33 | 38 | 18 | 5 | 5 | 1 |
| Food poisoning | 67 | 16 | 35 | 32 | 17 | 10 | 6 | 1 |
| Salmonella | 66 | 13 | 33 | 33 | 18 | 7 | 6 | 2 |
| Chemicals and fertilizers being used in food production | 46 | 21 | 20 | 26 | 32 | 11 | 10 | 1 |
| Genetically modified foods (split sample; *n*, 500) | 34 | 29 | 17 | 17 | 27 | 13 | 16 | 11 |
| Biotechnology being used in food production (split sample; *n*, 500) | 32 | 27 | 15 | 17 | 30 | 16 | 11 | 11 |
| Irradiation | 32 | 21 | 15 | 17 | 18 | 8 | 13 | 29 |
| Listeria | 25 | 14 | 14 | 11 | 15 | 6 | 8 | 47 |

[a] Interviewees were asked: 'I'm going to list some things people tell us they are concerned about when it comes to food safety. After each, tell me whether it is one of the things that worries you most about food safety, whether it worries you a great deal, somewhat, not too much or not at all' (*n*, 1000).

**Table 5.2.** Public attitudes toward various foods (from PIFB survey, 2001).[a]

| Food type | Very/somewhat favourable | Very/somewhat unfavourable | Don't know | Never heard of |
|---|---|---|---|---|
| GM food | 21 | 44 | 22 | 15 |
| Biotechnology used in food production | 25 | 30 | 25 | 19 |
| Irradiated foods | 17 | 25 | 22 | 36 |
| Organic foods | 64 | 15 | 15 | 5 |

[a] Interviewees were asked: 'Overall, do you have a very favourable, somewhat favourable, somewhat unfavourable or very unfavourable impression of?' (*n*, 1001).

**Table 5.3.** Levels of awareness of genetically modified foods (from PIFB surveys, 2003, 2004).[a]

| Response | 2003 | 2004 |
|---|---|---|
| Unsafe | 17 | 14 |
| Genes are altered/genetically modified | 14 | 11 |
| Controversial/much debate | 9 | 9 |
| Safe | 6 | 6 |
| Healthier/better than other foods | 5 | 5 |
| Opposition from others – especially other countries | 4 | 2 |
| Unnatural/not as good as natural food | 4 | 6 |
| General knowledge | 4 | 0 |
| Grow faster | 3 | 4 |
| Same as other foods | 2 | 2 |
| Reduce the cost of food | 1 | 3 |
| Immoral | 1 | 0 |
| Unregulated | 1 | 1 |
| Generally negative | 1 | 0 |
| Other | 5 | 16 |
| Don't know | 30 | 30 |
| Regulated by the FDA | – | 1 |

[a] Interviewees were asked: 'What have you heard about genetically modified foods? Please be as specific as possible' (Asked of those stating that they had heard a great deal, some or not too much about GM foods or foods derived from biotechnology. Multiple responses allowed).

introduction of GM foods to the food supply, many Americans opposed their introduction.

This is, after all, a relatively new technology and one that can easily be misinterpreted by consumers. Additionally, the fact that this technology is being applied to food makes it all the more controversial. Unlike other areas where new technology is applied, such as television, telephones and computers, food is something everyone must have to survive. Moreover, unlike medicine where the benefits of the technology accrue directly to the patient and the risks are borne by that patient, consumers perceive that they bear the food safety risks of consuming GM foods without receiving any direct benefit. Current foods derived from biotechnology may provide indirect benefits through cheaper food, fewer chemical applications in crop production and enhanced soil conservation, but consumers are unlikely to be aware of those benefits.

There has been a slight drop in numbers among those who oppose GM foods over the past several years. In 2001, the majority of Americans (58%) opposed them but, by 2003, 2004 and again in 2005, half or less (48, 47 and 50%, respectively) opposed their introduction (see Fig. 5.3). Again, this may be attributed to fewer negative news stories about GM foods now that the StarLink™ story has disappeared from the headlines.

It is important to note, however, that while opposition as a whole has dropped since 2001, the percentage of people who *strongly* oppose GM foods has remained stable at about one-third of the respondents. Through all four years, only one quarter of Americans favoured the introduction of GM foods and, by 2004 and 2005, an equal number were not sure of their opinion regarding GM foods (PIFB, 2001, 2003, 2004, 2005a).

The large percentage of respondents who were unsure in their answers can pose a problem in determining where their support is

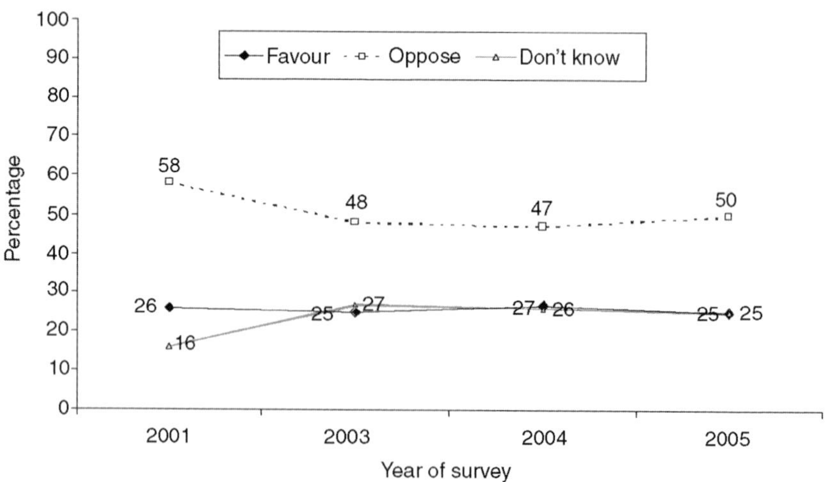


**Fig. 5.3.** Americans' approval of/opposition to GM food. Based on four PIFB surveys: 2001, 2003, 2004 and 2005 (*n*, 1000). The question asked was: 'Do you favour or oppose the introduction of GM food to the food supply?'

likely to lie. In their 2004 survey, Hallman *et al.* pressed those who did not give a definitive answer to a similar question (PIFB, 2000, 2003, 2004, 2005). They asked: 'How do you feel about the use of genetic modification to create plant-based food products. Do you approve, neither approve nor disapprove, disapprove or are you unsure of your opinion?' For those answering 'unsure' (38%) or 'neither approve nor disapprove' (11%), the authors asked to which direction they leaned. This resulted in a nearly even split between those who leaned toward approval (39%) and disapproval (36%). Twenty-five per cent still remained unsure. For the total sample, therefore, the addition of those leaning in one direction or the other moved approval from 28 to 47% and disapproval from 23 to 41% (Hallman *et al.*, 2004).

Looking at gender as a subgroup, men were less likely than women to oppose GM foods and more likely to consider them safe. Other surveys have shown similar results (Hossain *et al.*, 2002; Canadian Biotechnology Secretariat, 2005). Through all four years of the PIFB survey, about one-third of men favoured GM foods, while only one-fifth of women supported them. Between 2001 and 2005, however, the gender gap narrowed by ten percentage points for those opposed to the introduction of GM foods. In 2001, two-thirds of women opposed GM foods, while one-half of men opposed them. By 2005, slightly more than one-half of all women (53%) and slightly less than one-half of men (46%) opposed GM foods.

PIFB's poll results show that men reported having seen, read or heard about GM foods more (between 3 and 6 percentage points) than women every year but, unfortunately, the surveys do not reveal what type of articles each sex was reading.

One hypothesis for the narrowing gender gap could be the drop in the number of negative food safety articles in the years following StarLink™. In the PIFB 2005 survey, more than two-thirds of women considered themselves the primary shoppers for their households, as opposed to only two-fifths of men. Because primary shoppers are probably more likely to notice or pay attention to food safety articles, the subsequent drop in such articles may have allowed the gender gap to narrow to a more natural equilibrium. Additional survey results and a media analysis would be necessary to investigate this hypothesis further. During those same years, those who stated they did not know nor had an opinion on GM foods rose from 16 to 26% (PIFB, 2001, 2003, 2004, 2005a).

Given the question of how likely they would be to eat GM foods, a slight majority of Americans felt it was unlikely they would eat them (between 50 and 54% during 2001–2005), while a little more than one-third thought that they would eat GM foods (38–43% for the same period) (see Fig. 5.4).

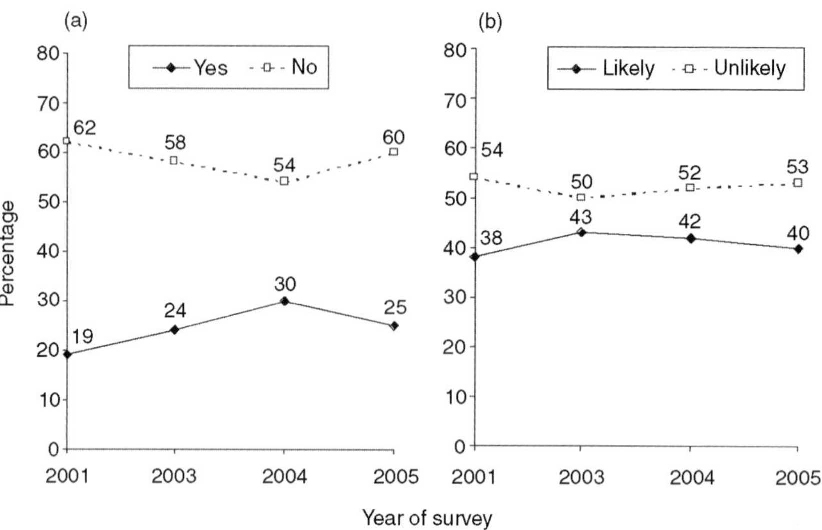


**Fig. 5.4.** Americans' consumption of GM food. Based on four PIFB surveys: 2001, 2003, 2004 and 2005 (*n*, 1000). The questions asked were: (a) 'Have you eaten GM food in the past?' and (b) 'Are you likely to eat GM food in the future?'

Since this is a question regarding individual choice and the numbers of consumers opposed to the introduction of GM foods is high, it is not surprising that many Americans felt it unlikely that they would eat GM foods.

The previous question was preceded by one asking: 'As far as you know, have you ever eaten genetically modified foods?' The answers to this question demonstrate several points. First, the majority of Americans continue to be unaware that GM foods are in the food supply. For the 2001–2005 timeframe, between 54 and 62% of Americans believed they had not eaten GM foods (PIFB, 2001, 2003, 2004, 2005a; Fig. 5.4). Knowledge about the presence of GM foods may be growing, however, because the number of those who believe they have eaten GM foods grew from 19% in 2001 to 30% in 2004 (Fig. 5.4). This trend is also seen in other polls (Hallman *et al.*, 2004). In 2005, this figure fell back to 25% (see Fig. 5.4), indicating that additional surveys would be necessary to verify a trend of increasing knowledge about eating GM foods (PIFB, 2005a). Finally, even those in the previous question who thought it was likely they would eat GM foods apparently viewed this as a future event, not as a current one. In 2005, 40% of consumers thought they would probably eat GM foods, but only 25% thought they had eaten them. In 2001 and 2003, the spread was even wider, with a 19-point difference in both years (PIFB, 2001, 2003; Fig. 5.4).

**Table 5.4.** Levels of public knowledge of genetics (from Hallman *et al.*, 2004).[a]

| Statement | Correct response (%) | Incorrect response (%) | Unsure response (%) |
|---|---|---|---|
| Ordinary tomatoes do not contain genes while GM tomatoes do | 40 | 9 | 51 |
| By eating a GM fruit, a person's genes could also become modified | 45 | 13 | 42 |
| The mother's genes determine whether a child is a girl | 57 | 12 | 31 |
| GM animals are always bigger than ordinary animals | 36 | 17 | 47 |
| It is not possible to transfer animal genes into plants | 30 | 18 | 52 |
| Tomatoes modified with genes from a catfish would probably taste fishy | 42 | 15 | 43 |
| Cloning produces genetically identical copies | 54 | 17 | 29 |
| More than half the human genes are identical to those of chimpanzees | 40 | 16 | 44 |
| Scientists sometimes genetically modify plants so that they cannot reproduce | 44 | 10 | 46 |
| Larger organisms have more genes | 38 | 14 | 48 |
| Most of the soybeans grown in the USA are a genetically modified variety | 27 | 11 | 62 |
| GM maize is required to be kept separate from non-GM maize | 12 | 39 | 49 |

[a] Interviewees were asked: 'For each of the following statements, please tell me whether you think it is true or false or are you not sure' (n, 600).

Further evidence of the disconnect between reality in the marketplace and public knowledge of GM foods can be gleaned from the PIFB 2001 poll and Hallman's 2004 poll (Hallman *et al.*, 2004). In 2001, only 14% of Americans thought that more than 50% of food in a typical American grocery store was genetically modified (PIFB, 2001). At the time, it was estimated that 70% or more of the processed food on American shelves contained genetically modified ingredients. Additionally, Hallman *et al.* found, in their 2004 survey, that even those who believed GM foods to be on the market had difficulty identifying which foods were genetically modified. Seventy-nine per cent of respondents thought that GM tomatoes were on the market when, in reality, GM tomatoes had not been on the market for 7 years. Moreover, 61% believed GM chickens were available, though GM chickens have never been on the market (Hallman *et al.*, 2004).

Finally, basic genetics knowledge and more specific knowledge of GM foods is very low among Americans (see Table 5.4). Hallman *et al.*

examined this issue by asking a series of basic questions and tracking whether consumers answered correctly or not. They found that less than 1% of respondents could answer all questions correctly, and 58% could not answer even half of the questions correctly (Hallman *et al.*, 2004).

## Regulatory Issues and GM Foods

It is difficult to tease out consumer opinion on complex policy questions when the public appear to know very little about the topic to begin with. One striking number to bear in mind when looking at polls asking policy questions is that, during the past 4 years, 57–65% of Americans polled by PIFB reported having seen, read or heard 'not too much' or 'nothing at all' about genetically modified foods. At the other end of the spectrum, only 9% or less stated they had seen, read or heard a great deal about these foods.

American public opinion, then, is still very much malleable and open to factors that could change it. For example, PIFB asked Americans whether GM foods were basically safe or unsafe. In each year polled (2001, 2003, 2004 and 2005), nearly one-third felt these foods were safe, a one-quarter felt they were unsafe and nearly two-fifths or more said they did not know or had no opinion (see Fig. 5.5). Americans were then told that more than half of the products at the grocery store had been produced using some form of biotechnology or genetic modification. Having this single piece of information altered many people's opinion about the safety of GM foods. After receiving this information, nearly one-half of Americans stated that GM foods were safe, an increase of 17 points or more depending on the year of survey.

On the other hand, it appears that those who felt GM foods were unsafe were not heavily influenced by this information. This category, like the category of those who opposed GM foods, changed only a little (a decrease of 2–4% depending on the year) over all four years. The majority of those who changed their minds regarding the safety of GM foods, then, came from the 'don't know/no opinion' category. Those stating originally that they 'didn't know or had no opinion' dropped by at least 13 points. This demonstrates the plasticity of American public opinion around this topic (PIFB, 2001, 2003, 2004, 2005a).

Given this malleability, asking questions of an uninformed public can be difficult and the answers one receives from such questions should always be viewed with a somewhat sceptical eye. It is easy to imagine that public opinion would change if 'they only knew what I know'.

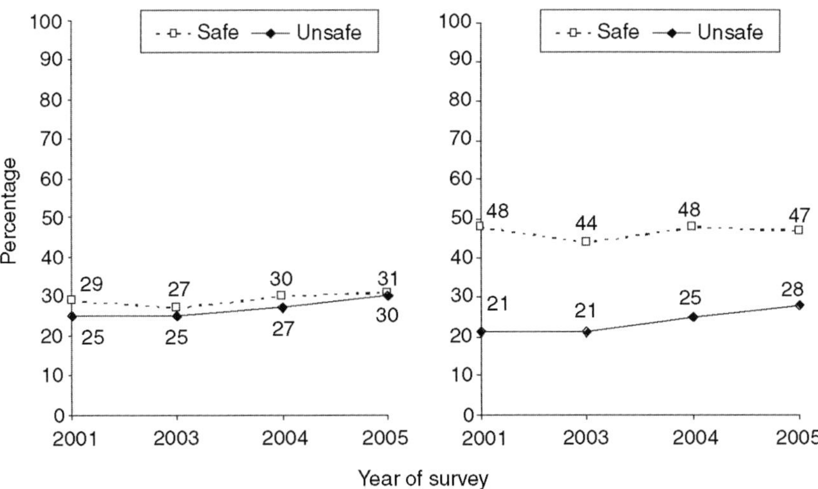


**Fig. 5.5.** Americans' attitudes to the safety of GM food. Based on four PIFB surveys: 2001, 2003, 2004 and 2005 (*n*, 1000). The questions asked were: (a) 'Are GM foods basically safe or unsafe?' and (b) 'Now, as you may know, more than half of processed products at the grocery store are produced using some form of biotechnology or genetic modification. Knowing this, is GM food basically safe or unsafe?' The term 'processed' referring to products was introduced only in the most recent of the four surveys.

Unfortunately, most polls are really only a snapshot of public opinion in time and provide little or no concrete information on the drivers for that opinion. To get at some of the underlying reasons for the public's opinion on agricultural biotechnology, PIFB has utilized lengthier and deeper polls, as well as focus groups. Even with these tools, it is only possible to scratch the surface of an issue like agricultural biotechnology. To truly understand public thought on GM foods and receive informed feedback, one would probably have to hold members of the public captive for a great deal of time while providing them with an in-depth education on agriculture, food production systems, ethics, regulatory policy, international trade, rural and international development and many other issues.

PIFB has tried through the years to ask policy questions from different angles in hopes of further elucidating what is driving public opinion. For example, in 2001, PIFB asked the public who they trusted for information on GM foods (see Fig. 5.6).

The source named as the most trustworthy was the Food and Drug Administration (FDA), with 83% of Americans stating that they trusted FDA a great deal (PIFB 2001). A similar question was asked in 2003 and, again, FDA was seen as the most trustworthy (PIFB, 2003).

In 2003 and 2004, PIFB explored the area of regulation of GM foods.

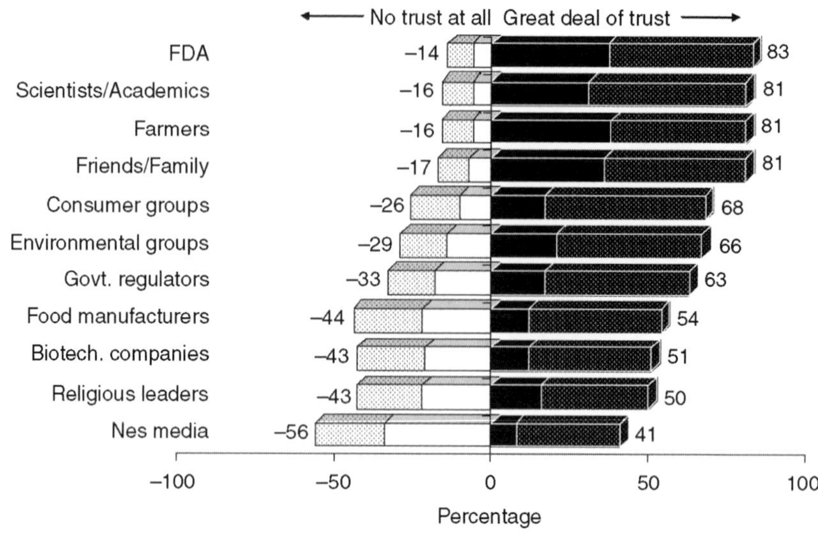


**Fig. 5.6.** Americans' trust in various sources of information about GM food. Based on a PIFB survey in 2001 (*n*, 1000).

Consumers were asked how much they knew about the federal regulation of GM foods. Not surprisingly, given how few Americans have heard, read or seen anything about GM foods, very few respondents reported knowing a great deal or even something about regulation of GM foods. Only 2 and 1% in 2003 and 2004, respectively, felt they knew a great deal about the regulatory system. This lack of knowledge, however, did not prevent Americans from giving their opinion about the amount of regulation required for GM foods. An average of 37.5% felt there was too little regulation, while 22% thought there was just the right amount (see Fig. 5.7). Only 9% responded that GM foods were over-regulated (PIFB 2003, 2004). This is probably a response on the part of Americans to regulations in general, given how little people indicated they knew about GM foods and regulation of those foods in these same polls.

In PIFB's 2003 poll, consumers were questioned on current and hypothetical FDA policies. Consumers were read a series of potential policy options and asked to state whether they agreed or disagreed with the statements. The responses to this question further demonstrated Americans' substantial faith in the FDA's ability to protect the food supply (PIFB, 2003).

As shown in Fig. 5.8, 89% of Americans would like to see a mandatory requirement for companies to submit data to the FDA and for the FDA to generate a determination of safety. Seventy-four per cent strongly supported this idea (PIFB, 2003). In the biotechnology

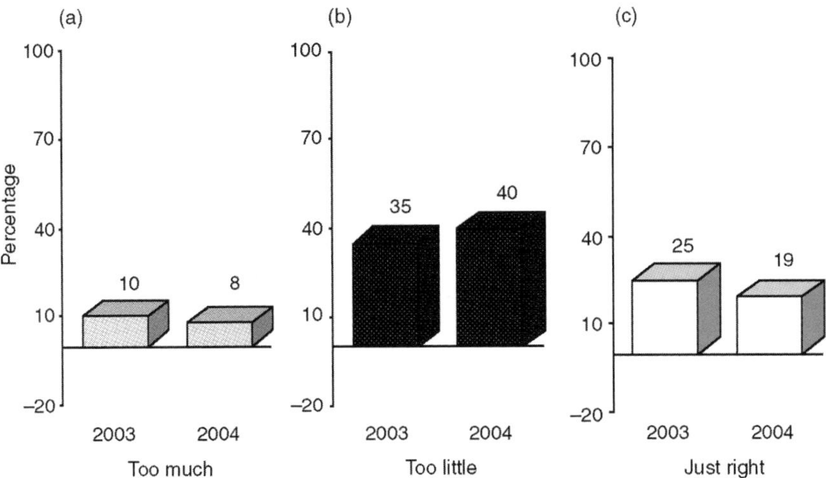


**Fig. 5.7.** Americans' opinions on the regulation of GM food. Based on two PIFB surveys: 2003 and 2004 (*n*, 1000). The questions asked were: 'Generally, is there (a) too much, (b) too little or (c) just the right amount of regulation of GM food?'

regulatory debate, the two ends of the spectrum are represented by those who wanted to ban the technology and those who argued for no government oversight at all. The majority of the American public, however, are not comfortable with either argument. Only one-third could agree with the statement that GM foods should be allowed on the market without review by FDA, and slightly less than one-third felt that GM foods should not be allowed to be sold even if the FDA found them to be safe (PIFB, 2003).

In 2004, PIFB probed more on questions involving FDA's policies regarding GM foods. Specifically, respondents were told: 'Currently, the FDA does not require companies to prove that conventional foods are safe before they can be sold on the market, and companies can bring new, conventionally produced foods to the market without prior approval from the FDA. The FDA can act to take a food off the market if there's a safety problem'. Respondents were then read a list of goals that had been suggested for the regulation of GM foods and asked whether they favoured or opposed those goals (see Fig. 5.9).

Figure 5.9 shows that the majority of Americans (80%) would like to see GM foods labelled. This feeling was also expressed routinely in focus groups. One Iowa woman stated: 'I think it's having the option. We don't have the option to eat or not to eat, we need to eat. So someone messing genetically with our food sources, that worries me. Technology as far as medications, inoculations – those I can opt in or out of, but I can't opt whether to eat or not to eat. So that's the part that concerns me, is the control' (PIFB, 2004).

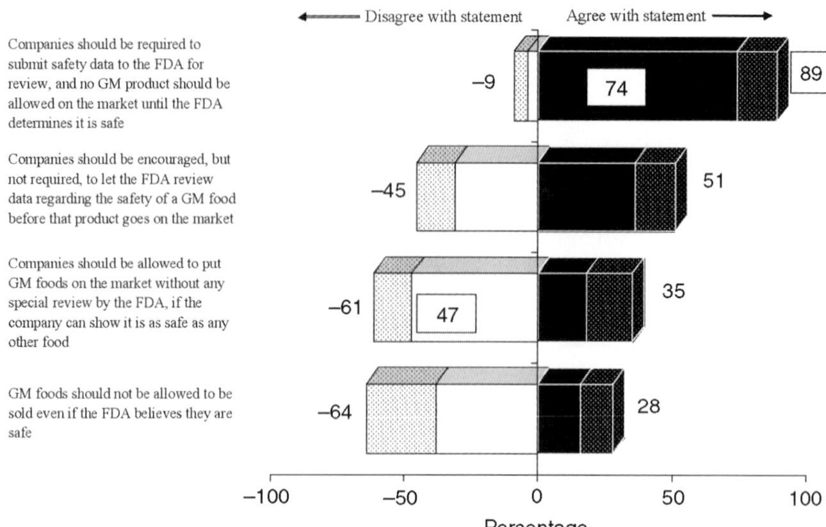


**Fig. 5.8.** Americans' opinions on the regulation of GM food by the FDA. Based on a PIFB survey in 2003 (*n*, 1000). Solid shading signifies stronger intensity of opinion.

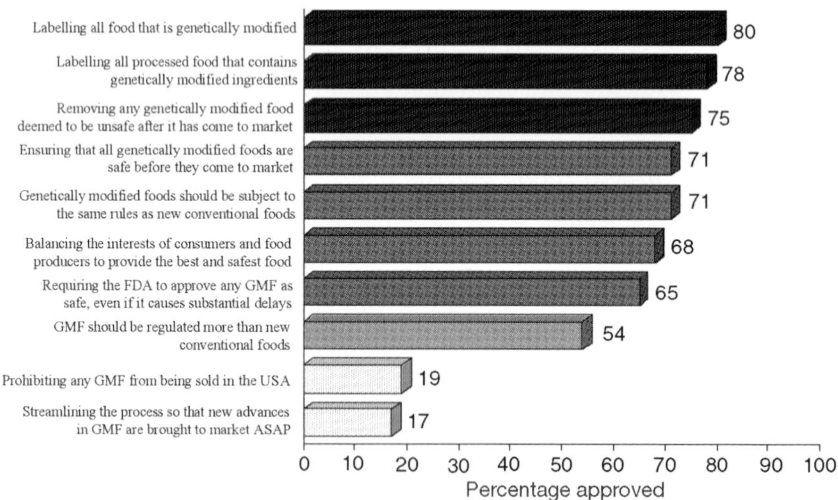


**Fig. 5.9.** Americans' opinions on various measures to improve the regulation of GM food. Based on a PIFB survey in 2004 (*n*, 1000).

The question of labelling GM foods is a difficult issue. When IFIC has asked consumers if there was any information not currently on food labels that they would like to see, only 1–2% named GM foods or biotechnology, demonstrating once again that this technology is not a top-of-mind concern for consumers (IFIC, 2005).

When prompted about labelling for specific reasons, consumers will routinely answer 'yes'. Hallman *et al.* (2004) found that 89% of Americans would prefer to have GM foods labelled after they were told that no laws currently required labelling. In this same survey, respondents were asked to rate, on a scale of 1–10, how important it was that food labels indicated whether foods were produced using certain methods or in certain localities. Using the means of those ratings it is possible to see that Americans rank the potential labels in order of importance as follows: (i) food grown with pesticides; (ii) food containing GM ingredients; (iii) food grown organically; (iv) country of origin; (v) food grown using traditional cross-breeding methods; (vi) food grown locally; and (vii) food grown in the USA (Hallman *et al.*, 2004).

The Center for Science in the Public Interest had a similar question in their 2001 survey. That survey showed that 76% of Americans would like food labelled where ingredients were from crops which had been sprayed with pesticides, 70% would like labels for crops that had been genetically engineered and 40% would also like food made from cross-bred maize to be labelled (CSPI, 2001).

The question then arises of whether the information is actually useful enough to the consumer that the consumer would, in fact, use it and pay the costs to cover it. CSPI explored this issue by asking respondents if they would be willing to pay for labelling if labelling increased the cost of their food. Three-fifths were not willing to raise their food budget by more than 0.1% per year, while less than one-third were willing to raise their food budget by 0.9% a year to cover the costs of labelling (CSPI, 2001). The labelling issue, then, poses problems for both policy-makers and surveys alike.

While the majority of Americans favour the labelling of GM foods, Fig. 5.9 demonstrates that the majority of Americans are also opposed to banning agricultural biotechnology. One woman in the Philadelphia focus group summed up the cautionary, but optimistic feelings of her group: 'Yeah, so at least you have the option, and given the fact that we have an economy that works the way it does, you can buy what you want, then you make the choice. I think you definitely have to have regulation, but I would not say ban it completely because maybe it would be better in some way that we can't conceive of just now as we sit here' (PIFB, 2004).

As demonstrated in Fig. 5.9, neither are Americans in favour of streamlining the regulatory process to speed products to the market. Only 17% supported that option, whereas 65% were in favour of a mandatory affirmative finding of safety from FDA, even if it caused substantial delays in bringing products to the market (PIFB, 2004).

In 2005, the PIFB poll delved into the area of regulating the importation of GM foods, as this topic is likely to grow in importance over the next decade as other countries become producers of GM crops. Four-fifths of those responding had not heard about GM food imports, and more than one-half reported they'd heard 'nothing at all' (PIFB, 2005a).

This response is not surprising, given that few GM food products are currently imported into the USA and the issue has received little media attention. The public, then, has had little chance to think about this issue. Nevertheless, respondents were willing to give their opinions about GM imports and how they should be regulated.

As shown in Fig. 5.10, 65% of Americans opposed the *importation* into the USA of foods derived from GM plants or animals, with a slight majority *strongly* opposing them: that is 15 percentage points higher than opposition expressed to the *introduction* of GM foods into the food supply. Additionally, Americans were more decisive about their opinions on this issue. Only 15% answered 'don't know' to the importation question, compared to 25% who answered 'don't know' to the introduction question (PIFB, 2005a).

To fully understand what is driving opinion on GM imports, additional surveys or focus group work would be necessary. One likely hypothesis, however, is that respondents may hold a negative view of imports of any kind, such that, when combined with their negative view of GM foods, they cannot help but form a very negative opinion of GM imports. An informal survey taken by Riviana Foods, Inc. (a rice-milling company based in Houston, Texas, USA), seems to support this idea. Between 7 and 31 October 2005, the company asked consumers who contacted them for any reason (*n*, 223) to answer two questions on GM rice: one related to imported rice and the other to domestically grown rice.

As Fig. 5.11 shows, consumers were much more willing to purchase GM rice grown domestically (40%) than that grown in China (21%) (Riviana, 2005, unpublished data).

## Animals *versus* Plants

Another area PIFB explored in 2003, 2004 and 2005 was the difference in public opinion between genetic modification of plants and animals. Americans are less aware of transgenic, or genetically engineered, animals than of their GM plant counterparts. Only 34% had heard of GM animals compared to 41% who had heard of GM

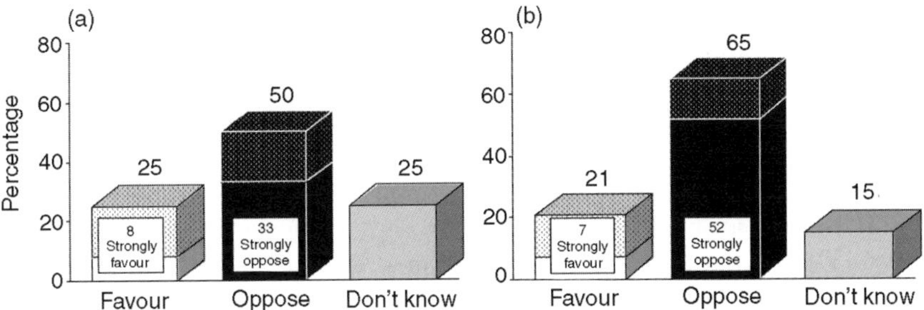


**Fig. 5.10.** Americans' opinions on the introduction and importation of GM food. Based on a PIFB survey in 2005 (*n*, 1000). The questions asked were: (a) 'Do you favour or oppose the introduction of GM food into the US food supply?' and (b) 'Do you favour or oppose the importation of food derived from GM plants or animals from other countries?' Solid shading signifies stronger intensity of opinion.

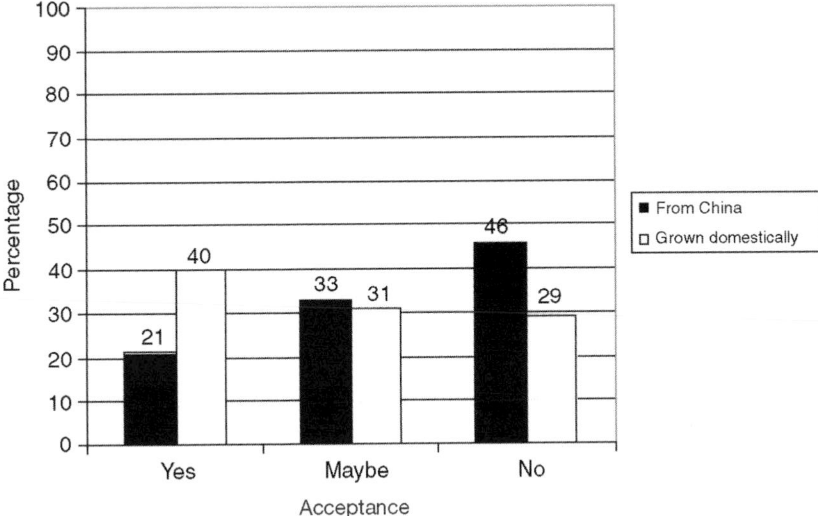


**Fig. 5.11.** Americans' opinions on GM rice, domestic and imported. Constructed from data presented by D.A. Coia of the USA Rice Federation at the Pew Initiative on Food and Biotechnology's policy dialogue: *Exploring Genetically Modified Imports: Implications for Domestic Policies*. Data collected by Riviana Foods, Inc., 7–31 October 2005, in a survey of 223 customers who had contacted the company for any reason. The questions asked were: (a) 'Next year, China will produce GM rice to increase yields and reduce the use of pesticides, making it more naturally resistant to insects. Some of that rice may find its way into international markets. If the FDA deemed this rice safe for human consumption, would you purchase it?' and (b) 'GM rice is not currently produced in the USA but, if it were, it could mean the use of fewer pesticides and a better-yielding rice crop. If the FDA deemed this rice safe for human consumption, would you purchase it?'

foods. This is not surprising, since GM animals – other than GloFish™ – have not yet been commercialized. Trouble may be brewing, however, for those who are considering commercializing GM animals, because the majority of Americans oppose GM research on animals.

Nearly 60% opposed such research (PIFB 2003, 2004, 2005a; Fig. 5.12). In 2003 and 2004, 46% were strongly opposed to it (PIFB, 2003, 2004). There was a drop in 2005 in those strongly opposed, however, as well as those strongly in favour of the technology (PIFB, 2005a). There is no clear explanation of this change, though additional studies may shed light on the subject. Hallman *et al.* found similar resistance to GM animals in their 2004 study.

American discomfort with animals is further demonstrated when they are asked to give their comfort level with respect to the genetic modification of types of organisms, with a scale of 10 being the most comfortable and 0 being the most uncomfortable (PIFB, 2003, 2004).

Americans were two times more comfortable with genetic modification of plants than with the modification of animals for other purposes (pets, racehorses, etc.) (see Fig. 5.13). They expressed slightly more comfort with the genetic modification of food animals, perhaps because they viewed those animals more distantly than other animals (PIFB, 2003, 2004). It has been suggested, and probably correctly, that the more people view an organism as closely related to

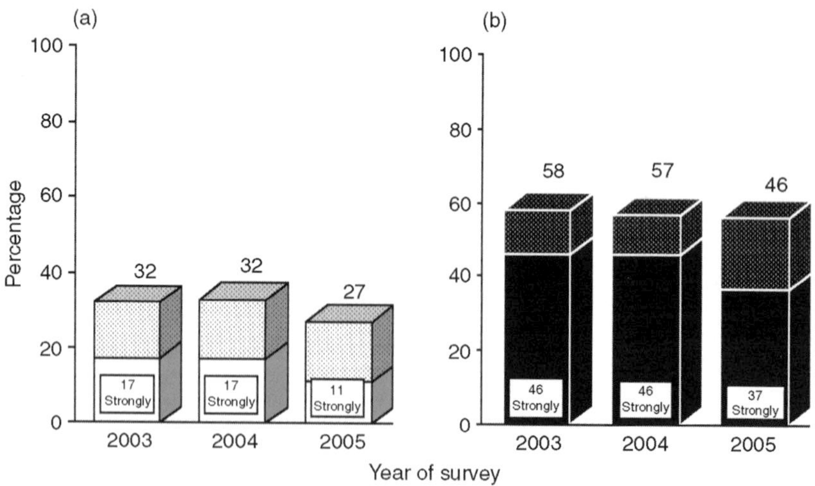


**Fig. 5.12.** Americans' opinions on GM research in animals. Based on three PIFB surveys: 2003, 2004 and 2005 (*n*, 1000). The question asked was: 'Do you (a) favour or (b) oppose GM research in animals?' Solid shading signifies stronger intensity of opinions.

themselves, the more likely they are to oppose the genetic modification of it.[2]

Animals also elicit strong ethical and moral feelings in people that plants and microbes do not. Again, this brings up a reality *versus* perception issue for biotechnology. Scientists often respond to the data in Fig. 5.13 with surprise, since they generally view the modification of animals as safer than the modification of microbes, because microbes can exchange DNA more easily and can be transported more easily.

The average person, however, may be more concerned about ethical and moral issues than about biological safety. A 2003 Gallup Poll of 1005 adults nationwide asked respondents: 'Regardless of whether or

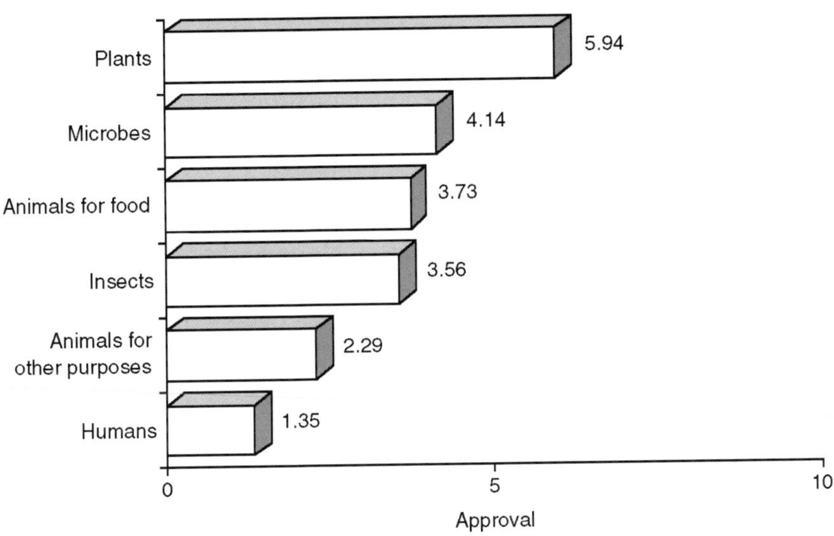


**Fig. 5.13.** Americans' opinions on genetic modification of various organisms. Based on two PIFB surveys: 2003 and 2004 (*n*, 1000). The figures represent how comfortable interviewees felt, on a scale of 0 (most uncomfortable) to 10 (most comfortable).

2 The exception to this idea seems to be insects, which may be rated with less comfort by the general public not because of ethical concerns, but very probaby from fear of release of giant bugs. PIFB has not researched this hypothesis. This is based on anecdotal information gathered from informal conversations with people during the development of PIFB's report 'Bugs in the System' and conference 'Biotech Bugs'. PIFB staff found that the first question asked by most people who knew nothing about the plans of scientists to prevent disease through the genetic modification of insects was: 'Are the scientists going to make them bigger?' This is usually said with some trepidation in their voice as well.

not you think it should be legal, for each one, please tell me whether you personally believe that, in general, it is morally acceptable or morally wrong.' The survey then listed 16 human acts, ranging from abortion, suicide and the death penalty, to buying and wearing clothing made of animal fur and having a baby outside of marriage.

Of the behaviours listed, cloning animals was less morally acceptable than all but four of the acts: suicide, cloning humans, polygamy, and married men and women having an affair. Sixty-eight per cent of Americans found animal cloning morally unacceptable (Gallup, 2003).

At PIFB's animals-related workshops, participants have identified several ethical reasons for their discomfort with cloning or genetically modifying animals (PIFB, 2002a, b). Those reasons included religious objections, a general unease or repulsion with the idea (sometimes called the 'yuck factor'), a fear that humans are likely to be cloned or modified next (sometimes referred to as the slippery slope) and a belief that this use of technology is somehow unnatural.

These moral and ethical arguments were clearly illuminated through PIFB's 2004 focus groups. Participants were given examples of likely applications of biotechnology to plants and animals and asked to rank them from those with which they felt most comfortable to those with which they were most uncomfortable. Some participants worried about upsetting the natural order of things.

For example, an Iowa man worried: 'The one I [ranked lowest] had at the bottom was "M" – to produce beef with less fat. Cattle – not to go against anybody who believes in creationism, but animals and people are very similar. When you affect one, you affect another.' A Philadelphia woman stated: 'I am not the great animal lover. I don't know, you are starting to change what God made. I am not even like this big God person. It is just like they are just changing things that should not be changed.'

Several participants worried about the slippery slope. A Philadelphia male participant argued: 'Why I'm so concerned [about animal biotechnology] is because you're one step from humans and you're getting closer to us.' A Philadelphia female participant echoed this concern, saying: 'Why do we need to do this and if we do this then – I think my other thing is if you start doing this in 2004, what happens in 2024? Do they start moving onto humans and stuff? It is just like it is that little step.'

Others were concerned about animal welfare. An Iowa man struggled with how to frame his concerns:

> I don't like the use of animals for testing, but, on the other hand, I realize it's a necessary thing that probably needs to take place, but I find it objectionable. I went back and forth on that, and I was battling myself on

that, because I'm also a hypocrite that will go and eat a McDonald's hamburger and not think twice about it. I would like to think that I had a bit more reverence for the life of [animals] ...

Other participants also took issue with the use of animals. A Des Moines woman stated: 'I just don't like messing around with other species that much. I know plants are species too, and I guess I have a different feeling about animals than I do plants' (PIFB, 2004).

To dig deeper into this issue, PIFB created a list of reasons to genetically modify a plant or animal and asked respondents to state whether it was a good or bad reason to genetically modify the organism.

Respondents clearly differentiated between plant products and animal products (see Fig. 5.14). All but one plant product was listed as a better reason for genetic modification than all the animal products. In particular, it should be noted that making affordable drugs by the use of plants was listed as one of the most favourable reasons for genetic modification, whereas making affordable drugs using animals was one of the least favourable reasons for genetic modification.

The split sample between the more general phrase 'to provide organs for transplant to humans' and the more explicit phrase 'to make it possible to transplant animal organs to humans' further reveals the negative reaction of respondents to using animals, even for something lifesaving (PIFB, 2003).

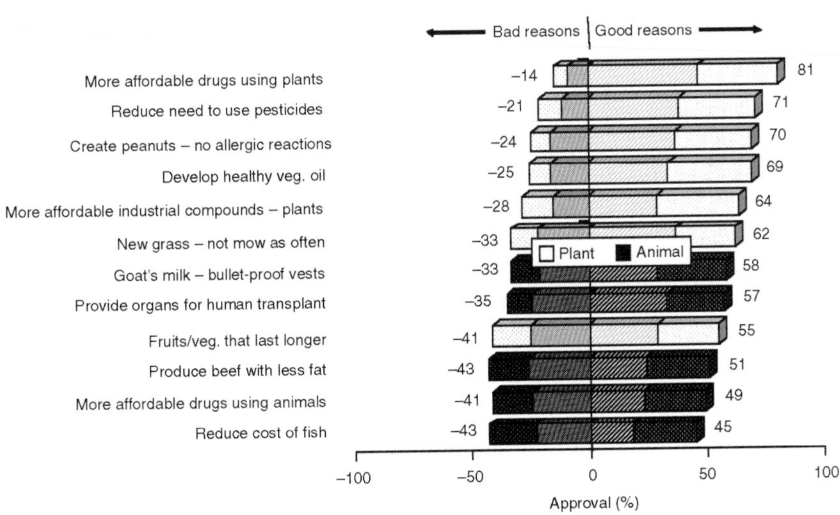


**Fig. 5.14.** Americans' opinions on the reasons for genetic modification. Based on a PIFB survey in 2003 (*n*, 1000). Diagonal shading signifies stronger intensity of opinion.

Besides pointing out the differences between how Americans view GM plant and animal applications, this question and another in 2004, which asked the importance of the application, illustrated that Americans favoured those applications which were seen to have the most benefit to the individual, their family or society.

For example, the majority of Americans favoured reducing the cost of food to reduce world hunger over simply reducing the cost of food, and few were in favour of simply reducing the cost of fish (see Fig. 5.15). Of course, all of the applications listed could be beneficial to some individuals or society, but respondents viewed these applications through their own life experiences and weighed some more heavily than others.

For example, many Americans may not understand food allergies and, thus, they are not likely to see the benefit in a non-allergic peanut. In the Des Moines women's 2004 focus group, however, one woman had a son who had a peanut allergy. Simply by explaining to the group what the impact of the allergy was on her family, she essentially swayed them into believing that this was an important application of biotechnology (PIFB, 2004).

Other focus group participants provide the reasoning behind why Americans ranked certain applications the way they did. Many ranked applications that affected health the highest. A Des Moines Man said: 'I tended to put near to top things that would either save lives or help

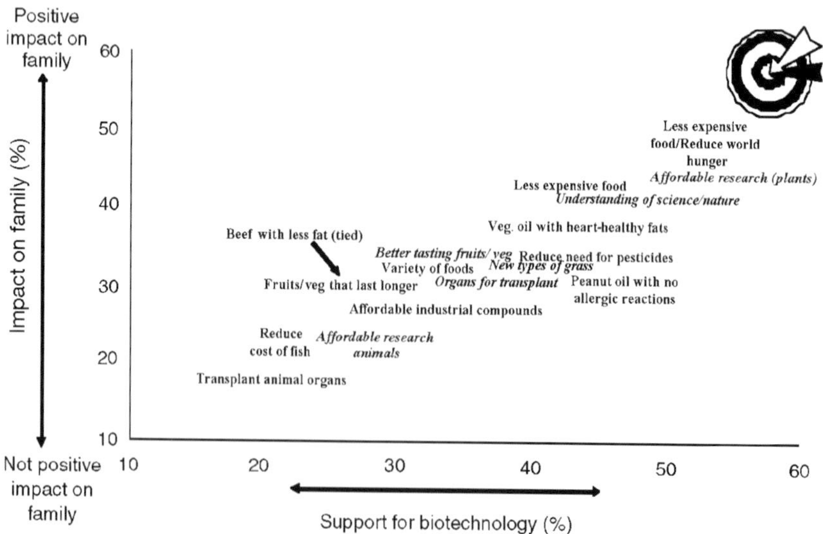


**Fig. 5.15.** Strategic map showing Americans' support for uses of biotechnology and beliefs that those uses might help them and their families. Based on a PIFB survey in 2004 (*n*, 1000).

us to stay more fit and healthy.' A Philadelphia woman went a bit further by stating: 'Things that could make our health better were important to me *versus* things that really don't affect my health a whole lot were less of a concern, like the cost of fish. I don't really care what the cost of fish is. How healthy it is, is another issue, but what it costs I don't care about' (PIFB, 2004).

Many ranked certain items based on safety concerns. One Philadelphia woman was concerned about the environmental safety of using plants to produce more affordable industrial components: 'The reason why I had that last is because more plastic means more pollution, which means more landfills and isn't plastic not biodegradable? I know they are recycling it but I just wonder if it is really that effective and stuff. That is why I did that, it was more of an environmental issue for me.'

One Iowa man was concerned about the issue of GM animals created to produce pharmaceuticals escaping: 'When you use a plant to bioengineer something, you put the plant in a [greenhouse], but when you bioengineer an animal, all it takes is one person to forget to put the lock on the cage. If an animal gets free, they're going to find another animal with the potential of passing on that trait, it's like a fan going out; one makes two, two makes four, four makes eight.' Several participants noted that they ranked new types of grasses that didn't need such frequent mowing higher, because they weren't food or medical items and thus were less likely to endanger anyone's health (PIFB, 2004).

Others found some items to be trivial uses of the technology. One Iowa man put it this way:

> I separated mine into two groups. I put them all out in front of me, and the ones I thought were important, I put off to the one side; the ones I thought were somewhat trivial, I put to the other side. Let's take the transplant one or some pharmaceutical ones. People that are in need of pharmaceuticals or surgical procedures ... probably don't have any choice in the matter. To come up with something that was low like to create fruits and vegetables which last longer – well, just eat your apple. Produce vegetable oils with heart healthy fats. Well, if you've got a problem, stay away from that stuff. A lot of these I think had more to do with behavioural issues than necessity.
>
> (PIFB, 2004)

In 2005, PIFB altered the list of specific applications which it asked respondents to consider by limiting it to animal-based applications and by adding several new choices, including pets, disease resistance and environmental protection. As in past surveys, respondents were asked to say whether they felt an application was a good reason or a bad reason for genetically modifying an animal.

As shown in Fig. 5.16, producing livestock resistant to diseases topped the list, with two-fifths of Americans favouring this application. It may be that headlines about avian flu near the time of the poll fielding and ongoing concerns about mad cow disease could have influenced this result (PIFB, 2005a). Furthermore, both of these applications have human health implications, which previous poll results have indicated is the most important reason for genetically modifying plants or animals (PIFB, 2003, 2004).

One test of this theory could be to utilize lesser-known diseases or diseases unlikely to affect human health. At the opposite end of the spectrum, Americans overwhelmingly rejected (79%) the idea of genetically modifying animals to produce novel pets (PIFB, 2005a). In the marketplace, however, the manufacturers of GloFish™, a novel pet, claimed brisk sales and were gearing up to bring additional GM fish to pet stores (GloFish, 2004).

In 2005, PIFB also asked survey respondents to react to policy proposals for regulating foods derived from GM animals. As Fig. 5.17 demonstrates, respondents favoured policies that balanced the interests of consumers and producers and that required government approval prior to sale, even if it caused substantial delays to the marketing of the food. As with previous PIFB polls on GM foods, the majority (63%) of consumers were not in favour of a ban on foods from GM animals, but they fully rejected (94%) the idea of allowing

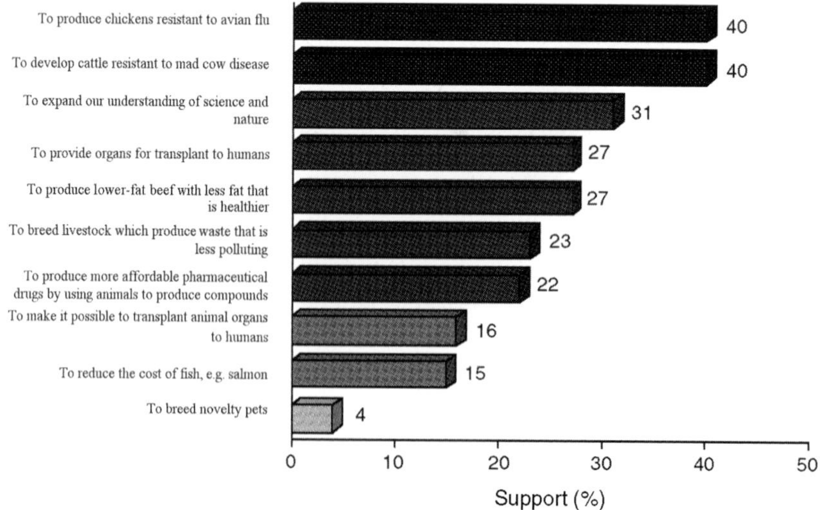


**Fig. 5.16.** Americans' reasons for supporting GM animals. Based on a PIFB survey in 2005 (*n*, 1000).

foods derived from GM animals to be sold as long as the developer believed them to be safe (PIFB, 2005a).

Finally, to complete its look at animal biotechnology, in 2005 PIFB asked a series of questions on cloning. The responses show that Americans claimed to have a much higher level of awareness of 'animal cloning' than of 'genetic modifications' of any sort (see Fig. 5.18). Additionally, two-thirds of Americans were uncomfortable with the idea of animal cloning, with nearly half feeling strongly uncomfortable (see Fig. 5.19).

Similar results were found in another 2005 poll (IFIC, 2005). These feelings against animal cloning were much stronger than those against genetic modification of plants or animals. Similarly, Americans were more concerned about the safety of foods derived from animal clones (43%) than of GM foods (30%) (PIFB, 2005a).

In the same timeframe as PIFB's 2005 poll, a poll funded by an animal-cloning company showed that nearly two-thirds of Americans would buy or consider buying meat and milk from the offspring of cloned animals if the FDA determined that the meat and milk from such animals was safe (KRC, 2005). On the other hand, fully one-third of respondents in the KRC poll indicated that they would never buy the products again. Furthermore, in a similarly worded poll, IFIC found that 57% were not likely to buy meat, milk or eggs from the offspring of cloned animals (IFIC, 2005).

Depending on the wording, then, the majority of consumers

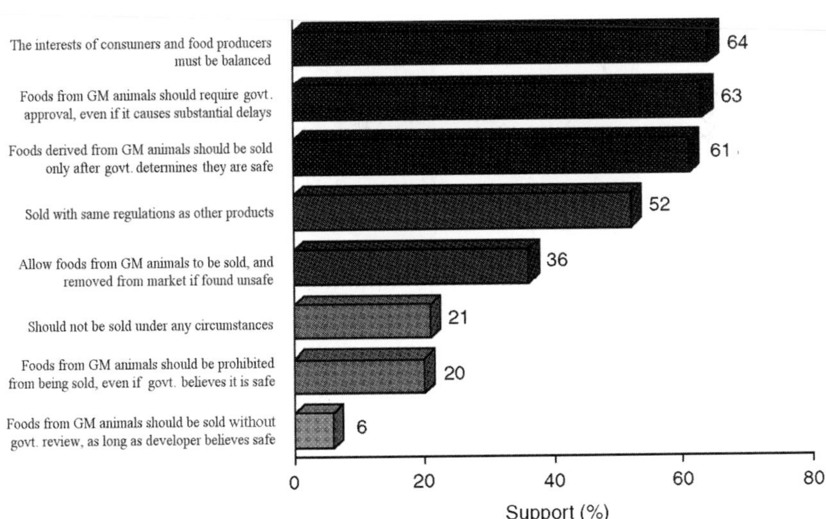


**Fig. 5.17.** Americans' opinions on the balancing of their interests with those of producers. Based on a PIFB survey in 2005 (*n*, 1000).

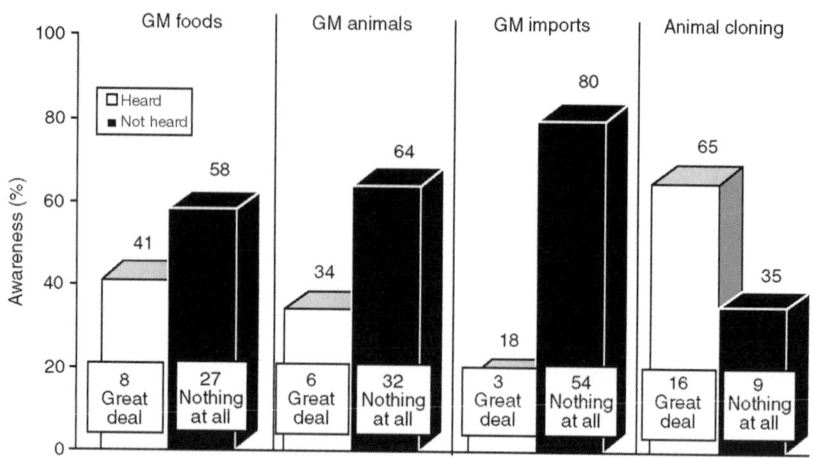


**Fig. 5.18.** Americans' awareness of animal cloning and genetic modification. Based on a PIFB survey in 2005 (*n*, 1000).

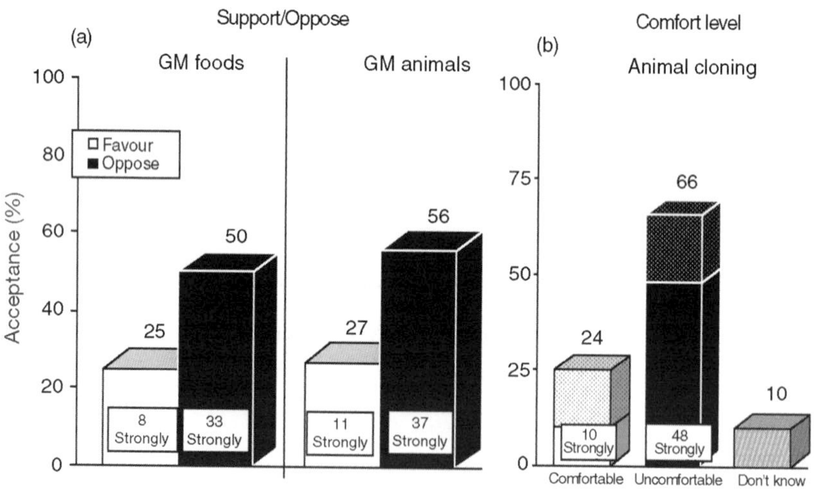


**Fig. 5.19.** Americans' opinions on (a) genetic modification of plants and animals and (b) animal cloning. Based on a PIFB survey in 2005 (*n*, 1000).

accepted or rejected products from the offspring of clones. Because polls are not good predictors of consumer behaviour, however, it will not be possible to determine how the entry of clones or their offspring might change American consumption of meat and milk until these products are allowed onto the market.

To understand the drivers behind public opinion of animal cloning,

PIFB asked respondents to explain their opinions through a series of either benefits or concerns depending on whether the respondent had been comfortable or uncomfortable with the technology.

As Fig. 20 shows, of those Americans who favoured cloning, they considered the most important benefits of cloning to be supporting scientific advancement, developing new medical treatments, lowering the price of food, helping small farmers and protecting endangered plants and animals (PIFB, 2005a). The benefits named lean towards helping society as a whole or protecting human health. For those concerned about animal cloning, Figure 5.21 illustrates the top three concerns: religious or ethical, safety and being personally uncomfortable with it (PIFB, 2005a).

To ascertain whether specific potential applications of cloning altered opinions of the technology, respondents were given a list of applications and asked whether they thought the application was a good or bad reason to clone an animal.

Though 66% of Americans said they were uncomfortable with cloning animals, the majority thought that cloning animals for biomedical research, livestock disease resistance and saving endangered species was a good idea (see Fig. 5.22). Americans were less enthusiastic about cloning to improve meat and milk traits, and they rejected the idea of cloning animals for pets.

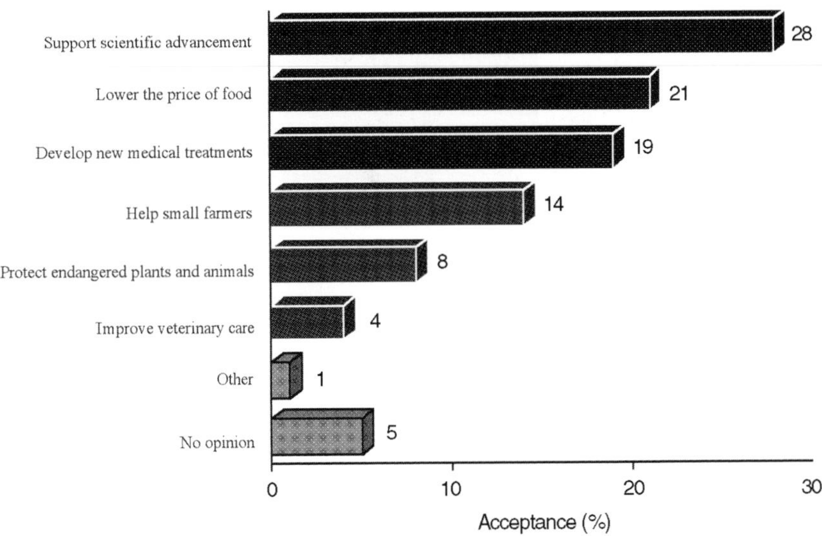


**Fig. 5.20.** Opinions on the benefits of cloning of those Americans who favour cloning. Based on a PIFB survey in 2005 (*n*, 242). The sample includes those who said that they *strongly* favoured or *somewhat* favoured cloning.

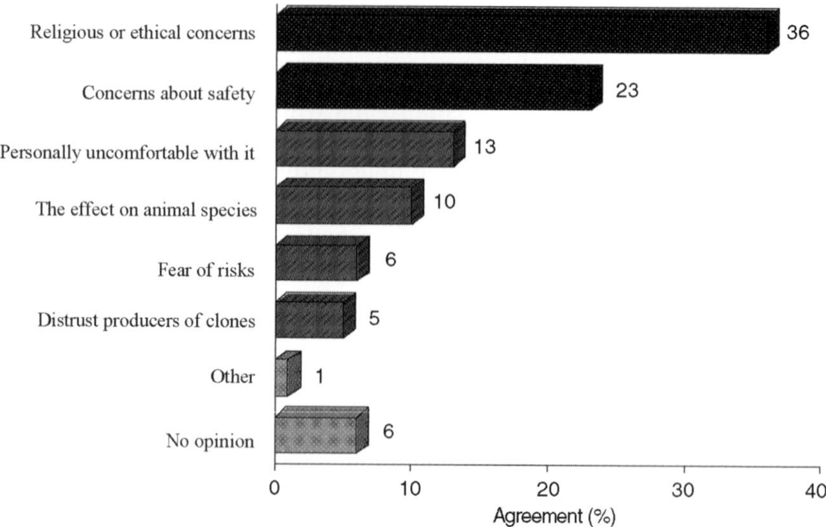


**Fig. 5.21.** Americans' concerns about animal cloning. Based on a PIFB survey in 2005 (*n*, 757). The sample includes those who said they *strongly* opposed and *somewhat* opposed cloning and those who *did not know* about cloning.

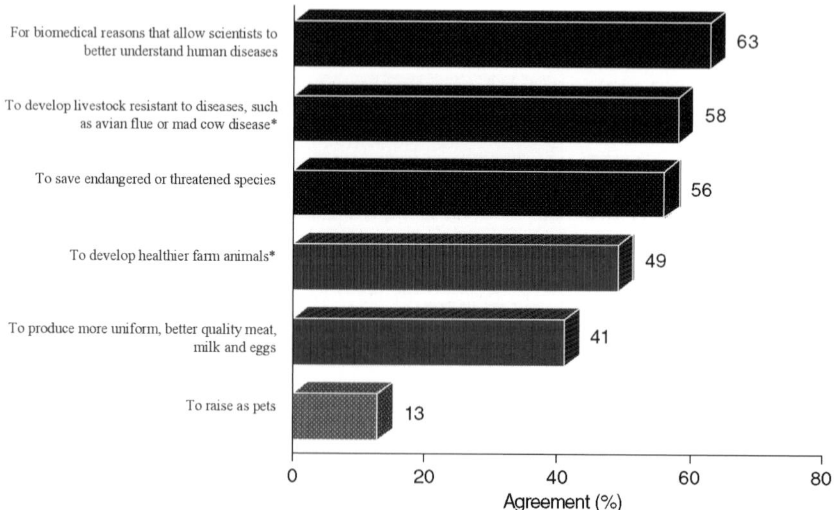


**Fig. 5.22.** Americans' accepted reasons for cloning animals. Based on a PIFB survey in 2005 (*n*, 1000). *Derived from a split sample (*n*, 500).

Finally, when asked whether they would like government regulators to consider moral and ethical considerations in addition to scientific evidence of risks and benefits when making regulatory decisions about cloned or genetically modified animals, or if regulators should focus strictly on the science evaluation, nearly two-thirds of respondents supported the idea of including ethical and moral considerations in the regulatory process, and more than half strongly supported this idea (see Fig. 5.23).

Among those who reported attending church only a few times a year or less (9%), 56% were in favour of including ethics, while 70% of those who attended church weekly (42%) felt ethics should be included. This result, however, begs the question of whose ethics respondents were considering when they answered. Providing respondents with a series of ethical choices in future surveys could provide clarifying answers as to where the public may draw the line between science-only-based decisions and the inclusion of ethical or moral considerations.

PIFB's results differ from the results published by Gaskell *et al.* In the latter study, respondents were asked two questions: (i) 'Should decisions about technology be left to the experts or based on the views of the public? and (ii) Should decisions be made on the basis of scientific evidence or on moral and ethical considerations?' From the

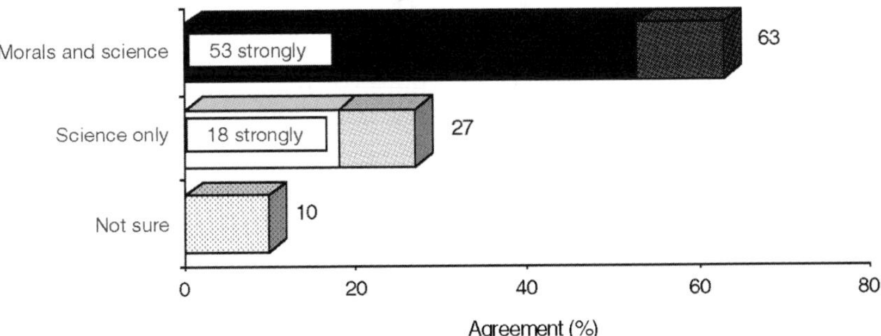


**Fig. 5.23.** Americans' opinions on incorporating moral and ethical considerations when regulating cloning. Based on a PIFB survey in 2005 (*n*, 1000). Interviewees were asked which of the two statements came closer to their point of view. Solid shading signifies stronger intensity of opinion.

answers, the authors demonstrated that two-thirds of Americans wanted science-based decision-making, and three-quarters wanted experts making the decision. Twenty-two per cent wanted experts to make decisions based on moral and ethical considerations, while 11% wanted science-based decisions made by the public (Gaskell *et al.*, 2005). It should be noted that these questions were asked in general and not directed specifically to cloned or GM animals.

Furthermore, as the authors point out, respondents were forced to choose between science and ethics (Gaskell *et al.*, 2005), whereas the PIFB poll allowed respondents to choose both in the one response (PIFB, 2005a). Nevertheless, the Gaskell study clearly shows that a large minority would like to see moral and ethical considerations included in decision-making and another minority would like more public participation (Gaskell *et al.*, 2005). Both issues are at the heart of the current debate over how cloned or genetically modified animals should be regulated in the USA (PIFB, 2002a, b, 2005b).

## Conclusion

Since 2001, Americans' knowledge of and opinions about GM food have changed relatively little. Most Americans have heard little about GM foods and awareness has tended to decrease since the highly publicized StarLink™ episode of 2000. Opposition to GM foods has decreased slightly, while support has remained stable, but not strong. Because Americans know very little about GM foods, their opinion about these foods is malleable and can change when given additional information. The safety of GM foods is not a 'top of mind' issue for most US consumers, but consumers still appear divided on the safety of GM food.

Given this, the future American public opinion of GM foods continues to remain open. If a positive, consumer-driven application were to appear, it could influence American perceptions of GM foods. On the other hand, a food or environmental safety incident involving GM plants or animals could push American opinion in a negative direction. Americans are looking towards the FDA as a trustworthy source of information about GM foods. They would like to see more regulation of GM foods, particularly by the FDA. With respect to specific applications, Americans oppose most animal-based applications, and are generally most supportive of those applications likely to provide themselves, their families or society in general with direct benefits.

## Acknowledgements

Both authors would like to acknowledge the support of the Pew Charitable Trusts. PIFB was established to be a source of balanced information to the public, media and policy-makers on the controversial topic of genetically modified foods. PIFB has sought to bring together stakeholders with an array of perspectives and opinions. PIFB has used a variety of tools to explore different aspects of the agricultural biotechnology debate, including conferences, workshops, white papers, issue briefs, stakeholder meetings, policy dialogues and public opinion research (see http://www.pewagbiotech.org).

## References

Canadian Biotechnology Secretariat (2005) *International Public Opinion Research on Emerging Technologies: Canada-US Survey Results URL.* http://www.biostrategy.gc.ca/CMFiles/E-POR-ET_200549QZS-5202005-3081.pdf

Center for Science in the Public Interest (2001) *National Opinion Poll on Labeling of Genetically Modified Foods.* http://www.cspinet.org/new/poll_gefoods.html

Gallup Poll (2003) *Polling Report.* http://www.pollingreport.com/values.htm

Gaskell, G., Einsiedel, E., Hallman, W., Hornig Priest, S., Jackson, J. and Olsthoorn, J. Social values and the governance of science. *Science* 310, 1908–1909.

GloFish (2004) *GloFish™ Fluorescent Zebra Fish now Available in Most US Markets; Unprecedented Consumer Demand Drives Pre-Market Sales.* http://www.glofish.com/pressreleases/pr.AvailableMostUSMarkets.pdf

Hallman, W.K., Hebden, W.C., Cuite, C.L. Aquino, H.L. and Lang, J.T. (2004) *Americans and GM Food: Knowledge, Opinion and Interest in 2004.* Publication number RR-1104-007. Food Policy Institute, Cook College, Rutgers – The State University of New Jersey, New Brunswick, New Jersey.

Harris Interactive (2004) *Genetically Modified Foods and Crops: Public Still Divided on Benefits and Risks.* http://www.harrisinteractive.com/harris_poll/index.asp?PID=478

Hossain, F.B., Onyango, A. Adelaja, B. Schilling, W. and Hallman, W. (2002) Uncovering Factors Influencing Public Perceptions of Food Biotechnology. *Food Policy Institute Working Paper* WP-0602-003, Food Policy Institute, Rurgers – The State University of New Jersey, New Brunswick, New Jersey.

International Food Information Council (IFIC) (2005) *US Consumer Attitudes Toward Food Biotechnology.* IFIC, Washington, DC.

KRC Research (2005) *Consumer Opinion on Animal Cloning.* http://www.clonesafety.org/documents/Cloning_use_by_farmers_pol_10_25_05.doc

Pew Initiative on Food and Biotechnology (PIFB) (2001) *Public Sentiment about Genetically Modified Food.* http://pewagbiotech.org/research/gmfood/

Pew Initiative on Food and Biotechnology (PIFB) (2002a) Proceedings from the workshop *Biotech in the Barnyard: Implications of Genetically Engineered Animals.* http://pewagbiotech.org/events/0924/proceedings1.pdf

Pew Initiative on Food and Biotechnology (PIFB) (2002b) Proceedings from the workshop *Animal Cloning and the Production of Food Products: Perspectives form the Food Chain.* http://pewagbiotech.org/events/0924/proceedings2.pdf

Pew Initiative on Food and Biotechnology (PIFB) (2003) *Public Sentiment about Genetically Modified Food.* http://pewagbiotech.org/research/2003update/

Pew Initiative on Food and Biotechnology (PIFB) (2004) *US Consumer Opinion Divided.* http://pewagbiotech.org/research/2004update/

Pew Initiative on Food and Biotechnology (PIFB) (2005a) *Public Sentiment about Genetically Modified Food.* http://pewagbiotech.org/research/2005update/

Pew Initiative on Food and Biotechnology (PIFB) (2005b) Proceedings from the workshop *Moral and Ethical Aspects of Genetically Engineered and Cloned Animals.* http://pewagbiotech.org/events/0124/proceedings.pdf

# 6

# Biotechnology and Consumer Information

J. Kolodinsky

*Department of Community Development and Applied Economics, University of Vermont, Burlington, Vermont, USA; e-mail: jane.kolodinsky@uvm.edu*

## Introduction

> Consumers are not intelligent enough to understand anything about agricultural biotechnology so we shouldn't even try to provide them with any information.

While I can't remember the exact quote, nor which scientist actually said it during their presentation at the Risk Communication for Scientists, Communicators and Administrators Conference in Florida in 2000, I vividly recall that the comment sparked conversation with colleagues who were sitting at my table. The threads of the conversations were similar. 'So we shouldn't provide consumers with any information about these new technologies because they are too stupid to understand?' Somehow, we all found this untenable.

But, we were *behavioural* scientists, not *natural* scientists. I had spent almost my entire career researching consumer information and here were scientists, presenting to a national audience, telling us that we shouldn't – no, *couldn't* – provide consumers with any information upon which to base their decision to purchase foods produced using new biotechnologies, specifically genetic modification.

But information is the key. The US economic system is one that more often than not touts 'let the marketplace decide'. On 3 January 2005, a Google search of this phrase revealed 993,000 hits. When narrowed down to genetically modified organisms (GMOs), over 4000 hits remained. Letting the marketplace decide means that where consumers choose to spend their dollars will determine the fate of a product, service or company. But, in order for that to happen, consumers need information to help them determine which goods and services meet their needs.

In the case of a good containing GM ingredients, information plays an even more important role because, for consumers, GMOs are credence goods as opposed to search or experience goods (Caswell and Mojduszka, 1996). With search goods, one can ascertain the quality of a product by using one's senses – before one even uses it. With experience goods, one can ascertain the quality after one uses the good. With credence goods, consumers depend on a means to verify the attribute.

To date, most genetically engineered agricultural products must be classified as credence goods to consumers at the retail level. For example, while bt maize may resist the maize borer (and farmers can see this benefit), the consumer at the supermarket only sees the same perfect ear of maize they have in the past. Because the bt gene is in the maize itself, and it does not affect the taste or appearance of the maize, the consumer cannot ascertain the bt attribute. The only way a consumer can know about the bt attribute is through information. In fact, we often see government regulation in the case of credence goods (Giannakas, 2002; Fulton and Giannakas, 2004).

Of course, consumers are always threatened with information overload. If there are ten product choices and each product has three characteristics that aren't necessarily the same for all products, it quickly becomes clear that it is almost impossible for consumers to sort out and compare every possible combination. *Consumer Reports Magazine* helps for some products and services, but even they narrow down the characteristics to a few basic attributes.

So what do consumers do in reality? They narrow down their choice sets to a reasonable number. This is called their evoked set. Consumers then determine whether any or all of the products in that set meet their needs. This can be done using several different choice rules that are explained in any number of consumer behaviour textbooks. The important point is that consumers must have at least basic information as to whether or not a product has a particular attribute that impacts a decision to purchase.

The attention-getting comment that opens this chapter is only one in a series of events that strengthens the argument that the agricultural biotechnology issue is a consumer information issue.

## The Emergence of Consumer Information Needs Surrounding Biotechnology

The beginning of the twenty-first century wasn't the beginning of the consumer information story: it is actually somewhere in the middle of a continuing saga. In the early 1990s, research was taking place regarding what effect, if any, the use of rBST had on cows. rBST is a genetically engineered version of the hormone Bovine Somatotropin (bST) that is produced naturally in the pituitary gland in cattle (Food Commission, undated; rBSTfacts, undated).

In the 1930s, it was discovered that BST taken from slaughtered cows could significantly increase milk production when injected into lactating (milk-producing) cows (Azimov and Krouze, 1937). Since bST could only be taken from slaughtered animals in small amounts, it was too expensive to be used by dairy farmers. In 1973, the development of recombinant DNA technology enabled large quantities of rBST to be produced at a relatively low cost. Using recombinant DNA techniques, scientists can identify and remove a particular gene, and inject that gene into a bacterium that creates copies of itself with the new DNA included (Scott, 1995). In November 1993, the Food and Drug Administration (FDA) ruled rBST safe for use in milk production.

While natural science research was taking place, consumer research was in its infancy. There were many questions to answer regarding consumers and rBST. What do consumers know about this? How do they know whether the milk they buy is produced using rBST? Do they *want* milk produced using it? Do they *care*?

About the time I decided that this would be a great, cutting-edge research project, several faculty members at my institution were called together by an administrator. We were all told not to speak to the press about anything related to rBST research on campus. I had never been approached by the press. My aim on the road to tenure was to write academic articles for prestigious academic publications. I didn't understand. What needed to be hidden from the press? I came away from the meeting even more convinced that rBST might just be as much a consumer information issue as a natural science issue.

In the 1990s, milk was still basically a commodity. There were fat variations including skim, one per cent, two per cent and whole; other attributes included store *versus* name brand. But there were few other decisions a consumer had to make except for size of container. By 1993, things were changing. Now, consumers had to decide whether to purchase milk that was produced using rBST. But how would they know unless the container provided that information?

Several researchers, mostly at universities in 'dairy states', were asking this and similar questions. Questions related to whether consumers would accept milk produced with rBST and the range of new biotechnologies that would follow came first (Douthitt, 1990 a, b; Slusher, 1991; Busch, 1992; Fox, 1995; Grobe *et al.*, 1996 a, b, c).

In Vermont, we began by adding questions to an annual poll that collected information about Vermonters' perceptions, attitudes and behaviours toward rBST; later, questions were added concerning the food-related use of the more general GMO attribute (Center for Rural Studies, 1990–2004; Kolodinsky *et al.*, 1997, 1998; Wang *et al.*, 1997). Regardless of the study, where it was conducted or the year, it was clear that consumers wanted information about how their food was produced. In 1994 in Vermont, half of consumers wanted an informational label on their milk. By 1995 that number had increased to 90% (see Table 6.1). Douthitt (1990b) reported that, in 1990, a majority of Wisconsin consumers desired an informational label.

Vermont did initiate a mandatory labelling law in 1994. This law required milk that might have been produced using rBST to have a blue sticker referring to a sign in the store that notified consumers of the meaning of the 'blue dot'. The State was almost immediately sued by several dairy groups and Monsanto, the producer of the patented Posilac, the brand name for rBST[1].

While the law was upheld in District court, a three-judge panel of the US Court of Appeals for the 2nd Circuit ruled in New York in 1995 that the producer's 'right not to reveal how they produce their product (free speech) was more important than a consumer's right to

**Table 6.1.** Demand for labelling of food.

| Study | Year | Support for labelling (%) |
|---|---|---|
| Wisconsin Poll | 1990 | 68[a] |
| Vermonter Poll | 1994 | 50[a] |
| Vermonter Poll | 1995 | 90[a] |
| Harris | 2000 | 86[b] |
| Pew | 2001 | 75[c] |

[a] Labelling of milk.
[b] Want government to require labelling of GM foods.
[c] Want to know if product contains GMOs.

1 The International Dairy Foods Association, The Milk Industry Foundation, The International Ice Cream Association, The National Cheese Institute, The Grocery Manufacturers of America, Inc. and The National Food Processors' Association – all trade, lobbying and promotional corporations – filed suit asserting that the Vermont statute was unconstitutional; they were joined by the Monsanto Corporation as *Friend of the Court* (Grossman, 1996).

know how a product was produced'. In fact, consumers' rights to know were dismissed as 'consumer curiosity' (US Court of Appeals, 1996). It wasn't until 2001 that voluntary labelling guidelines were produced by the Food and Drug Administration (FDA, 2001). These guidelines were much broader than simply labelling milk produced using rBST.

## More Research was Needed

By the late 1990s, it was becoming increasingly clear that consumer research was important. The literature showed a huge range in the percentage of consumers who were supportive and/or accepting of biotechnology. The numbers ranged from 25 to 70% (Hoban, 1996, 1998; Hoban and Katic, 1998; Shanahan *et al.*, 2001). A review of the literature seemed to reveal two things: (i) whether a study was funded by an industry group, a general media group or an advocacy group (either pro- or anti- biotechnology) seemed to be related to the results; and (ii) how questions were worded seemed to make a big difference (Kolodinsky *et al.*, 2004).

The review by Shanahan *et al.* (2001) provides some examples. General media polls, including Pew (1999) and Harris (2000) asked: 'Do you think the Food and Drug Administration should require labelling on all fruits, vegetables or foods that have been genetically altered, or don't you think labelling is necessary?' Eighty-four and 89% of a representative sample of Americans answered affirmatively. The point here is that no information was provided, just the questions. If respondents aren't familiar with the concepts, they might over-react.

A poll sponsored by the International Food Information Council[2] (IFIC, 1999, 2000) – an industry-supported group – asked: 'Simply labelling products as containing biotech ingredients does not provide enough information for consumers. It would be better for food manufacturers, the government, health professionals and others to provide more details through toll-free phone numbers, brochures and websites. Please tell me whether you: (i) strongly agree; (ii) somewhat

[2] IFIC is supported primarily by the broad-based food, beverage and agricultural industries. IFIC's purpose is to bridge the gap between science and communications by collecting and disseminating scientific information on food safety, nutrition and health and by working with an extensive roster of scientific experts and through partnerships to help translate research into understandable and useful information for opinion leaders and, ultimately, for customers. These groups find the IFIC reservoir of scentific and health data a valuable and easily accessed resource http://www.ific.org/about/index.cfm, accessed 4 January 2005).

agree; (iii) somewhat disagree; (iv) strongly disagree; or (v) don't know about the statement.'

These studies found that only 12% of a representative sample of American consumers *disagreed* with the statement. This question suffers from a fatal flaw that is outlined in any elementary research methods book: when formulating survey questions, avoid leading or loaded questions. These are 'a major source of bias in question wording' (Zikmund, 2003, p. 336).

The Pew Initiative on Food and Biotechnology has conducted yearly polls of Americans since 2001. They asked a neutral question: 'Do you favour or oppose the introduction of genetically modified foods into the US food supply?' In 2001, slightly more than half of consumers were opposed; by 2003, slightly less than half were opposed. Their conclusion? About half of US consumers remain opposed to genetically modified foods (Pew, 2003). In 1999 and 2000, Gallup found the same result (Shanahan *et al.*, 2001).

The Food Marketing Institute contracted with an academic at North Carolina State University to conduct its consumer research. Results of these studies were published as industry reports (FMI, 1996a), in agricultural economics-related journals (Hoban, 1997) and in food industry-related journals (Hoban, 1997; Hoban and Katic, 1998). The questions that were asked focused on the *benefits* of biotechnology, such as being able to 'resist insect damage' or being 'better tasting or fresher'. A summary paper notes: 'Regardless of how we measure consumer perceptions, the surveys described (above) document that between two-thirds and three-quarters of American respondents are positive about plant biotechnology' (Hoban, 1998, http://www.agbioforum.org/v1n1/v1n1a02-hoban.htm). In 2001, the Food Policy Institute at Rutgers University, New Jersey, reported similar findings for specific benefits related to plant biotechnology applications, and much lower acceptance for animal applications (Hallman *et al.*, 2002).

Research using a sample of Vermont residents examined whether the wording of questions decreased support for GMOs (Kolodinsky *et al.*, 2004). Respondents were randomly provided with either a statement of only the benefits of a GMO or a statement that provided both risks and benefits (see Table 6.2 for exact wordings of statements). The level of difference in their degree of support was calculated. Question wording *did* make a difference. Almost 90% of respondents reported reduced support for GM when both risks and benefits were provided compared with only benefits. The biggest impact of changes in question wording were on those respondents who reported that they planned to purchase products with GM ingredients. These consumers were more likely to report a decrease in support for the technology.

**Table 6.2.** Positive and balanced wordings of statements given in Kolodinsky *et al.*, 2004.

| Positive statement | Balanced statement |
|---|---|
| Crops that can resist pests with smaller amounts of pesticide | Crops that can resist pests with smaller amounts of pesticide, but may potentially be harmful to animals that were not the intended target, or may result in pesticide-resistant insects |
| Crops that are resistant to certain herbicides | Crops that are herbicide resistant; but may pass these traits to their weed relatives |
| Crops that are more nutritious | Crops that are more nutritious, but may potentially cause people to ignore broader social issues, such as poverty, which may be causing malnutrition |
| Crops that have received genetic material from completely unrelated species | Crops that have received genetic material from completely unrelated species, but may potentially cause allergic reactions in humans |

One conclusion regarding the above summary of consumer research is for consumers of research to beware. The basic rules of consumer research are upheld: always ask: (i) Who did the study?; (ii) How were the questions asked?; (iii) Are the results logical?; (iv) Are the implications forthright?; and (v) Are there any biases in the study? As will soon become apparent, we will also conclude that communicators have built campaigns around research results that, at best, found consumer support for biotechnology split 50/50. A variety of organizations have filled the *communication* need. Yet it remains questionable as to whether or not the *information* needs of consumers have been met.

## Consumer Communication Campaigns in the Age of the Sound Byte

After examining literally hundreds of consumer-oriented messages about agricultural biotechnology while conducting a literature review, two conclusions can be reached:

**1.** Anti- and pro-biotech groups have used propaganda-like techniques to further their own position – this in many ways confuses rather than informs the public.
**2.** Much of the debate regarding consumer information has focused on normative arguments rather than on fact.

Anti- and pro-biotechnology groups have used propaganda-like techniques to further their own position and this confuses rather than informs the public. The Funk and Wagnalls New Standard Dictionary

(1956) defines propaganda as ‘effort directed systematically toward the gaining of public support for an opinion or a course of action’. Figure 6.1 shows several images that appear on a variety of websites and in advertising campaigns. Clearly, the intent of these images is to provoke a certain emotional (not intellectual) response from consumers.

The International Food Information Council (IFIC) – whose mission is to communicate science-based information on food safety and nutrition to health and nutrition professionals, educators, journalists, government officials and others providing information to consumers – developed a media kit and included a list of ‘words to use’ and ‘words to lose’ with regard to genetic modification of food. Words to lose include (not a complete list) genetically modified organism, GMO, change, genes, altered, scientists and technology. Words to use include agricultural biotechnology, bounty, crops, earth, farmer, fruits and grandparents.

Early on in the information conversation, *The Economist* (I’m modified ..., 1998) reported that, in Germany, several large food

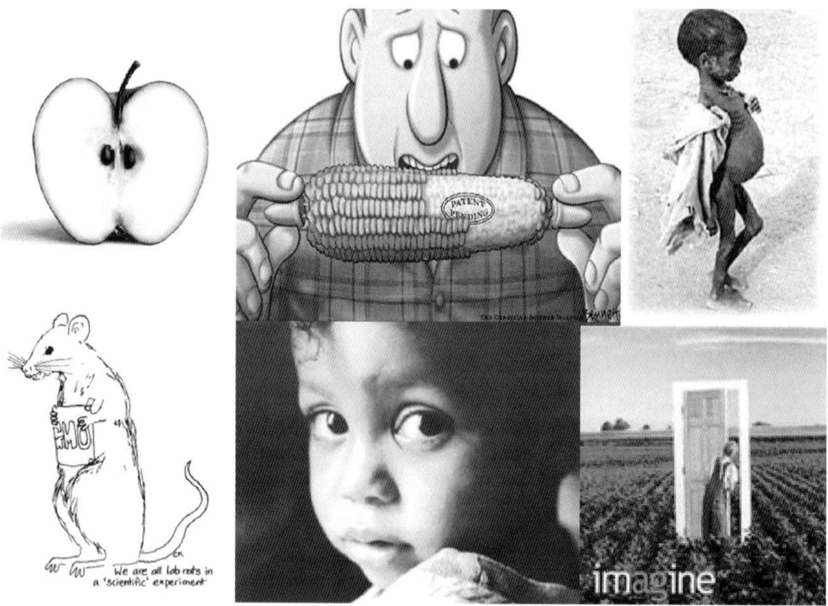


**Fig. 6.1.** Media images associated with agricultural biotechnology. Web sites: apple, http://www.agilent.com/.../2002aug14_biotechnology.html; maize, http://www.duke.edu/.../biotech/environmental.html; rat, http://www.ramshorn.ca; child’s face, http://www.wlf.org/communicating/advocacy/hunger.htm; imagine, http://www.monsanto.co.jp

producers – including Monsanto – had developed a plan to include leafleting, website development and materials for schools. In Austria, Novartis planned 'fireside chats'. On the other hand, the Center for Food Safety (Center for Food Safety, undated) had also developed a leafleting campaign, warning consumers of the potential dangers of genetically engineered food. The issue of swaying public opinion *rather* than providing consumers with unbiased information on which to base purchasing decisions appears to be at the forefront here, regardless of whether the information source is pro- or anti-biotechnology. Recently, these headlines appeared on the Monsanto website:

- Genetics, Not Cannibalism; Report: Biotech Corn in Africa Can Relieve Hunger (17 May 2005).
- Schmeiser's Rhetoric; Unjust And Irrational Fears; Long On Speculation; Reason To Fight Back; Unconscionable Acts Of Greenpeace (19 May 2005) (http://monsanto.com/monsanto/layout/sci_tech/ag_biotech/default.asp).
- Dupont's biotechnology section of their website is titled: 'Biotechnology is the Science of Miracles' (DuPont, 2000; http://www2.dupont.com/Biotechnology/en_US/science_knowledge/index.html).

The above discussion leads to the second point: much of the debate regarding consumer information has focused on normative arguments rather than on fact. Of the more than 150 sources reviewed for this chapter – including newspaper articles, journal articles, websites, reports for funding agencies and news releases – about 40% appear to be scientifically based. The other approximately 60% used normatively based arguments. None of the sources appear to be conclusive on either side of the issue.

Consumers need unbiased information, presented in an unbiased manner, if the marketplace is going to be the battleground for determining whether genetically modified organisms will be the wave of the future (Streiffer and Rubel, 2003). Unfortunately, the reliance on propaganda and normative arguments to sway public opinion continues to limit the amount of information available to consumers.

## Why Not Just Label?

One of the major impediments toward providing consumers with adequate information on which to make choices about whether to purchase genetically modified food is the stalemate that continues between those who feel there needs to be more information and those who believe the only important information for consumers is that

GMOs are safe according to current FDA, EPA and USDA regulations – and therefore require no labelling. The President of the Kansas Farm Bureau reported, at a hearing before the Subcommittee on Risk Management, Research, and Specialty Crops (1999), that: 'Many consumer concerns and the views to allow imports of biotechnology products are an outgrowth of inadequate understanding or an unwillingness to comply with the scientific findings of the regulatory systems within these countries' (p. 32).

It has also been asserted that providing more information – even scientific – is useless and destructive, and will 'frighten consumers away from a new technology that has the potential to lend infinite variety and choice to their diet' (Browning, 1993). Others have asserted that providing scientific information on labels could erode public confidence (Miller and Huttner, 1995; Carter and Gruère, 2003).

Safety, however, is only one of the myriad of characteristics of food upon which consumers base their purchase decisions (Douthitt, 1991; Caswell and Mojduszka, 1996; Caswell, 1998). Caswell (1998) reported that: 'Regulators and consumers may care about process attributes for a number of reasons. First, there may be concerns about the impact of use of the process on the final quality attributes of consumer-ready products. Second, the process may have impacts on nutrition, the environment, animal welfare, worker safety, or other important attributes.'

In fact, safety is almost a 'sunk cost' characteristic: if it isn't safe, it doesn't enter into the consumer's evoked set in the first place. There are many reasons why a consumer might be interested in having information on which to base choice, including: (i) economic issues related to small-scale agriculture (Marion and Willis, 1990; Kolodinsky *et al.*, 1997, 1998); (ii) 'interference' in the natural order of things (Douthitt, 1991); (iii) perceived risk and unintended consequences as yet not uncovered by current standards (Douthitt, 1991; Suddeth, 1993; Pollack, 2004; Committee on ..., 2004); (iv) fairness about who derives the benefits from purchase of these goods, business or consumers (Busch, 1992); (v) values concerning food and its social significance (Busch, 1992; Thompson, 1997); and (vi) consumers' general lack of trust in science (Thompson, 1997).

By discounting the attributes by which consumers make their decisions, regardless of whether producers and some scientific groups believe that the only important characteristic is safety as defined by the current regulatory process, the market's ability to decide whether genetically modified foods will succeed is denied.

Several groups have taken an anti-labelling position. Many of these have stated opinions in opposition to increased consumer information in the form of labels, either regulated or voluntary. A partial list of these groups include: American Seed Trade Association (ASTA), Council for Agricultural Science and Technology (CAST), Kansas Farm Bureau, American Farm Bureau Federation, American Crop Protection Association, National Corn Growers' Association (NCGA), American Soybean Association, Biotechnology Industry Association, Consortium for Plant Biotechnology Research, National Cotton Council, American Apparel Manufacturers' Association, American Cotton Shippers' Association, American Textile Marketing Association, National Cotton Council of America and the National Cotton Seed Products Association.

These companies have a huge amount of research and development capital invested. If GMOs fail in the final consumer market, none of this capital can be recouped. Thus, there is a push toward getting GMOs to the market as quickly as possible (see, for example, Jacobs, 1999). It has also been asserted that the current regulations to test whether a product is safe aid producers of GMOs in getting their products to the final market. For example, Burstein (1997) asserted: 'Monsanto and other industry giants love EPA regulation. It adds another stamp of approval to their products, and it squeezes out smaller companies that can't afford the time and money the regulatory process demands.'

And 'EPA regulations already favour industry. The Federal Insecticide, Fungicide, and Rodenticide Act (FIFRA) directs the EPA to take the economic impact of a product into consideration during the approval process' (http://www.motherjones.com/news/feature/1997/01/burstein.html)'.

Einsiedel (2000) demonstrates that the struggles between meeting the needs of all parties have made the labelling issue an extremely demanding one in the case of biotechnology. In the USA, labelling guidelines have been in place, but not without controversy. Consumers have made choices for several years, but the marketplace has experienced mishaps in following the labelling guidelines.

The industry itself has asserted that it would most likely embrace labelling once aspects of GMOs that revealed direct consumer benefits were introduced. For example, currently, the benefits of GMOs flow to the producer of both the GMO and perhaps the farmer who uses these products. rBST may result in greater milk production for an individual farmer, leading to higher revenues for him or her, but to no discernable benefit for consumers: they do not see lower milk prices. Bt maize results in lower current use of pesticides for a farmer, but the price of maize is not likely to change for the final consumer. On the other hand, if a GMO affords a consumer a higher level of vitamins per serving of fruit, for example, a label that touts this wonder will probably be embraced.

Regardless, labelling remains voluntary and we continue to see only a pocketful of labelled products – all on products that do not contain GMOs. And the FDA seems to be 'cracking down' on misuse of GMO-free claims (see Kolodinsky, 2005 for an overview).

## Consumer Information and Biotechnology in 2004/2005

Buzz and consumer research about biotechnology seems to be calming down as we approach the second half of the first decade in the twenty-first century. GMOs are quietly continuing to enter the marketplace. Table 6.3 summarizes results from four national polls that were conducted in 2005: (i) Gallup asked questions with positive undertones; (ii) The International Food Information Council (IFIC), with its industry ties, continued to put a positive spin on its questions; (iii) The Pew Initiative on Food and Biotechnology (PIFB) attempted to remain neutral; and (iv) The University of Vermont (UVM) survey continued to focus on information and labelling, asking questions in a neutral manner (see Shanahan *et al.*, 2001).

In 2005, Gallup didn't provide a definition of genetic engineering, but couched the issue as a 'new scientific technique' and lumped genetic engineering and genetic modification of food in with medicines. In 2005, about 40% of Americans reported following news on these issues somewhat or very closely, about the same level since 2001 (Gallup, 2005). Also in the Gallup results, it is shown that 45% of Americans remain opposed to the use of biotechnology in agriculture and food production, down a couple of percentage points (it is not revealed in the results whether this is a statistically significant decline).

The 2005 IFIC study is clearly industry oriented. This survey explored biotechnology initially with open-ended questions. That is, they did not specifically ask respondents about biotechnology in this early section, but rather asked whether an individual had avoided or eaten less of any food items. Not surprisingly, the largest response was a decrease in sugar/carbohydrates (given the Atkins diet fad of 2004/05). Biotechnology was not mentioned. The survey then moved into specific biotechnology questions, using positively worded questions, including the words 'new scientific techniques' and 'improve crop plants.' They found that 70% of Americans had heard of biotechnology and only 34% knew that these types of foods were on supermarket shelves.

The labelling question did not ask specifically about GM foods. Instead, respondents were told about FDA policy in simple terms ('the FDA requires labelling when a food is produced under certain conditions: when biotechnology's use introduces an allergen or when it

**Table 6.3.** GMO Opposition to labelling items compared across multiple polls.

| Study | Year | Poll item | Results |
|---|---|---|---|
| Gallup | 2005 | As you may know, some food products and medicines are being developed using new scientific techniques. The general area is called 'biotechnology' and includes tools such as genetic engineering and genetic modification of food. Overall, would you say you strongly support, moderately support, moderately oppose or strongly oppose the use of biotechnology in agriculture and food production? | 45% opposed use of biotech in agriculture and food production |
| IFIC | 2005 | The FDA requires labelling when a food is produced under certain conditions: when biotechnology's use introduces an allergen or when it substantially changes the food's nutritional content, like vitamins or fat, or its composition. Otherwise special labelling is not required. Would you say you support of oppose this policy of the FDA? | 55% supported this policy |
| IFIC | 2005 | As you may know, some food products and medicines are being developed with the help of new scientific techniques. The general area is called 'biotechnology' and includes tools such as genetic engineering. Biotechnology is also being used to improve crop plants. | 70% had heard of biotech 34% knew these foods at supermarkets |
| Pew | 2004 | Currently, the FDA does not require companies to prove that conventional foods are safe before they can be sold on the market, and companies can bring new conventionally produced foods to the market without prior approval from the FDA. The FDA can act to take a food off the market if there's a safety problem. Now I'm going to read you some goals that people have set out for the regulation of genetically modified foods. After each I'd like you to tell me whether you favour or oppose that goal: labelling all food that is genetically modified | 92% supported labelling |
| Pew | 2004 | Now I'm going to list some life forms which can be genetically modified. After each, please tell me how comfortable you are with genetic modification of that specific life form. Use a scale of zero to ten where zero means that you are very uncomfortable with genetic modifications of that life form and ten means that you are very comfortable with genetic modification of that life form. | Plants: mean = 5.94 Animals used for food sources (including cattle, fish and shrimp): mean = 3.73 |
| UVM | 2005 | Should products containing GMOs be labelled, products that do not contain GMOs be labelled, both products be labelled or neither products be labelled? | 92% supported some kind of labelling policy |
| UVM | 2005 | For the purposes of this survey, we have defined GMOs as a form of biotechnology where scientists selectively and deliberately move genes from one organism (like a plant, animal or micro-organism) to a different type of organism to achieve certain desired characteristics that would not occur in nature or traditional breeding. Are you very supportive, supportive, neutral, opposed or very opposed to the use of GMOs in commercially available food products? | 43% opposed or very opposed |

substantially changes the food's nutritional content, like vitamins or fat, or its composition. Otherwise, special labelling is not required. Would you say you support or oppose this policy of the FDA?'). The responses were that 55% indicated they supported this policy (IFIC, 2005).

Pew's September 2004 study found that consumers remained split on whether they believed GM foods to be safe (30% safe/27% unsafe/others unsure). And, although consumers did not know a great deal about the regulatory process (84%), they supported a strong regulatory system for GM foods, with 85% indicating they wanted regulators to ensure GM foods were safe before they came to market. Ninety-two per cent supported GM labelling (Pew, 2004). Pew also found that about half of respondents remained opposed to genetic modification, but were most accepting of plant modification.

A University of Vermont 2005 national survey found that half of respondents had heard of the term genetically modified organism (or GMO), and a little over 40% believed that less than 50% of processed foods contained GM ingredients. Only 20% knew that more than 70% of processed foods contained GM ingredients (see Cornell Cooperative Extension, undated). Ninety-two per cent of respondents thought there should be some GM labelling policy, either labelling all foods, only foods containing GM ingredients or only foods that were GM-free. This is the same as the Pew findings, but very different from the IFIC findings (by 92 percentage points!). With regard to support for GM foods, 17% of respondents were supportive, one-third were neutral and 43% were opposed. These findings are also on a par with the Pew study.

The University of Vermont study also asked about information search behaviour. Only 7% of respondents chose the option: 'I seek information on GMOs', while a little over 40% chose the option stating: 'I pay attention to information on GMOs if it catches my eye.'

So, where are we in terms of information about biotechnology in 2006? We continue to face persistent information challenges, yet consumers do enjoy the benefits of new labelling guidelines. The greatest remaining challenge in the biotechnology information debate is the continuing confusion of research results. As was found by Kolodinsky *et al.* (2004), the wording of questions matters and researchers can load their questions to obtain the desired results. The surveys using more neutral questions found consumers evenly divided about GM food acceptance. Consumer information search also remains challenged by the source and credibility of the information available. There is no standard for the provision of information about GM foods and consumers must have the skills to wade through the 'muck' of claims and statements made about biotechnology by groups supporting and opposing its use. Even the scientific literature does not find consensus on many consumer-related aspects of biotechnology.

The greatest consumer benefit, however, has been the voluntary labelling of GM foods. For the small percentage of consumers who care deeply about the issue, they now have the ability to find GM-free foods through this labelling programme. The FDA has issued guidelines and appears to be enforcing them (see FDA, 2001; Kolodinsky, 2005).

Markets for agricultural products have witnessed a definitive industry response to consumer demand regarding biotechnology attributes. Immediately following the 1993 approval allowing the use of synthetic rBST on dairy cows, several independent dairies and small processors made steps to create a niche market for 'rBST-free.' Cedar Grove Cheese of Plaine, Wisonsin, emerged immediately to distinguish their product as 'rBGH-free' (another terminology for the synthetic bovine growth hormone (Petrak, 2005). In the Northeast, Oakhurst Dairy of Maine did the same in 1994. By 2005 we had seen continued expansion of new dairy brands nationwide that used labels to distinguish that their milk was produced without the use of synthetic growth hormones (Ruff, 2006).

The most recent news for 2006, however, indicates an industry response in the Northeast that takes on a regional significance. Two of the largest Northeast milk processors: HP Hood and Dean Foods announced in autumn 2006 that they will no longer accept milk that has been produced using synthetic hormones (Mohl, 2006). Amidst documented shortfall in the supply of organic milk, this move indicates that the industry is responding to a demonstrated consumer demand for hormone free dairy products. 'Let the marketplace decide' appears to be an appropriate policy tool when consumers are given information on which to base their decisions.

## References

Asimov, G. and Krouze, N. (1937) The lactogenic preparations from the anterior pituitary gland and the increase in milk yield from cows. *Journal of Dairy Science* 20, 289–306.

Browning, G. (1993) Food fight. *National Journal* 25, 2658–2661

Burstein, R. (1997) Paid protection. *Mother Jones* 42. Retrieved from http://www.motherjones.com/news/feature/1997/01/burstein.html

Busch, L. (1992) Biotechnology: consumer concerns about risks and values. *Food Technology* 45, 96–101.

Carter, C.A. and Gruère, G.P. (2003) Mandatory labelling of genetically modified foods: does it really provide consumer choice? *AgBioForum* 6, 68–70.

Caswell, J. (1998) How labelling of safety and process attributes affects markets for food. *Agricultural and Resource Economics Review* 27, 151–158.

Caswell, J. and Mojduszka, E. (1996) Using informational labelling to influence the market for quality in food products. *American Journal of Agricultural Economics* 78, 1248–1253.

Center for Food Safety (undated) *Fact Sheet on the Risks of Genetically Engineered Food.* Available from CFS, 666 Pennsylvania Ave., Suite 302, Washington, DC 20003.

Center for Rural Studies (1990, 1993, 1994, 1995, 2004) *Results of the Annual Vermonter Opinion Poll.* University of Vermont, Burlington, Vermont.

Committee on the Biological Confinement of Genetically Engineered Organisms, National Research Council (2004) *Biological Confinement of Genetically Engineered Organisms.* National Academies Press, Washington, DC.

Cornell Cooperative Extension (undated) *GE Foods in the Market.* Genetically engineered organisms public education project (http://www.geo-pie.cornell.edu/crops/eating/html, accessed 18 September 2006).

Douthitt, R. (1990a) *Wisconsin Consumer Attitudes Regarding Acceptance of Food-related Biotechnology* 1st DPLS edn, 1997. College of Letters and Science Survey Center, Madison, Wisconsin.

Douthitt, R. (1990b) *Wisconsin Consumers' Attitudes toward Bovine Somatotropin (BST) and Dairy Product Labeling.* Retrieved from http://www.mindfully.org/GE/Consumer-Surveys-Labelling-GMOs.htm

Douthitt, R. (1991) *Biotechnology and Consumer Choice in the Market Place: Should there be Mandatory Product Labeling? A Case Study of Bovine Somatotropin and Wisconsin Dairy Products.* American Council on Consumer Interests, Ames, Iowa, pp. 97–104.

DuPont (2000) *RandD in Support of Growth – the Never Ending Transformation.* DuPont Online (retrieved from http://www.dupont.com/corp/science/growth.html).

Einsiedel, E. (2000) Consumers and GM food labels: providing information or sowing confusion? *AgBioForum* 3, 231–235.

FDA (2001) *Guidance for Industry: Voluntary Labeling Indicating Whether Foods Have or Have Not Been Developed Using Bioengineering.* FDA, Washington, DC.

Food Commission (undated) *Food Issues: Have we gone too far in Manipulating Nature?* (retrieved from http://www.foodcomm.org.uk/genetics.htm).

Food Marketing Institute (FMI) (1996) *Trends in Europe: Consumer Attitudes and the Supermarket.* Food Marketing Institute, Washington, DC.

Fox, J.A. (1995) Determinants of consumer acceptability of Bovine Somatotropin. *Review of Agricultural Economics* 17, 51–62.

Fulton, M. and Giannakas, K. (2004) Inserting GM products into the food chain: The market and welfare effects of different labelling and regulatory regimes. *American Journal of Agricultural Economics* 86, 42.

Funk, C. (1956) *New Practical Standard Dictionary of the English Language, 2* (M–Z). Funk and Wagnalls, New York.

Gallup Poll (2005) *Social Series: Consumption Habits.* Gallup Poll, 7–10 July 2005, Princeton, New Jersey.

Giannakas, K. (2002) Information asymmetries and consumption decisions in organic food product markets. *Canadian Journal of Agricultural Economics* 50, 35.

Grobe, D., Douthitt, R. and Zepeda, L. (1996a) *Exploring Consumers' Risk*

*Perception of Recombinant Bovine Growth Hormone and Recombinant Porcine Growth Hormone by Income and Gender: a Focus Group.* Institute for Research on Poverty, Madison, Wisconsin.

Grobe, D., Douthitt, R. and Zepeda, L. (1996b) *Measuring Consumer Knowledge and Risk Perceptions of Food-related Biotechnologies.* Institute for Research on Poverty, Madison, Wisconsin.

Grobe, D., Douthitt, R. and Zepeda, L. (1996c) A model of consumers' risk perceptions toward recombinant Bovine Growth Hormone (rBGH): the impact of risk characteristics. *Consumer Interests Annual* 42, 395–422.

Grossman (1996) Retrieved 3 January 2005 from http://www.ratical.org/corporations/rBSTlabelling.html

Hallman, W.K., Adelaja, A.O., Schilling, B.J. and Lang, J.T. (2002) *Public Perceptions of Genetically Modified Food: Americans Know Not What they Eat.* Food Policy Institute, Rutgers University, New Brunswick, New Jersey (retrieved from http://www.foodpolicyinstitute.org/docs/reports/Public%20Perceptions%20of%20Genetically%20Modified%20Foods.pdf).

Harris (2000) Results from a poll conducted on consumer's opinions on the labelling of genetically modified foods (retrieved from http://www.harrisinteractive.com/harris_poll/index.asp?PID=96).

Hoban, T.J. (1996) Trends in consumer attitudes towards biotechnology. *Journal of Food Distribution Research* 27, 1–10.

Hoban, T.J. (1997) Consumer acceptance of biotechnology: an international perspective. *Nature Biotechnology* 15, 232–234.

Hoban, T.J. (1998) Trends in consumer attitudes about agricultural biotechnology. *AgBioForum* 1, 3–7 (http://www.agbioforum.org/v1n1/v1n1a02-hoban.htm).

Hoban, T.J. and Katic, L.D. (1998) American consumer views on biotechnology. *Cereal Foods World* 43, 20–22.

International Food Information Council (IFIC) (1999, 2000) *Functional Foods: Attitudinal Research* (retrieved 4 January 2005 from http://www.ific.org/about/index.cfm).

International Food Information Council (IFIC) (2005) *Food Biotechnology Not a Top-of-the-Mind Concern for American Consumers* (retrieved from http://www.ific.org/research/biotechres03.cfm).

Jacobs, P. (1999) Protest may mow down trend to alter crops. In: *LA Times* (retrieved from http://www.latimes.com/class/employ/healthcare/19991005/t000089515.html).

Kolodinsky, J. (2005) Affect or information? Labeling policy and consumer valuation of rBST-free and organic characteristics of milk. In: *American Council on Consumer Interests* (retrieved from http://www.consumerinterests.org/files/public/Kolodinsky_AffectorInformationLabelingPolicyandConsumerValuation.pdf).

Kolodinsky, J., Conner, D. and Wang, Q. (1997) Who gets it right? Consumer experience with mandatory labelling of dairy products containing rBST. *Consumer Interests Annual* 43, 96–101.

Kolodinsky, J., Conner, D. and Wang, Q. (1998) rBST labelling and notification: lessons from Vermont. *Choices* (3rd quarter), 38–40.

Kolodinsky, J., DeSisto, T. and Narsana, R. (2004) Influences of question wording on levels of support for genetically modified organisms. *International Journal of Consumer Studies* 28, 154–167.

Local North Dakota dairy offers region's first hormone-free milk. (9 Aug, 2005). PR Newswire US. (retrieved from http://www.lexisnexis.com/universe).

Marion, B. and Willis, R. (1990) A prospective assessment of the impacts of Bovine Somatotropin: a case study of Wisconsin. *American Journal of Agricultural Economics* 72, 326–336.

Miller, H. and Huttner, S. (1995) Food produced with new biotechnology: can labelling be anti-consumer? *Journal of Public Policy and Marketing* 14, 330–333.

Mohl, Bruce (25 Sept, 2006). 2 dairies to end use of artificial hormones; hope to compete with organic milk. Boston Globe pp. A1. (retrieved from http://www.lexisnexis.com/universe).

Petrak, Lynn (October 2005). Growth chart: the industry debate over rBST continues, as individual processors and consumers make their choices at the dairy case. Gale Group, 188 n. 10 Pg. 44. (retrieved from http://www.lexisnexis.com/universe).

Pew Initiative on Food and Biotechnology (PIFB) (2001, 2003, 2004) *Public Sentiment about Genetically Modified Food* (retrieved from http://pewagbiotech.org/research).

Pollack, A. (2004) *No Foolproof Way Is Seen to Contain Altered Genes* (retrieved from http://online.sfsu.edu/~rone/GEessays/nocontainment.htm).

rBSTFacts (undated) Retreived from http://www.rbstfacts.org/

Ruff, Joe (28 July, 2006). Dairy firm sees a niche for hormone-free milk. Omaha World-Herald p. 01D. (retrieved from http://www.lexisnexis.com/universe).

Scott, D. (1995) *Bst Fact Sheet.* US Food and Drug Administration Center for Food Safety and Applied Nutrition (http://www.cfsan.fda.gov/~ear/CORBST.html)

Shanahan, J., Sheufele, D. and Lee, E. (2001) The polls trends: attitudes about agricultural biotechnology and genetically modified organisms. *Public Opinion Quarterly* 65, 267–281.

Slusher, B. (1991) *Consumer Acceptance of Food Production Innovations: an Empirical Focus on Biotechnology and BST.* American Council on Consumer Interests, Ames, Iowa, pp. 105–116.

Streiffer, R. and Rubel, A. (2003) Choice *versus* autonomy in the GM food labelling debate: comment. *AgBioForum* 6, 141–142.

Suddeth, M.A. (1993) Genetically engineered foods: fears and facts: an interview with FDA's Jim Maryanski, *FDA Consumer* 27, 10–14 (retrieved 9 September 2005 from http://www.fda.gov/bbs/topics/CONSUMER/CON00191.html).

*The Economist* (1998) I'm modified, buy me. *The Economist* 346, 60.

Thompson, P. (1997) Food biotechnology's challenge to cultural integrity and individual consent. *Hastings Center Report* 27, 34–38.

USA Court of Appeals for the Second Circuit (1996) Docket No. 95-7819 (retrieved from http://laws.lp.findlaw.com/2nd/957819.html).

Wang, Q., Halbrendt, C., Kolodinsky, J. and Schmidt, F. (1997) Willingness to pay for rBST-free milk; a two-limit Tobit model analysis. *Applied Economic Letters* 4, 619–621.

Zikmund, W. (2003) *Business Research Methods.* Thomson Southwestern, Mason, Ohio.

7

# What do Brazilians think about Transgenics?

L. Massarani[1*] and I. de Castro Moreira[2**]

[1]*Museu da Vida/Casa de Oswaldo Cuz, Fundação Oswaldo Cruz, Rio de Janeiro, Brazil;* [2]*Department of Physics and Program of History of Science, Techniques and Epistemology, Federal University of Rio de Janeiro\Praia do Flamento 200, Rio de Janeiro, Brazil;*
*e-mail: *lumassa@coc.fiocruz.br; **icmoreira@uol.com.br*

## Introduction

In Brazil, there has been considerable controversy over genetically modified (GM) crops and food. It is the fifth largest country in the world with a land area of approximately 8.5 million km$^2$, and agriculture is a key sector of the economy of the country. Since 1998, attempts to produce GM crops on a commercial scale have been made, but growing and selling GM crops was prohibited until March 2005, when the so-called biosafety law rendered them legal (Massarani, 2005).

In 2003, the controversies surrounding GM crops were especially significant due to the fact that, in February, it was found that a major proportion of Brazilian soy crops were transgenic due to illegal planting in southern states. Shortly after an announcement was made to maintain the ban on GM crops, the government decided to allow the sale of GM soy for animal and human consumption, sparking protest within environmental sectors of the government and from environmental groups. The federal government argued that the decision was taken after being faced with an important social and economic problem involving thousands of millions of tons of transgenic soy and small farmers who did not want their crops destroyed (Massarani, 2003). The decision was initially limited to the 2003 harvest but, later, special permission was given for selling GM soy in subsequent harvests.

The economic dimension of the issue cannot be ignored. According to the Embrapa's (Brazilian Agricultural Research Corporation) website,

 *The Media, the Public and Agricultural Biotechnology* (ed. D. Brossard *et al.*)

Brazil was the second biggest producer of soy in the world in 2004, producing 50 million tons of the grain – equivalent to 25% of the world production (http://www.cnpso.embrapa.br/ index.php?op_page=22andcod_pai=16, accessed 28 September 2005). Before the approval of the commercial growing of GM crops in Brazil, the country was the only one of the three biggest exporters of soy (USA, Argentina and Brazil) in which it was not allowed commercially to grow GM crops, warranting room for selling the grain in markets that had restrictions on the consumption of transgenic food (Menasche, 2003, unpublished PhD thesis).

Due to its very large territory and the different historical processes through which the five regions of the country have developed, Brazil has manifested different reactions toward GM crops. In Rio Grande do Sul (the southern state), Olívio Dutra, governor during the period 1998–2002, publicly stated several times his objective of keeping the state free of transgenics (Menasche, 2003, unpublished PhD thesis). However, due to smuggled seed from Argentina and the lack of vigilance at the federal level, about 90% of the 95,000 soy producers in Rio Grande do Sul are now growing GM soy, according to Carlos Sperotto, president of the state's Agriculture Federation (Massarani, 2004a).

But, in the neighbouring state of Santa Catarina, governor Roberto Requião had ordered that the port of Paranaguá, in his state, be closed to genetically modified organism (GMO) crops, and also threatened to close highways to trucks hauling them (Rohter, 2004), raising the question as to whether state legislation can overrule federal legislation.

Although some attempts to grow GM cotton and maize have been made in Brazil, soy is the main GM crop in this country; GM crops are in less of a reality, at least at such a massive scale as in the southern, north-eastern and south-eastern regions, where soy is not a main issue for the local economy. In the Amazon states, soy has increasingly been occupying more agricultural land, but data are controversial on whether some of it is, in fact, transgenic.

## Previous Studies in Brazil

Despite the significant economic and social impact of the GM crops and food in Brazil – and the big debate around the theme – little has been done in Brazil to analyse people's attitudes towards genetics, biotechnology and GM crops. One of the rare relevant studies investigating public attitudes and opinions towards GM crops and food was carried out by Menasche (2003, unpublished PhD thesis), who analysed the social representations of consumers and farmers from the state of Rio Grande do Sul, the region in which the smuggled seeds were detected. The analysed data were obtained through ethnographic research developed among farmers from two rural regions and through in-depth interviews with urban people in Porto Alegre, the capital of the state.

Menasche observed that transgenic food is perceived by the interviewed consumers as being included in a list of contemporary fears, being related to cloning, radiation, mad cow disease, mutation, malformation of fetuses and cancer. But, even though those interviewed rejected transgenic foods, this does not necessarily mean that they don't buy them in the supermarket, showing that their behaviour can be different from what was explicit in the discourse. The author also observed that the main fear of GM crops among farmers who do not grow transgenic crops is not related to the potential impact on the environment or human or animal health, but to the loss of control in the production process.

Brugnerotto (2003, unpublished PhD thesis) focused on the perceptions that urban and rural populations from 28 municipalities of Alto Vale do Itajaí, in the state of Santa Catarina, had toward agrotoxics and modern biotechnology. Defining his sample, he developed simple random samples and interviewed 541 people (57.5% urban; 42.5% rural). According to the results, 47.1% of the interviewees reported that they knew what a transgenic plant was. When questioned as to whether they would eat transgenic food, the figures were: 75.5% of the consumers and 91.5% of the farmers replied in the negative.

Some specific small surveys have been done. For example, in the newspaper *Zero Hora*, published in Rio Grande do Sul in the Grande Porto Alegre (an area including the capital, Porto Alegre, and the surrounding area), 418 people over 18 years old were interviewed by the Federal University of Rio Grande do Sul on transgenics (*Zero Hora*, 1999). The results were: 66% had never heard about transgenics; 71.8% would never buy oil made by transgenic soy; 54% would give up buying Argentinian chocolate or American fried potatoes including transgenic ingredients; 60.9% would pay more for a GM-free product. However, 64.4% supported the view that those who illegally grew transgenic soy should not be punished. Half of the interviewees felt that Rio Grande do Sul should be a transgenic-free zone; 58.3% were pro-moratorium and 95.2% supported that research on transgenics should be done.

A newspaper in Rio de Janeiro, *Jornal do Brasil*, also carried out a quick and simple poll on this issue, according to which 69% of the respondents affirmed that the commercialization of transgenic food should not be allowed in Brazil; the results were published in that journal on 11 July 2000. Furthermore, the Brazilian Society of Genetics came into the debate, publishing a document in its website discussion some advantages and disadvantages of GM crops: 66.5% were against transgenics.

At the national level, three surveys were conducted by the Brazilian Institute of Public and Statistical Opinion (IBOPE) in 2001, 2002 and 2003 (IBOPE, 2001, 2002, 2003). In the next section, we will focus on the results of the 2003 poll. Afterwards, we will discuss some qualitative results obtained by our research group on a study case carried out among young people in Rio de Janeiro, in which young people's attitudes towards modern genetics were investigated (Massarani, 2001, unpublished PhD thesis; Massarani and Moreira, 2005).[1]

## National Results

The national 2003 poll exploring public opinion on GM food was carried out by IBOPE through interviews with questionnaires. Two thousand urban Brazilians across the country were interviewed; the estimated error margin is +/–2.2%.

The survey showed that 63% of Brazilians had never heard of genetically modified (GM) food. But, after having been given an explanation of what GM organisms were, 74% preferred to consume GM-free products, while 13% supported the use of transgenic food.[2]

Those with higher incomes were slightly more likely to reject GM food than were those with lower ones: 77% belonging to classes A and B (upper economical classes) rejected GM food, compared to 78% in class C and 71% in classes D and E. Similarly, the higher the level of education, the higher the percentage of people who preferred non-transgenic food; for example, 84% among those with university-level education opposed GM food; the result for those at the lowest education level was 70%.

People aged 16–24 years old – the age group of the study case that

[1] See also Guivant (2005), who gives an overview on transgenics and public perception of science in Brazil.

[2] It should be observed, however, that exact definition is tricky: the following was used here: An organism is called transgenic, or genetically modified, when a change is made in its DNA, that is, where the characteristics of a living being are stored. Through genetic engineering, genes are retrieved from a vegetal or animal species and transferred to another. These new genes are submitted to a kind of reprogramming, and it might be capable of producing a new kind of substance, different from the original organism.

we will discuss later in this paper – were those who exhibited the lowest percentage of rejection toward transgenics. In this group, 72% preferred non-transgenic food; meanwhile the other results were: 77% (25–34 years old), 76% (35–49) and 70% (> 49).

This poll also showed that a large majority (92%) thought that food with genetically modified ingredients should be identified as such on the label.[3]

Support for food labelling increased with higher levels of education and income. The group that presented the lowest percentage supporting the labelling were those > 49 (85%).

The survey also found that 73% of the respondents were of the opinion that transgenic crops should be prohibited until all the important questions concerning the risks involved had been adequately addressed. On this item, the rejection index is lower for those in the two categories of lower education level (71%); meanwhile, the other groups exhibit a similar value).

For all the above-mentioned items, the results of the IBOPE survey in 2001 and 2002 were similar to those of the subsequent year. However, some differences can be identified. For example, there was a significant increase in those that affirmed to having already heard about transgenic food: 37% in 2002 to 63% 1 year later. An increase was also observed in the group of people that would prefer to eat non-transgenic food (71% in 2002 to 74% in 2003).

The 2002 IBOPE survey included new questions, which showed that 38% of interviewees were concerned about the market implications, believing that Brazil could have difficulties in exporting GM food. More than half considered that GM foods could damage the environment, and nearly two-thirds thought that they could pose a hazard to human health. From the total, 60% agreed that Brazil would have difficulties in exporting transgenic products since they were rejected by consumers from several countries; 17% disagreed and 23% didn't know/didn't answer.

The poll also asked whether the interviewees would agree that: (i) production would be greater with transgenic crops in comparison to traditional agriculture (42% agreed, 35% disagreed and 23% didn't

3 All human and animal food sold in Brazil that contains more than 1% of genetically modified (GM) ingredients is supposed to be labelled under a law that came into force on April 2004 (see Massarani, 2004b), even before GM crops are allowed into the country. Paradoxically, however, this new law does not require products containing the 2003 GM soy to be labelled, which had a special licence. This is not the first time that Brazil has legislated on the labelling of GM food: the government of ex-president Fernando Henrique Cardoso introduced a law that products with more than 4% GM ingredients should be labelled. However, neither of these laws was ever put into practice.

know/didn't answer); (ii) production costs would be lower (31% agreed, 32% disagree and 27% didn't know/didn't answer); and (iii) transgenic food would help alleviate world hunger (28% agreed, 55% disagreed and 18% didn't know/didn't answer).

## Young People's Attitudes

In 2001, we investigated young people's attitudes towards modern genetics and biotechnology.[4]

We studied high school students of nine public and private schools in Rio de Janeiro, through questionnaires and focus groups (Massarani, 2001, unpublished PhD thesis; Massarani and Moreira, 2005). Questionnaires were used to explore students' current attitudes towards biotechnology, while focus groups were employed to analyse *how* they thought and *why* they thought that way, i.e. how their knowledge and opinions had been shaped by the different contexts of which they were part. Brazil has a large population, and there are significant regional differences and large social inequalities in many aspects – including scientific education and access to information. Our case study investigated only high school students from a single metropolitan area (Rio de Janeiro). Hence, the attitudes of these students will not necessarily reflect the attitudes of young people in the country as a whole.

The questionnaire was, in part, based on the Eurobarometer 1996 survey (Durant *et al.*, 1998). Students were asked to what extent they agreed or disagreed with the following:

- Using biotechnology in the production of foods, for example to increase size and protein level, or to change the taste.[5]
- Taking genes from plant species and transferring them to crops, to render them more resistant to insect pests.

Questions were asked about the applications with respect to: (i) their actual use; (ii) their benefits to society; (iii) the risks involved in their application; (iv) their moral acceptability; and (v) whether they should be encouraged. The following response options were provided: 'strongly agree', 'somewhat agree', 'somewhat disagree', 'strongly disagree' and 'don't know'.

[4] For those interested in similar studies in other countries, see Macer *et al.* (1994) and Cavanagh *et al.* (2005).

[5] The original question in the Eurobarometer study was: 'Use modern biotechnology in the production of foods, for example to make them higher in protein, keep longer or change the taste.'

We analysed 610 questionnaires filled out by students in the second grade of secondary school (370 from a public school and 240 from a private school). Our study incorporated different kinds of schools in Rio de Janeiro, namely: (i) three schools that were funded by the government, did not charge fees and were located in poor areas; (ii) two schools that were government-funded, did not charge fees but were located in socially and economically more privileged areas, although students came from a variety of economic backgrounds; (iii) three private schools that charged monthly fees; they were located in socially and economically privileged areas and concentrated on students of higher purchasing power; and (iv) one school that, despite being situated in a poorer area, charged a monthly fee and, hence, mainly accommodated students from the local elite. The questionnaires included closed questions as well as space for additional comments.

In the qualitative phase of the research, we subjected some of the questionnaire data to a more in-depth analysis by means of focus groups. This allowed us to assess the students' perceptions of the relative importance of the different arguments presented to them and of the issues raised by the students themselves in the additional comments section of the questionnaires.

We organized six focus groups, each of which consisted of about five students from the same school.[6] The schools were selected in such a way that students from different social and economic backgrounds within the city would be represented. There was a broad interest in participating in the focus groups. The participants were chosen randomly among the students that had indicated their willingness to participate. However, we cannot rule out that those students that agreed to participate in the focus groups were more interested in science issues than the average student in their respective schools was. Yet, it should be stressed that a broad interest in modern genetic issues was also observed in most of the questionnaires.

A synthesis of the main results of the questionnaire data is presented in Table 7.1.

While GM food was seen to be useful by two-thirds of the students, more than three-quarters of the respondents viewed this application as risky and about one-third did not think it morally acceptable. Applications to crops were more likely to be encouraged by about two-thirds, since a smaller majority saw this application as risky (63%) and less than one-quarter saw GM crops as not morally acceptable.

To carry out a cross-analysis with the results obtained by IBOPE, we will focus on the 2001 poll, since this was carried out in the same

6 All discussions were tape-recorded and fully transcribed.

**Table 7.1.** High school students' attitudes towards GM crops and food.[a]

| | Utility | | Risk | | Moral acceptability | | Encouragement | |
|---|---|---|---|---|---|---|---|---|
| Item | Agree (%) | Disagree (%) | Agree (%) | Disagree (%) | Agree (%) | Disagree (%) | Agree (%) | Disagree (%) |
| GM food | 66 | 29 | 78 | 16 | 55 | 35 | 56 | 32 |
| GM crops | 81 | 14 | 63 | 26 | 65 | 24 | 65 | 23 |

[a] The responses 'strongly agree' and 'somewhat agree' have been combined as 'agree'; likewise, 'strongly disagree' and 'somewhat disagree' are now 'agree'.

period of our study. According to this poll, among all age groups, people between the ages of 16 and 24 – i.e. those in the same age range as the students in our study – were the group that showed the lowest rejection rate towards transgenics. In this age group, 68% preferred non-transgenic products compared to 79% for 25–34-year-olds, 76% for 35–49-year-olds and 72% for those 50 years of age and older. As far as labelling of GM food is concerned, the only group that showed a lower approval rate than the people under 25 years of age were the over-50s, with 83%; for all other age groups, including the one consisting of young people, the figures were in the 93–94% range (IBOPE 2001).

In the same poll, most of the people under 25 years of age (67%) believed that transgenic crops should be prohibited until all the important questions surrounding the risks involved were adequately addressed. For the other age groups, the numbers were 69% (25–34), 70% (35–49) and 62% (50 and over).

Among the students in our study, 66% agreed that the use of biotechnology in food was socially useful, for example by increasing size or protein content or by changing the flavour; in the case of using biotechnology for increased pest resistance in cultivated plants, the percentage increased to 81%. However, there was a strong perception that there were risks associated with this technology: 78% of respondents for the first application and 63% for the latter. With regard to the first application, 56% supported its encouragement, while the approval rate was 65% for the second one.

The results of our own research corroborate the finding in the IBOPE polls that, in general, young people were more open to receiving technological innovations than were the other age groups. Both studies show that a significant number of young people thought genetic engineering of food useful and to be encouraged, but that risks were involved. The students would prefer that transgenic crops not be grown until these risks could be adequately addressed. Furthermore, they would prefer non-transgenic food but, if this food was to be put on the market, they thought it should be labelled.

In the focus groups, transgenic food was one subject in which students showed great interest. They discussed – sometimes on their own initiative – environmental, economical and social aspects related to the issue, as well as the need for extensive research as a prerequisite for releasing transgenic food into the marketplace. Environmental implications, associated economic interests and the importance of labelling were widely mentioned.

All focus groups mentioned the risks of transgenic technologies, but many advantages were also brought up, such as: (i) increased productivity (with a consequent reduction in price); (ii) elimination of world hunger; (iii) increased resistance of food to pests and spoiling; (iv) production of food with enhanced taste and appearance; and (v) production of food containing ingredients that protect against disease. Those students that neither saw advantages nor condemned transgenic technologies argued that GM products would do you no good, but they would not harm you either: they are as unhealthy and unnatural as most currently available foods that are produced with agrochemical substances.

The following disadvantages of transgenic food were identified: (i) unpredictable health risks, resulting from the fact that it is not natural or from the fact that we do not know what is being changed; (ii) possible environmental imbalances; (iii) possible deterioration of the quality of life for small farmers, due to a predominance of large companies; and (iv) the belief that it would not solve the world hunger problem, since this can be solved only by the implementation of economic and political changes that benefit developing nations, rather than through technological means.

However, it was clear in the focus groups that some students did not understand that each genetically modified crop or food had been designed with a specific objective in mind – e.g. therapeutic properties, increased protein concentration, increased tolerance for herbicides, increased resistance to rotting or to insect pests – and different objectives often require different techniques. Some of them seemed to perceive GM food as being part of one broad, single category. Another misconception was that all genetically modified crops are sterile, supposedly forcing farmers to buy new seeds for every crop.

Considering the several applications of the new biotechnology, both the analysis of the questionnaires and of the focus groups showed that most students were able to make clear distinctions between the different purposes of each technique used in genetic manipulation. This was reflected in their assessment of the utility, the potential risks and the need for encouragement of the different techniques. For example, there is a general positioning against the use of the genetic selection of embryos for aesthetic purposes and against genetic tests that might induce discrimination among people (by the labour market or by insurance companies, for example). In other cases, especially those concerning medical use, the students' attitudes were more favorable, although they were aware that risks remain an important issue.

## Final Considerations

An intriguing issue to be faced in the future is to compare the history of introduction of GM crops in Argentina and Brazil. These neighbouring countries, with historical similarities, have very different trajectories in this respect. Genetically modified soy was introduced to Argentina in the mid-1990s, with no significant resistance by the Argentinian public. In Brazil, the first attempts were recorded in 1997 but, since then, a significant resistance from some Brazilian groups was seen and a series of legal disputes has blocked such attempts (Menasche, 2000).

At present, almost all the soy produced in Argentina is transgenic and the country is the second largest producer and exporter of genetically modified seeds, following the USA (Polino and Fazio, 2005). However, a survey shows that 61% of Argentinians are unaware that the country is producing transgenic soy. Half of the interviewees affirmed that they would prefer to pay more for a non-transgenic food (Secyt, 2004).

One aspect that deserves our attention is the relationship between the level of education on the one hand and the acceptance of transgenic food on the other. Some researchers have suggested that people do not accept transgenic food because they do not have a sufficient educational background to understand the benefits of new technologies (e.g. Oda, and Soares, 2001). Yet, in Brazil the IBOPE surveys show that the simplistic assumption that there is a direct link between the degree of education and the degree of approval for transgenic food is not sustainable.

Durant *et al.* (1998) have also linked education level to the degree of approval of transgenic food, but in a different way. Their results, based on studies carried out in 17 European countries, suggest that people with higher knowledge levels are more likely to express a definite opinion about biotechnology. However, this does not necessarily imply a *positive* opinion. As these authors explain: 'Knowledge is one of the resources that contribute to the formation of opinions, but those opinions may be positive or negative' (Durant *et al.*, 1998, p. 200).

Despite the differences between the methodologies utilized in the studies mentioned in this article, it is clear that there is a high level of rejection of transgenics in Brazil. This result is also corroborated by a Citizens Jury held in 2001, designed to give voice to the opinions of poor Brazilian citizens in the north-east of the country – and organized by the Brazilian non-governmental organizations ActionAid and Esplar – who rejected the proposal to introduce genetically modified organisms into Brazil (Toni and Braun, 2001).

In this sense, an issue that calls attention from the results from national poll, corroborated by further studies that our group has carried out, is the fact that the governmental decisions ignored public opinion that transgenic crops should be prohibited until the important questions concerning the risks involved had been adequately addressed.

An episode symbolic of this is the announcement made by the federal government that special permission had been given to illegal soy planted in southern states only a couple of months after the results of the 2002 IBOPE survey were disseminated, with a huge impact in the mass media. Later, as mentioned before, new special licences were given for GM soy in every new harvest and the biosafety legislation was approved, including permission to grow and sell GM crops.

Also of concern is that the fact the labelling legislation was also ignored, although a large majority of the population exhibited the point of view that food with genetically modified ingredients should be identified as such on the label. These episodes show that the Brazilian government has not been hearing their citizens' voices.

## References

Cavanagh, H., Hood, J. and Wilkinson, J. (2005) Riverina high school students' views of biotechnology. *Electronic Journal of Biotechnology* 8 (2) (available at: http://www.ejbiotechnology.info/content/vol8/issue2/full/1/index.html, accessed 13 September 2006).

Durant, J., Bauer, M. and Gaskell, G. (eds) (1998) *Biotechnology in the Public Sphere*. Science Museum, London.

Guivant, J. (2006) Transgênicos e percepção pública da ciência no Brasil. *Ambiente e Sociedade* 9 (1), 81–103.

IBOPE (2001) Pesquisa de Opinião Pública sobre Transgênicos (http://www.greenpeace.com.br/transgenicos/pdf/pesquisaIBOPE_agosto2001.pdf, accessed 20 September 2005).

IBOPE (2002) Pesquisa de Opinião Pública sobre Transgênicos (http://www.idec.org.br/files/pesquisa_transgenicos.pdf, accessed 20 September 2005).

IBOPE (2003) Pesquisa de Opinião Pública sobre Transgênicos (http://www.greenpeace.org.br/transgenicos/pdf/pesquisaIBOPE_2003.pdf, accessed 28 September 2005).

Macer, D., Asada, Y., Akiyama, S. and Tsuzuki, M. (1994) Bioethics in high schools in Australia, Japan and New Zealand. In: Macer, D. (ed.) *Bioethics for the People by the People.* Eubios Ethics Institute, New Zealand, pp. 177–186.

Massarani, L. (2003). Brazil to Allow Sale of Illegally Grown GM Food (http://www.scidev.net/News/index.cfm?fuseaction=readNewsanditemid=357andlange=1, accessed 1 April 2003).

Massarani, L. (2004a) Brazil Labels GM Food. SciDev.Net (http://www.scidev.net/News/index.cfm?fuseaction=readNewsanditemid=1329andlanguage=1, accessed 6 April 2004).

Massarani, L. (2004b) Brazil Delays GM Crops and Cloning Bill (http://www.scidev.net/News/index.cfm?fuseaction=readNewsanditemid=1618andlanguage=1, accessed 6 September 2005).

Massarani, L. (2005) Brazil Says 'Yes' to GM Crops and Stem Cell Research (http://www.scidev.net/content/news/eng/brazil-says-yes-to-gm-crops-and-stem-cell-research.cfm, accessed 26 September 2005).

Massarani, L. and Moreira, I.C. (2005) Attitudes towards genetics: a case study among Brazilian high school students. *Public Understanding of Science* 14 (2), 201–212.

Menasche, R. (2000) Uma cronologia a partir de recortes de jornais. *História, Ciências and Saúde – Manguinhos* 7 (2), 523–540.

Oda, L. and Soares, B. (2001) Biotecnologia no Brasil. Aceitabilidade pública e desenvolvimento econômico. *Parcerias Estratégicas* 10, 162–173.

Polino, C. and Fazio, M.E. (2005) La opinión pública de los argentinos sobre los organismos genéticamente modificados (OGM). *El caso de la soja transgénica* {In press}.

Rohter, L. (2004) Planting-time soy quandary for Brazil. *New York Times,* 13 October (http://www.nytimes.com/2004/10/13/business/worldbusiness/13seed, accessed 13 October 2004).

Secyt/Secretaría de Ciencia, Tecnología e Innovación Productiva – Ministerio de Educación, Ciencia y Tecnología (2004) *Los Argentinos y su Visión de la Ciencia e la Tecnología.* Primera Encuesta Nacional de Percepción Pública de la Ciencia, Buenos Aires

Toni, A. and Braun, von J. (2001) Poor citizens decide on the introduction of GMOs in Brazil. *Biotechnology and Development Monitor* 47, 7–9.

*Zero Hora* (1999) 36 (12.534), 39.

# II Theoretical Perspectives

8

# Where do Science Debates Come From? Understanding Attention Cycles and Framing

M.C. Nisbet[1*] and M. Huge[2**]

[1]*School of Communication, American University, Washington, DC, USA;*
[2]*School of Communication, The Ohio State University, Columbus, Ohio, USA;*
*e-mail: *nisbetmc@gmail.com; **huge.8@osu.edu*

## Introduction

In the summer of 2002, when operatives from a coalition of environmental groups sent samples of Taco Bell shells to a laboratory to be checked for contamination by StarLink™ maize, they hoped that news of a positive test result might catalyse a major media spectacle, focusing critical attention on the central assumptions of plant biotechnology regulation. Since the early 1980s, a few interest groups had been calling attention to perceived systemic-level problems in the monitoring and successful segregation of plant biotechnology products but, despite extensive efforts, these groups had little success in changing policy.

Yet, a series of key federal policy decisions in the late 1980s and early 1990s had successfully limited official debate about the technology to a narrow range of short-term health and environmental factors. Out of bounds for serious consideration in regulation were uncertainties about long-term environmental or health risks, or calculations of social, ethical or economic impacts. What made the possible mixing of StarLink maize with traditional maize varieties especially useful as a focusing event was StarLink's lack of approval for human consumption. The environmental groups claimed that if consumers ingested StarLink maize, genetically engineered to express a special protein, they risked severe allergic reactions.

After the test results on the taco shells came back positive for traces of StarLink, the coalition of groups calling themselves Genetically Engineered Food Alert contacted a political reporter at the

*Washington Post* specializing in science regulation (for more details on this strategy, see Rodemeyer and Jones, 2002). In a 18 September second-page *Washington Post* story announcing the discovery, Larry Bohlen of Friends of the Earth was quoted as voicing the following warning: 'This maize is absolutely not supposed to be in our food, but an independent lab. found it there anyway. This shows a major regulatory failure and raises some real human health concerns.'[1] On 22 September, after confirming the presence of StarLink in their taco shells, Kraft Foods issued a nationwide recall. The company's decision drew the attention of *The New York Times* to the event for the first time in a 23 September article fronting the business section of the paper.[2]

News of the StarLink discovery triggered quakes across the food industry that would play out for the next 2 years. Other food companies, including Safeway and Kellogg's, would be forced to recall millions of boxes of taco shells and similar products. Aventis, developer of StarLink, would fire its chief executives in crop science and put the division up for sale, with estimates of $100 million in losses (Taylor and Tick, 2001).

In late December 2000, when news reports revealed that the EPA knew as early as 1997 that StarLink maize had contaminated the human food supply, one possible interpretation by journalists was that of a major political cover-up, complete with the drama of possible congressional hearings. As Marion Nestle (2004) suggests, journalists could have easily run with a scandal storyline. What did Aventis and the EPA know and when did they know it? And why did it take a coalition of environmental groups to draw attention to the public health risk rather than industry or regulators?

As we review in this chapter, however, major news organizations did not react to the issue as a revelation worthy of the scandal label, assigning coverage to the politics desk and the front page. Instead, the press characterized the controversy predominantly from an industry and regulatory angle, with coverage delegated to business and science reporters, an editorial decision consistent with several decades of news coverage of the technology.

Our findings indicate that the StarLink affair generated a historic spike in press attention to plant biotechnology, but even compared to

1 The following day, the announcement was also covered on page one of the Life section of *USA Today*, in a news brief in the *Christian Science Monitor*, in the foreign section of *The Guardian*, in a wire story by *Agence France Press* and in brief mentions on CBS *Evening News* and on *Fox News Special Report* with Brit Hume.

2 The same day, news stories were also carried by the *Washington Post* (page A09), the *Associated Press* and CNN *Saturday Morning*.

other science- and food-related issues at the time, plant biotechnology received only modest attention. In part, the press strategy employed by Genetically Engineered Food Alert was hindered by its timing, coming during the heat of a historically tight Presidential race, with any subsequent political fallout from StarLink lost in the competing noise of the controversy over the disputed Florida vote count.

Not surprisingly, polls indicate that the American public's concern over plant biotechnology remained minimal, and that the event had little or no impact on collective public attention (Shanahan *et al.*, 2001), a finding that other chapters in this volume confirm. Moreover, as we will review, history shows that regulatory change in response was incremental at most. The StarLink affair became just another event where the press played an important role in constraining the scope of controversy in the USA surrounding plant biotechnology.

## Controlling Attention and Framing the Policy Debate

The efforts of Genetically Engineered Food Alert to alter the direction of biotechnology policymaking by boosting the level of media attention and by morphing the image of the issue in the press are time-honoured tactics employed by operatives across the policy spectrum. As part of the power game of politics, advocates routinely attempt to control media attention to an issue while simultaneously defining an issue in favourable terms. Yet, beyond these political playbook tactics, can social science research offer us a more systematic understanding of how press lobbying activities and strategies can be applied across issues and time, shaping the trajectory of policymaking? And how exactly do these media-policy interactions account for the relative absence of wider controversy in the USA over plant biotechnology?

As a starting point, we turn to work by the economist Anthony Downs (Downs, 1972), who articulates in a provocative way what appears to be an essential element about the mysterious nature of collective attention when it comes to policy disputes. According to Downs, an issue like plant biotechnology rests in a pre-problem stage until a triggering event catapults it into public attention. This triggering process is often followed by a period of public concern and collective enthusiasm to solve the problem. But, according to Downs, policymakers and the media inevitably exhaust dramatic elements of the issue that are needed to sustain interest, and new issues in the attention pipeline take its place.

Once the issue has passed through the attention cycle, it remains on average more likely to receive future attention than other issues that might have been left behind in the primeval soup of pre-discovery. The rise in attention leads to the creation of institutional arrangements to solve the problem, and these institutional arrangements persist long after initial attention subsides. Also according to Downs, important aspects of the issue may become attached to a separate issue that later comes to dominate attention.

Sociologists Steve Hilgartner and Charles Bosk (Hilgartner and Bosk, 1988) add to Downs by arguing that levels of attention to a problem are a function not of objective conditions alone, but are determined by a social contest to define the nature and importance of issues. Political operatives select a particular interpretation of an issue from a plurality of realities, and the interpretation that comes to dominate public discourse has profound implications for the future life cycle of the issue, for the interest groups involved and for policy decisions.

Moreover, this interpretative struggle occurs across various 'public arenas', social environments such as the news media or various political institutions where issues compete for attention. These arenas have limited carrying capacities, meaning that they can pay attention to only a limited number of problems at any given time. Consequently, unless carrying capacity increases, the rise of one issue to agenda status means that another issue is likely to be bumped from consideration. Competition means that there are a few very successful problems that achieve widespread 'celebrity' status and attention, a few other moderately successful issues and an overwhelming number of less successful issues.

As we will see in the case of plant biotechnology, when an issue shifts through Downs' upward swing in attention, a historic peak in the issue's level of attention does not necessarily mean that the topic achieves celebrity status and dominates the media agenda. Rather, its importance is contingent on a number of principles of selection that we outline, as well as the status of other competing issues at the time.

Building on these two explanations to understand news coverage of plant biotechnology and its connection to the policy debate, we present in this chapter several important underlying social mechanisms that drive cycles of media attention and definition to policy issues, what we call a 'a model of mediated issue development'.

In Fig. 8.1, we sketch the cycle of attention outlined by Downs, and then highlight as underlying mechanisms: (i) the type of policy venue where debate takes place or is centred; (ii) the media lobbying activities of competing strategic actors as they attempt to interpret or 'frame' the issue advantageously; (iii) the tendency for different types

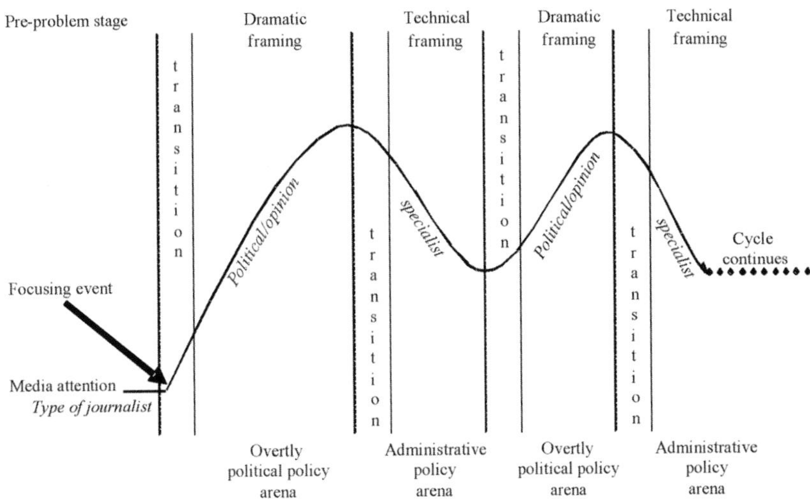


**Fig. 8.1.** Model of mediated issue development.

of journalist to depend heavily on shared news values and norms to narrate the policy world; and (iv) the context relative to other competing issues.

Some issues achieve celebrity status as they go through these cycles, but other issues – even in their peak years of attention – still rest relatively modestly on the overall media agenda. An important part of this process is the shift in coverage across news beats, from specialist journalists such as science writers and business reporters to political writers and general assignment reporters, and the spread in attention across the opinion pages. It may be useful for the reader to refer back to this figure as an aid and heuristic as these mechanisms are further explained throughout the chapter.

To demonstrate these mechanisms, we identify the major stages in policy development specific to plant biotech., and then measure patterns in media attention to – and definition of – the issue at the *New York Times* and *Washington Post* from the earliest mention of the issue in 1978 through to the end of 2004.

## Competition and mobilization of bias across policy venues and arenas

As previously mentioned, a few key principles and mechanisms underlie the model presented in this chapter. First, advocates and officials competing to shape American policy operate within a political system with multiple venues as access points. For example, on the issue of plant biotechnology, influence can be gained within administrative agencies such as the Environmental Protection Agency (EPA), the USA Department of Agriculture (USDA), the Food and Drug Administration (FDA) or an independent scientific advisory body such as the National Academy of Sciences (NAS). At other times, influence over decision-making might occur within more overtly political arenas such as Congress or the White House.

The political system is an uneven playing field. Each policy venue holds certain biases, and tends to favour certain actors and arguments over others. Power and influence turns on the number of actors and interest groups that become involved in a policy matter, a phenomena identified as 'the scope of participation' (Schnattschneider, 1960). For those actors advantaged by *status quo* decision-making on an issue within any single policy venue, it is in their best interest to limit the scope of participation, since adding new seats to the bargaining table might disrupt the balance of power. For the disadvantaged, it is almost always wise to try to expand the scope of participation, since recruiting new players to the conflict increases the potential for a change in power (Cobb and Elder, 1983).

New participants do not join the conflict randomly or distribute themselves equally across sides. Instead, they are lured by the way parties in the conflict manipulate the image of the issue (Rochefort and Cobb, 1994). Therefore the ability to advantageously define an issue 'is the supreme instrument of power' (Schnattschneider, 1960). We will return to issue definition and framing in a subsequent section.

## Administrative and overtly political policy arenas

Typically in administrative venues, the scope of participation is limited, with just a few actors given access and input to decision-makers. These venues are characterized by consensus: things happen incrementally and scant attention follows. In overtly political arenas the scope of participation is much wider, with a diversity of actors granted access and input. In these venues, consensus usually erodes, conflict can be high, there is the potential for non-incremental change and, when things happen, heavier media and public attention follows (Maynard-Moody, 1992, 1995). The contests within each of these policy venues also vary in nature. When the site of decision-making moves to an overtly political context such as Congress, the possible intensity of conflict and scope of participation expands.

In the case of emerging technologies such as plant biotechnology, administrative policy arenas typically prioritize access and input from industry and the scientific community, enabling mostly insular decision-making by administrators, scientists and independently constituted scientific advisory boards. Decisions are often to the exclusion of the general public or other interests, and arguments based in scientific and technical terms are typically the most persuasive.[3]

In contrast to administrative contexts, overtly political arenas such as Congress and the White House are open to a greater diversity of interest group involvement and are a more pluralistic policy domain. In these venues, consumer groups, citizen activists, and environmentalists might all hold either equal or greater sway over decision making. The scientific community holds less authority, and dramatic arguments centered on ethics or social concerns often win out over an emphasis on scientific evidence or cost-benefit calculations. After some deliberation, however, and a temporary solution adopted, overtly political arenas ultimately prove inadequate for handling certain technical decisions, and policy is delegated back to the administrative context for further formulation and implementation (Maynard-Moody, 1992, 1995).[4]

3 We do not mean to suggest that decisions within administrative arenas go uncontested. A glance at the news over the past few years will reveal multiple claims of 'politicization of science' emanating from various camps in society (Mooney, 2005). Such claims are not new. For several decades, across issues, environmentalists, consumer groups, religious advocates and scientists have all disputed the selective application of expertise within administrative arenas, alleging, for example, that peer-review rosters are stacked, that contrary evidence is ignored or that non-scientific perspectives are left unconsidered (Nelkin, 1975, 1994). In particular, high-stakes regulatory decisions about health or the environment often involve complex questions about emerging and uncertain science, with various actors competing to interpret evidence in ways that favour their preferred policies (Jasanoff, 1987). In these cases, language serves as the key tool in constructing authority and drawing boundaries between legitimate and illegimate participants or claims. Despite the contested nature of administrative arenas, our central point, however, is that in comparison to overtly political policy venues such as Congress or the White House, decision-making when centred in these regulatory contexts remains predominantly defined in highly technical terms, attracts less attention, with the scope of participation limited.

4 As in the case of administrative arenas, there are exceptions to these general characteristics. Whereas administrative arenas might be favourable territory for policy monopolies, limited participation can also occur within Congress, especially when an issue falls under the jurisdiction of just a single committee, and is never taken up by a diversity of committees or debated on the open floor (Baumgartner and Jones, 1994). Indeed, a recent policy study of biotechnology concluded that, in comparison to medical applications such as stem cell research, attention to plant biotechnology within Congress has been distributed across a much narrower range of committees, generating fewer hearings. The focus of these hearings was predominantly on benefits and was overwhelmingly positive, with most testimony coming from industry members (Sheingate, 2006).

## Framing

Examining the issues of nuclear energy, tobacco and pesticides, past research finds striking similarities in the links between issue definition, media attention and policy venues. In the case of each issue, its early history was characterized by positive image-making and enthusiasm for creating institutional arrangements that would further market development. Eventually, however, these pro-industry policy monopolies were broken up by opponents who successfully redefined the issue in provocative and negative ways, and who shifted decision-making away from administrative arenas to more overtly political contexts (Bosso, 1987; Weart, 1988; Baumgartner and Jones, 1994).

With each of these issues, did changes in the nature of media discourse parallel changes in the nature of 'real world' conditions? Did nuclear energy and tobacco, for example, become inherently more risky across decades, leading eventually to the demise of their favoured policy status? Both the policy definition and social problems literature emphasize that understanding shifts of power on an issue depends on analysing the relationship between the approximately 'objective' conditions surrounding an issue and how their perceived conditions are defined by various actors (Hilgartner and Bosk, 1988). The emphasis on social construction introduces framing as a second key mechanism underlying the issue attention cycle.

A 'frame' is a central organizing idea or story line to a controversy that provides meaning to an unfolding of a series of events, suggesting what the controversy is about and the essence of an issue (Gamson and Modigliani, 1989). Frames are 'thought organizers', devices for packaging complex issues in persuasive ways by focusing on certain interpretations over others, suggesting what is relevant about an issue and what should be ignored. Framing occurs at the policy level, the media level and/or at the public level (Scheufele, 1999). At the media level, 'frames may best be viewed as an abstract principle, tool or schemata of interpretation that work through media texts to structure social meaning' (Reese, 2001, p. 14). By giving more weight to some dimensions of a controversy than others, the frames in news coverage help guide policymaker and citizen evaluations about the causes and consequences of an issue, and what should be done (Ferree *et al.*, 2002).

Frames as general organizing devices should not be confused with specific policy positions. Individuals can disagree on an issue but share the same interpretative frame. Each frame as an organizing device for arguments and interpretations is 'valence neutral,' meaning that it can take pro-, anti- and neutral positions, though one position might be in more common use than others (Gamson and Modigliani, 1989; Ferree *et al.*, 2002). Consider the ethics/morality frame on which we elaborate later in the chapter. This interpretation could be applied to package plant biotechnology as 'playing God in the Garden', and violating the natural order of things, therefore leading to negative attributions about the issue. But the ethics frame could also be used to package plant biotechnology in a promotional light, emphasizing the moral duty to pursue a 'gene revolution' that could 'end world hunger'.

By linking framing to the type of policy arena where debate takes place and how much attention an issue receives, in this chapter we integrate research on framing with related work from political science in problem definition. When issues are debated within administrative contexts, as a way to limit wider involvement, advantaged actors typically frame an issue in narrowly technical terms using referential symbols that are devoid of emotional content (Edelman, 1964; Nelkin, 1992; Bennett, 2001).

Framing strategy has important implications for how much attention an issue receives, since frames that emphasize the technical dimension of an issue have less 'symbolic weight', 'potency' and/or 'urgency', and are therefore an effective way to limit conflict expansion and subsequent attention to an issue (Cobb and Elder, 1983). As part of the strategic effort to shift an issue away from an administrative policy arena to an overtly political arena, disadvantaged actors try to expand the scope of participation and intensity of conflict by employing frames that evoke emotionally charged condensational symbols (Edelman, 1964; Nelkin, 1992; Bennett, 2001). Again, in an important factor connecting issue attention and definition, frames emphasizing dramatic dimensions of an issue – including ethics, morality, conflict and uncertainty – are the type of 'symbolically urgent' tactics that help drive conflict expansion and wider public concern (Cobb and Elder, 1983).

Strategic actors understand that any single issue competes with many others for attention, and these actors actively work to frame an issue in ways that either deflect or attract attention, and persuade others to support their side. For political operatives, 'framing an issue is therefore a strategic means to attract more supporters, to mobilize collective actions, to expand actors' realm of influences and to increase their chances of winning' (Pan and Kosicki, 2001, p. 40).

Of particular relevance to the current study, previous research has pinpointed the shift in framing of an issue from technical terms to dramatic terms as a key element in promoting the scope of participation around science-related controversies (Nelkin, 1992). In sum, as Sheingate (2006) notes in his policy study of biotechnology, and as we detail in this chapter, how plant genetic engineering has been defined both reflects and shapes where the issue has been decided, by whom and with what outcomes.

## News narratives

Theorizing relative to news narratives is based in a journalist-centred perspective. Across a range of issues, media researchers have linked Downsian-like cycles or 'waves' in media attention to journalists' need for narrative structure. According to this line of research, journalists are more likely to produce heavy coverage of an issue if they can locate themselves within a continuing story with a beginning, middle and end, ascertaining newsworthy moments and thereby assigning larger meaning to passing events.

Issues that receive greatest media attention are those that are most easily dramatized or narratized (Fishman, 1978, 1980; Cook, 1996, 1998; McComas and Shanahan, 1999; Lawrence, 2000; Bennett, 2001). The frames adopted by journalists are closely linked to this need for a narrative structure, explaining in part journalists' preferences for framing public life as if it were simply a matter of political competition and conflict (Patterson, 1994; Capella and Jamieson, 1997). This journalist-derived interpretation, labelled by scholars as 'strategy framing', emphasizes who is ahead or behind, and the episodic day-to-day tactics employed by strategic actors to gain an advantage. These frames generate drama by focusing on conflict, with an expectation of 'winners' and 'losers', and personalizing the news by focusing on individual battles between specific actors (Bennett, 2001).

As Lawrence (2000) describes, the use of the strategy frame is context dependent, and it is closely linked to the amount of media attention given an issue. If journalists need to be able to visualize events as part of a broader storyline, then policymaking becomes most newsworthy when there is clear conflict leading up to eventual promise of resolution, a scenario that Cook (1996) describes as 'conflict with movement'. When policymaking appears quagmired, with no movement on legislation or decision on the horizon, the strategy frame is less likely to be applied by journalists, and media coverage is unlikely to build in anticipation of a final outcome.

## Transfer across news beats and specialist journalists

Of importance to understanding cycles of attention and media framing, often an issue will transcend various 'news beats.' For example, as an issue shifts from the science or industry sectors into overtly political arenas, coverage by science writers or business journalists may give way to coverage by political and general assignment reporters. This transfer of an issue across coverage domains from specialist journalists to political reporters helps explain in part a rise in media attention, and the concomitant rise in emphasis of the strategy frame. Conversely, a shift in beats away from political reporters and back to specialist journalists helps explain a drop in media attention, and a return to more technical frames.

Coverage of science has traditionally been associated with the science writer, specialist journalists who often view themselves as conduits between scientists and the public, with the goal of effectively communicating a scientist's results so that the public can have a better appreciation and understanding of the science topic or subject. Given this orientation, science journalists have been described as most likely to define an issue as scientists do, using scientific and technical frames (Nelkin, 1995). Even science journalists, however, require elements of drama to write a news story. Scientific uncertainty, in this case, serves as a dramatic device (McComas and Shanahan, 1999).[5] Very little research exists on the coverage tendencies of business writers when it comes to technology but, given their beat, we might expect an emphasis on economic and market development and the surrounding policy context.

In contrast to these types of specialist correspondents, political journalists specialize in the technical matters of the political game (Kepplinger, 1992, 1995; Hallin, 1994; Lawrence, 2000), meaning that political journalists are much more likely to rely on the strategy frame as a way of interpreting the complexities of many public issues (Nisbet, 2004; Mooney and Nisbet, 2005).

In light of these differences in reporting style and preferred frames across categories of correspondent, the shift in news beats and media definition has important implications for the amount of attention an issue receives. Any topic can become 'politically relevant', and rise into the coverage domain of the political reporter with dramatic politically oriented frames replacing technically oriented frames. Therefore, due to the greater number of political reporters in comparison to science writers, and due to the greater amount of space given political coverage, when an issue becomes defined as politically relevant the potential for volume of coverage about the topic increases (Kepplinger, 1992, 1995).

5 Science writers have also often been accused of downplaying differences in opinion across disciplines about the impacts and risks of biotechnology. In her analysis of biotech coverage, Priest (2001) observed that science writers often rely heavily on the voices of university-based plant biotechnologists who define risk narrowly in terms of short-term threats to human health or the environment, while leaving out views from other disciplines, such as ecologists who might perceive risk in terms of the impacts on the ecosystem or social scientists who might discuss social, economic and ethical risks. Indeed, surveys of university scientists and social scientists reflect the diversity in opinion about biotechnology that occurs outside of the discipline of plant genetics, including contrary views that are likely to go unreported if science writers focus narrowly on plant scientists as their sources (Priest and Gillespie, 2000; Lyson, 2001).

Moreover, when an issue is successfully defined as politically relevant, it is easier for political reporters to fit the issue into a narrative structure with a clear beginning to the controversy, with resolution of the conflict typically marked by legislative passage or other government action, conditions that conform to Cook's (1996) 'conflict with movement'.

In contrast, when the issue remains defined by technical frames, it is much more difficult for a science writer or business reporter to fit the issue into a narrative structure. Scientific research has a perpetually moving goal line, with one discovery leading to the next discovery, and applications or significant breakthroughs often decades in the future.

Similarly, market growth and developments are less easily defined in terms of conflict, with fewer identifiable climactic events around which to build coverage in anticipation of a resolution to the narrative. As a result, science writers and business reporters are less likely than political journalists to be able to produce news consistently about an issue, instead relying more on routine channel opportunities such as a newly released scientific study, press conference, official report, stock price fluctuation, new product introduction or company merger in order to file a story.

Ethical and moral frames are often difficult for all types of journalists to include in news narratives. Although reporters may occasionally take time to contextualize the ethics debate within news coverage, it remains challenging to think of news pegs or hooks for covering the ethical dimension of a policy debate, unless it is connected to some type of routine channel source such as the release of an ethics commission report. Coverage of ethics may also interfere with journalists' preference for the appearance of impartiality.

Instead, editors are likely to delegate coverage of ethics to the opinion pages. Actors and sources connected to the debate generate fresh material in the form of letters, Op-eds, and arranged editorial meetings that strategically emphasize the ethical side of an issue, and editors are happy to take advantage of the material, covering this dramatic dimension while insulating themselves partly from direct criticism. Therefore, as an issue moves to overtly political arenas, and sources start lobbying the opinion pages of media outlets with submissions framed in ethical/moral terms, the prominence of the ethical/moral frame rises in coverage and the total volume of coverage (defined as the sum of both news and opinion articles) devoted to the topic increases.

## Policy Background on Plant Biotechnology

In this section and in Box 8.1, drawing upon a number of selected policy studies on the topic, we identify key stages of policy development specific to plant biotechnology. Though we do not want to imply that the complex nature of the issue should be understood in a linear fashion, in analysing and understanding 25 years of news coverage of plant biotechnology, it is useful to break down and compare our news media findings across roughly identifiable historical stages and their key events since, as we have explained, media coverage is likely to both reflect and shape the policy debate.

As we detail in Box 8.1, a number of early US policy decisions addressed regulatory uncertainty by narrowly defining questions about

**Box 8.1.** Stages of USA Policy Development on USA Plant Biotechnology.

| Stage | Policy arenas | Key events |
|---|---|---|
| Managing regulatory uncertainty, and walling off participation, 1975–1987 | NIH, WH OS and TP, EPA, FDA, USDA, limited Congress | Fearing Congress might enact legislation regulating recombinant DNA research, in 1975, scientists meeting at Asilomar, Calif. ask for regulatory oversight of recombinant DNA research from the NIH, with proposals for experiments evaluated by peer-review. Later, in 1980, the USA Supreme Court, in a 5–4 decision, concludes that companies could patent the products of biotechnology, a decision that helps catalyse biotech's commercialization. Across the early 1980s, uncertainty over the adequacy of the NIH peer-review scheme grows in reaction to a chain of events, including requests for field trials, jurisdictional challenges from the EPA, Federal lawsuits, local protests in California, and House inquiry hearings (Jasanoff, 2005; Sheingate, 2005). In 1986, the Reagan administration, guided by a preference for deregulation and fearing that Congress might move to pass special biotech legislation, issues an Inter-Agency Coordinating Framework. By applying an emphasis on the 'substantial equivalence' of biotech end products, rather than the process by which they were made, the Coordinating Framework creates no new regulation, and assigns authority to the FDA, EPA, and USDA. |
| Early market development and regulatory precedents, 1988 to 1994 | FDA, USDA, EPA | In 1988, the FDA begins review of rBST, a bio-engineered hormone that increases milk production in cows, and issues final approval of rBST derived milk in 1993, requiring no special labelling. Setting an important precedent for later plant biotech decisions, the FDA rules that concerns over socio-economic impacts cannot be used as grounds for deciding approval or labelling. Based on the 1986 Framework, the only considerations that matter are environmental and health risks (Nestle, 2004; |

**Box 8.1.** *Continued.*

| Stage | Policy arenas | Key events |
|---|---|---|
| | | Jasanoff, 2005). Applying the same logic, the FDA approves recombinant chymosin in 1990, an enzyme used in cheese making. Similarly, in 1994, the Flavr Savr tomato is approved by the FDA without the requirement of labelling (Nestle, 2004). The FDA formalizes its rules on GM food in 1992, rules that do not require pre-market approval. The USDA relaxes its rules on field trials, stipulating only prior notification. |
| Consolidation and investor optimism, 1995 to 1997 | EPA, USDA, FDA | The mid-1990s feature heavy investor speculation in biotech companies, especially Monsanto, as the EPA approves BT pesticide producing crops, and Monsanto introduces its trademark Roundup Ready soy. Monsanto launches a global PR campaign to hype the benefits of plant biotechnology while aggressively buying up smaller biotech companies, a move that consolidates its ownership of important patents, and boosts its stock price from \$10 to nearly \$60 (Leiss, 2000). In 1997, British scientists announce the birth of the cloned sheep Dolly, galvanizing worldwide attention. Yet, the overwhelming focus is on human genetic engineering, with little spill over in the USA to plant biotechnology (Priest, 2001; Nisbet and Lewenstein, 2002). |
| Trade conflict and social protest, 1998 to 1999 | EPA, FDA, USDA, USA State Department, limited Congress | The EU establishes in 1998 an elaborate system for labelling and tracing GM products, resulting in a *de facto* moratorium on USA imports. In several major European countries, a wider scope of participation in policy decisions that includes consumer, labor, and environmental groups sets the precedent for EU regulation that rejects the American emphasis on substantial equivalence of GM products, and defines genetic modification as a process with unique social implications (Jasanoff, 2005). In 1999, a series of high profile focusing events turns European public attention to potential safety risks of plant biotechnology. Events include protests led by anti-globalization leaders in France, statements of opposition from England's Prince Charles, and the claims of a British scientist who announces on television that rats fed GM potato suffered adverse health effects (Jasanoff, 2005). In the USA, the publication of a letter to *Nature* in May 1999 reports that Monarch butterflies fed GM maize pollen in the lab subsequently died. Environmental groups press their case about the ecological risks of plant biotechnology. During the |

| | | |
|---|---|---|
| | | summer of 1999, protests outside the World Trade Organization (WTO) meetings in Seattle include GM food opponents, and protests also occur at FDA public comment hearings on labelling in several cities (Nisbet and Lewenstein, 2002). The trade conflict with Europe and the emerging social protests take their toll on the plant biotech industry. By December 1999, with its share price at $38, Monsanto merges with Pharmacia and Upjohn, spinning off its plant biotechnology operations, and the Monsanto name (Leiss, 2000). In Congress, Rep. Dennis Kucinich introduces a series of failed bills proposed each year through 2005 that would require labelling and increase regulatory testing and review of GM products (Becker, 2005). |
| Food contamination 2000 to 2002 | EPA, FDA, USDA, USA State Dept, limited Congress | (*The dominant event of this period, the contamination of food products by StarLink maize, is detailed in the opening to this chapter.*) In early 2000, the Cartegena Biosafety Protocol is adopted by 176 countries. The Protocol, differing substantially from American policy definitions of biotechnology, advocates a 'precautionary approach' to risk assessment, the labelling of GMO shipments, and the inclusion of economic impacts in trade decisions (Segarra, 2000). Later in 2000, researchers publish the genome for Golden Rice, a GM variety designed to combat Vitamin A deficiency in less developed countries. In 2001, a report in *Nature* details gene flow from GM varieties to native Maize in remote areas of Mexico. An industry PR campaign attacks both the researchers and the science, leading to critical letters and an editorial note in *Nature* questioning the scientific basis for publishing the original paper (Nestle, 2004; Jasanoff, 2005). |
| Moral diplomacy and agency review, 2003 to 2004 | EPA, FDA, USDA, USA State Dept, limited Congress, limited White House | In 2003, the USA initiates a case before the WTO contending that the EU moratorium on imports has not only blocked trade but also fueled unwarranted concerns about GM food globally. In a rare instance where plant biotech is directly addressed by a USA President, George W. Bush, against the backdrop of Anglo-Euro tensions over Iraq, emphasizes the perceived moral imperative of winning the trade war: 'For the sake of a continent threatened by famine, I urge the European governments to end their opposition to biotechnology.' In spring 2004, the EU breaks the moratorium by approving a few biotech products, though leaving 30 products still embargoed (Becker, 2005). While trade disputes fester, in 2004, the National Academy of Sciences (NAS) reconfirms the |

**Box 8.1.** *Continued.*

| Stage | Policy arenas | Key events |
|---|---|---|
| | | legitimacy of product-based regulation of biotech food, but signals a possible shift by advocating that modified foods be evaluated on a case-by-case basis, and that the ability of scientists to predict adverse long term consequences remains limited. A separate NAS 2004 report cites studies that some GM organisms can cross-breed with traditional crops, and urges developers to take measures to prevent cross-breeding (Becker, 2005). |

genetic engineering in technical ways. Dominant definitions turned on either short-term threats to the environment and human health or were in line with market logic that prioritized commercialization over social or ethical considerations. By considering only short-term technological impacts, these early policy decisions rarely, if ever, considered ethical or social questions. In each case, the cognitive authority of science (and sometimes the market) was evoked to legitimate the decision and to undermine the claims of biotechnology opponents (Priest, 2001; Jasanoff, 2005).

As part of this technical framing of the debate, these early policy decisions had strong feedback and reinforcing effects in limiting the scope of participation in future policymaking to scientists, Federal agencies and industry (Sheingate, 2005), insulating key decisions within primarily administrative arenas, and walling-off input from environmental, consumer or other social groups. In Europe, where there was a much wider and pluralistic scope of participation in decision-making, policy heavily weighed social and economic considerations and defined biotechnology as a unique process requiring special regulatory attention (Jasanoff, 2005).

Starting in the late 1990s, the mobilization of biotech opponents in the USA around the publication of several scientific studies and the StarLink Affair did little to change US regulatory policy. Decision-making remained insulated within mostly administrative arenas, with minimal attention from Congress or the White House. These events, however, did slow the growth of industry, impacting the economic fortunes of several industry leaders.

## Media analysis

In applying the principles of our model to understanding the role of the media in the plant biotechnology debate, we were interested not only in media attention but also in its relationship to socially constructed meanings and interpretations. Therefore, we chose to compare our evaluation of policy development against trends and indicators compiled from a quantitative content analysis of coverage at the *New York Times* and the *Washington Post*, where we focused on frames and the type of journalist assigned coverage as key content features.

In subsequent studies of this topic and other tests of our model, we hope that this type of analysis can be complemented by interviews and surveys of journalists and sources. This choice to focus on the elite national newspapers of record complements what other media analysts have observed: stories tend to spread vertically within the news hierarchy, with editors at regional news outlets often deferring to elite newspapers and newswires to set the news agenda.

We ran a Lexis-Nexis keyword search using a comprehensive search string to reach a best approximation of the total population of articles in the *Washington Post* and *New York Times*.[6] During analysis, we discarded articles that were not substantially related to plant biotechnology, were duplicates or were non-articles such as content summaries for a newspaper edition, resulting in a final combined population of 767 news and opinion articles. We developed the coding instrument across a period of several months. As key features

6 Key words used to identify plant biotechnology-related articles included 'plant biotech'. or 'plant biotechnology' or 'crop biotechnology' or 'crop biotech'. or 'food biotech'. or 'food biotechnology' or 'ag biotech'. or 'agricultural biotechnology' or 'genetically modified food' or 'genetically modified crop' or 'genetically modified agriculture' or 'genetically engineered food' or 'genetically engineered crop' or 'genetically engineered agriculture' or 'frankenfood' or 'GM food' or 'GM crop' or 'GM agriculture' or 'GMO' or 'genetically modified organism' or 'transgenic crop' or 'transgenic agriculture' or 'transgenic food' or 'genetically altered crop' or 'bioengineered food' or 'bioengineered crop' or 'bioengineered agriculture' or 'genetically engineered maize' or 'genetically engineered soy' or 'genetically engineered cotton' or 'genetically engineered potato' or 'genetically engineered tomato' or 'genetically engineered rice' or 'genetically engineered bacteria' or 'genetically engineered microbe' 'genetically engineered organism' or 'genetically modified maize' or 'genetically modified soy' or 'genetically modified cotton' or 'genetically modified potato' or 'genetically modified tomato' or 'genetically modified rice' or 'genetically modified bacteria' or 'genetically modified microbe' or 'transgenic maize' or 'transgenic soy' or 'transgenic cotton' or 'transgenic potato' or 'transgenic tomato' or 'transgenic rice' or 'transgenic bacteria' or 'transgenic microbe' or 'transgenic organism' or 'genetically altered food' or 'genetically altered agriculture' or 'genetically altered maize' or 'genetically altered soy' or 'genetically altered cotton' or 'genetically altered potato' or 'genetically altered tomato' or 'genetically altered rice' or 'genetically altered bacteria' or 'genetically altered microbe' or 'genetically altered organism' or 'genetically modified plant' or 'genetically engineered plant' or 'transgenic plant' or 'GM maize' or 'GM soy' or 'GM cotton' or 'GM potato' or 'GM tomato' or 'GM rice' or 'GM bacteria' or 'GM microbe' or 'GM organism'.

of content, we identified a typology of relevant frames by reviewing congressional testimony, official government reports and websites, as well as articles in a diversity of newspapers and magazines.

Guiding this process, we relied on previous studies of frames in coverage of politics and in coverage of technical controversies (Gamson and Modigliani, 1989; Capella and Jamieson, 1997; Durant *et al.*, 1998; McComas and Shanahan, 1999; Patterson, 2001; Nisbet and Lewenstein, 2002; Nisbet *et al.*, 2003), with the assumption that, although some frames may be issue- or domain-specific, other frames such as the strategy frame are generalizable across issues. We further developed the validity of the frame typology in a series of pilot studies that the authors, as the two coders in the project, used to familiarize and train themselves in applying the frames to print coverage. These nine frames are outlined in Box 8.2.

Some frames, including strategy/conflict, ethics/morality, scientific uncertainty and public engagement, have stronger elements of drama and emotion than other frames, such as the release of a new scientific

**Box 8.2.** Framing typology for coverage of plant biotechnology.

***More technical frames***

NEW RESEARCH (RSCH)
Focus on new research released, discovery announced, new medical or scientific application announced, clinical trial results announced (e.g. government study, scientific journal article, scientific meeting paper, science-by-press-conference).

SCIENTIFIC BACKGROUND (SBKD)
General scientific, technical or medical background of the issue (e.g. description of previous research, recap of 'known' results and findings, description of potential agricultural or medical applications/uses).

POLICY AND/OR REGULATORY BACKGROUND (POLICY)
Focus on regulatory rules for plant biotech./framework for regulation/ jurisdiction or oversight over research and market regulation. Includes regulatory approval and oversight in for field testing, field application and market introduction. Focus on rules, enforcement and technical details of labelling and consumer disclosure. Includes international trade agreements, European or other national/trade zone policy or regulation related to ag. biotech.

MARKET/ECONOMIC PROSPECTS OR INTERNATIONAL COMPETITIVENESS (MARKET)
Focus on international trade, imports/exports, agricultural commodity prices, company market share, stock prices, company mergers and takeovers, overall growth or health of industry, financial health of farmers, reaction of investors, development/introduction of products for market, implications for domestic economy, global competitiveness and free/fair trade.

PATENTING, PROPERTY RIGHTS, OWNERSHIP AND ACCESS (PATENT)
Focus on ownership and control of new research, control and ownership of seeds or field and market products, patenting/patent approval of new crop

strains or discussion of national, international, or cross-national property rights. Also, international agreements, such as the specifics of WTO rules.

***More dramatic frames***

ETHICS AND/OR MORALITY (ETHICS)
Focus on the ethics of GM agricultural practice, focus on environmental values, emphasis on ethics of tampering with nature or 'playing God' or 'Frankenfood'. Focus on traditional/indigenous perspectives or values, discussion of impeding scientific/medical or social progress, emphasis on 'hope' and solution to world hunger, malnutrition or production of breakthrough medications/treatments.

SCIENTIFIC UNCERTAINTY (UNCERTAINTY)
Includes focus on the 'precautionary principle', definition of environmental and human health risks or moving ahead in the face of unknown risks and benefits. Includes emphasis on contesting the results of field trials or human health trials, uncertainty about the ability to reliably sort in harvesting and processing non-GMO and GMO crops, or ensure that food products contain no GMO products; or criticism of scientific claims of opponents, dismissing as not legitimate or 'sound science'.

POLITICAL STRATEGY AND/OR CONFLICT (STRAT)
Focus on the strategy, actions or deliberations of political figures, Presidential administrations, members of Congress, other elected federal or state officials, government agencies. Includes the lobbying of interest groups and the tactics of strategic actors. Focus here is not on specifics or context of policy, but rather on who is ahead or who is behind in the political conflict and their tactics for gaining an advantage. Can apply to contexts outside of the USA

PUBLIC ENGAGEMENT/EDUCATION (ENGAGE)
Focus on poll results, reporting of public opinion statistics, reference to public/consumer 'support', 'awareness', 'concern', 'education', 'demands', 'backlash', etc. or general reference to 'public opinion', 'public sentiment' or the 'battle over' public opinion. Focus on informing the public as a way to either ease their concerns or to raise alarm. Besides poll results, also includes focus on reaction or opinion specifically from an 'average man on the street' or a non-expert or local community leader. Also include emphasis on personal narrative or testimonial of a farmer, citizen, consumer or activist.

study, scientific background, policy/regulatory background, market or economic development or patenting/property rights. These latter frames tend to be more technical and contextual. Adopting a frame operationalization scheme used by Nisbet *et al.* (2003), each frame was coded as 'not present (0)', 'present (1)' or 'outstanding focus/appearing in the lead (2)'. This scoring system allows us to calculate and display a mean score for each frame across years, rendering a relative indicator of frame prominence. We tested our inter-coder agreement on a 20% probability sample of the population of articles.

Using Krippendorf's alpha (Krippendorf, 1980), we reached a reliability for each variable in the content analysis that was 0.80 or higher (at this level of consistency in interpretation, it is expected that a separate team of coders would arrive at similar results). After establishing reliability, we then moved forward to code the rest of the articles in the population.

We also wanted to record the type of journalist authoring an article. At major papers such as the *New York Times* and *Washington Post*, reporters can be segmented into science journalists, political/general assignment reporters, foreign correspondents, business reporters, wire services, style or arts reporters and other journalists, as well as guest opinion authors, regular columnists, letter-writers and in-house editorials.

For each article appearing in the *Washington Post* and *New York Times*, author names were recorded during coding. Later, each recorded author was categorized by specialty or type of opinion article.[6] In some cases, the specialty or beat of the reporter could be identified by their byline (i.e. science desk, national desk, metro desk, foreign desk, style desk, etc.), and at other times the reporter's speciality was identified by making reference to the news organization's website, or through Web searches that offered background information about the journalist, though not all authors could be categorized.[7]

[7] Authors or departments coded as science writers included Allan Coukell, Andrew Revkin, Boyce Rensberger, Carol Kaesuk Yoon, Gina Kolata, Gordon Graff, Harold M. Schmeck Jr., Health (Desk), Henry Fountain, Jane Brody, Keith Schneider, Richard D. Lyons, Warren E. Leary, William Claiborne, William Stevens, Rick Weiss, Malcolm Gladwell or Science Desk. Authors or departments coded as business writers included Andrew Pollack, B.J. Feder, Bloomberg News, Business Desk, Charles L.P. Fainweather, David Barboza, Edward Wyatt, Floyd Nannis, Justin Gillis, K. Schneider, Kurt Eichenwald, Larry Rohter, M. Lacey, Melody Peterson, Nell Henderson, Patrick J. Lyons, Paul Blustein, S. Rai, Sabra Chartand, Sandra Sugarwara, Business Desk, Sanna Siwolop, Stephanie Strom, Steven Pearlstein, Suzanne Kapner and C.H. Deutsch. David Barboza and Justin Gillis have specialized in covering the biotechnology industry, contributing industry news with a heavy technical and scientific background emphasis. Authors or departments coded as foreign correspondents included Alan Cowell, Anthony DePalma, Christopher Marquis, Craig Smith, Craig Timberg, Daniel Williams, Donald G. McNeil, Edmund Andrews, Elaine Sciolino, Elizabeth Becker, H.E. Cauvin, John Burgess, Joseph Gregory, Joseph Kahn, K. Tolbert, Lizette Alvarez, Michael Spector, Nora Bousrany, Business Desk, Roger Cohen, Sophia Kishkovsky, Stephen Buckley, Steven Erlanger, Suzanne Daley, T.R. Reid, Tony Smith, Warren Hoge, William Drozdiak, Foreign Desk or International Desk. Authors or departments coded as political/national/or general assignment reporters included Elizabeth Olson, Ernesto Londono, Judith Lederman, Judy Sarasohn, M. Cooper, M. Tolchin, Marc Kaufman, Nodine Brozan, National Desk, Shankar Vedantam or Metro Desk. Kaufman, given his experience and background as a political reporter, was classified in this category rather than as a science or business writer since, in reviewing the range of his coverage of issues other than biotech, his coverage spans the political and science beat, specializing in the politics of science. Authors or departments coded as wire services included Associated Press, Agence-France and Reuters. Authors or desks coded as style or food writers include Adrian Higgins and Marian Burros.

## Media trends and policy development

In Fig. 8.2, we see a classic Downsian pattern to media attention. A single first mention of plant biotechnology-related applications appears in a news article detailing the conclusions of a government report on agricultural industrialization appearing in the *Washington Post* in 1978. Mention of plant biotechnology did not occur again until the appearance of a single article on the biotech. industry in the *New York Times* in 1983, and then a series of articles in the *Times* and *Post* in 1984 detailing challenges to the NIH's decision to allow field tests of GM organisms.

Media attention then slowly, but only marginally, increased across the 1980s, despite the growing uncertainty over regulation and significant events, including requests for field tests, lawsuits and localized protests in California. Starting in 1986, there was a first, but very small, upward swing in media attention to 26 articles, increasing to 31 in 1987. During this rise in attention, the Reagan administration announced its Inter-Agency Framework, the business pages ran features on proposed new plant biotech products and news reports focused on separate cases of unapproved releases of GM organisms.

After the 1986 Inter-Agency Framework successfully insulated decision-making within administrative policy arenas, media attention subsided to fewer than 20 articles annually across most of the 1990s.

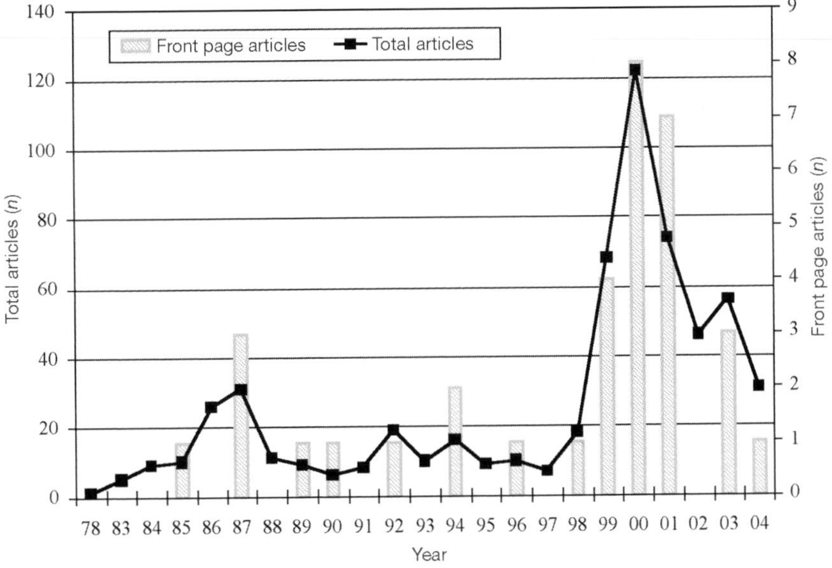


**Fig. 8.2.** Media attention to plant biotechnology.

The low level of media attention is in contrast to the key events that occurred during this period, including the precedent-setting FDA approvals of rBST, chymosin and the FlavrSavr tomato in the early 1990s, the rise of Monsanto during the mid-1990s and continued scientific research and technological development.

We interpret these findings as consistent with the conclusions of previous studies, which argued that low levels of media attention to the Inter-Agency Framework decision in the late 1980s (Nisbet and Lewenstein, 2002) and the lack of attention from the national press to the rBST approval process (Priest, 2001) served as important episodes of 'non-decision making'. In these cases, without a media spotlight on plant biotechnology, industry and scientists were better able to manage the scope of participation, keeping decision-making behind closed doors and away from wider social input.

McInerney *et al.* (2004) cite the Monarch announcement in 1999 as providing an important impetus for journalists to devote greater attention to the plant biotechnology debate, and to actively question the safety of GM agriculture. Based on interviews with Greenpeace and Union of Concerned Scientists staff members, Nisbet and Lewenstein (2002) conclude that the Monarch study may have been potentially important because it was perceived as catalysing increased commitment to the issue from large-membership environmental groups such as the Sierra Club. Indeed, in one content analysis, the Monarch announcement was found to result in a sharp increase in the reporting of risks about plant biotechnology (Marks *et al.*, 2003). Jasanoff, however, observes that the media, in covering the Monarch study, missed an opportunity to contextualize the findings within a wider debate over the ecological issues linked to plant biotechnology, and whether or not regulators were addressing these questions (quoted in Rodemeyer and Jones, 2002).

In terms of the StarLink affair, as we detailed at the opening of this chapter, biotech. critics had hoped that the discovery would serve as an important catalysing event, a potentially 'turn-the-tables on industry-focusing moment'. However, as we detailed in Box 8.1 and at the beginning of this chapter, though the Monarch study and the StarLink affair in combination with other events may have impacted the fortunes of the biotech. industry, it has done little to change the nature of policy regulation.

One likely reason is that, despite Downsian upswings in attention in 1987 and 2000 and the potential of many other focusing events to generate widespread media attention, plant biotechnology has never achieved 'celebrity' status as a topic, and has always rested relatively modestly on the overall media agenda, even during its peak year of attention in 2000. We reach this conclusion after examining two key indicators. First, in Fig. 8.2, as indicated by the limited number of front page articles devoted to the issue, plant biotechnology has never been given major agenda priority by the two elite newspapers, with a historic high of only eight and seven front page articles in 2000 and 2001, respectively.

In another indicator, in Box 8.3, we ran article frequencies from the combined coverage appearing in the two elite newspapers for

**Box 8.3.** Topical media attention, 2000: issue or topic with number of related *New York Times* and *Washington Post* articles for 2000.

Gun control, 1000+
Presidential election, 1000+
Elian Gonzalez, 774
Rising oil prices, 735
Microsoft anti-trust case, 702
Slobodan Milosevic, 596
Firestone tyre safety, 463
Nuclear energy, 403
Michael Jordan, 377
Broadband Internet, 373
Human Genome Project, 324
AOL Time Warner merger, 295
Dotcom crash, 267
Super Bowl, 262
Climate change or global warming, 234
Napster, 221
Beatles or Rolling Stones, 213
West Nile Virus, 207
Endangered species, 206
Y2K, 196
Space shuttle, 168
Food borne illnesses, *E. Coli* or *Salmonella*, 158
**Plant biotechnology, 155**
Campaign finance reform, 143
Kursk submarine, 132
Steven Spielberg, 120
Alternative medicine, 116
Florida tobacco case, 108
USS Cole, 108
Monica Lewinsky, 100
EPA air pollution regulations, 92
Mad cow disease, 91
Stem cell research, 76
Assisted suicide, 72
Welfare reform, 61
High-definition TV, 52
Lyme disease, 36
Nanotechnology, 33
Animal or human cloning, 32
Alien abductions, UFOs or Roswell, NM, 24
Shooting of 6-yr-old in Michigan, 17
Kansas evolution debate, 11

Source: Lexis-Nexis Universe index of *New York Times* and *Washington Post* for 2000. The issues chosen for comparison with agricultural biotechnology are derived in part from the top ten national news stories as indicated by the end-of-the-year Associated Press poll of newspaper editors, including: (i) the Presidential election; (ii) Elian Gonzalez; (iii) USS Cole attack; (iv) oil prices; (v) Firestone tyre safety; (vi) Microsoft anti-trust case; (vii) Human Genome Project; (viii) the Year 2000 and Y2K; (ix) removal of Slobodan Milosevic from power; and (x) jury verdict in Florida tobacco case.[a] In a separate survey of business editors, the Associated Press found that the StarLink maize affair was ranked the 14th top business story of 2000.[b] Other issues, either related to science and technology, politics, business or popular culture, reflect the authors' estimation of major 2000 events or trends.

[a] Evans, Mark (2000) 2000's top stories: Election, Elian. Associated Press, 22 December 2000.
[b] Geller, Adam (2000) Stock market tumble leads list of top AP business stories for 2000. Associated Press, 26 December.

various topics covered in 2000, the peak year historically for media attention to plant biotechnology.

Although the issue received greater or equal attention in the elite press than several events that the media considered top stories – such as the Florida tobacco case and the terrorist attack on the USS Cole – other technological or scientific topics received far greater attention, including broadband Internet, nuclear energy, West Nile virus, climate change, the Human Genome Project, Y2K and the Firestone tyre safety controversy. Even other food-related issues, such as Salmonella or E-Coli poisoning, received relatively similar levels of attention to plant biotechnology. The issue was dwarfed in coverage by attention to celebrity issues in 2000, including Elian Gonzalez, rising oil prices, the Microsoft anti-trust case, the Dotcom crash, gun control and, most notably, the 2000 Presidential election.

## Issue definitions and frames constraining media attention?

If relatively low levels of press attention have helped limit the scope and intensity of debate over plant biotechnology, according to the model we have outlined in this chapter it is likely that plant biotechnology's more commonplace status as a media issue is paralleled by a consistent appearance of technical frames in coverage. In Fig. 8.3, displaying the relative prominence of technical frames across the outlined stages of policy development, more thematic backgrounders emphasizing the specifics of policy and regulation have been dominant interpretations, appearing on average as a frame in every article (mean prominence $\leq 1.0$) in four of the six policy stages, with the heaviest emphasis during the first stage ($M = 1.48$).

A second technical emphasis on market development and economic competitiveness has also been dominant in coverage, peaking in 1995–1997 during the meteoric rise of Monsanto ($M = 1.21$). Across the history of plant biotechnology, an emphasis in coverage on patenting and property rights has been extremely rare, despite the importance of this dimension as a key feature driving technological development and investment (the frame peaks at 0.24).

The two technical frames dealing with the science of plant biotechnology are less prominent than either the policy or economic interpretations. In this case, in the two earliest stages when media coverage was minimal, an emphasis on scientific background was stronger ($M = 0.63$ and $M = 0.68$, respectively) but, as of 1998–1999, when media attention rose, the prominence of this technical frame was relatively low ($M = 0.35$). Somewhat surprisingly, across the history of plant biotechnology there have been very few articles framed around

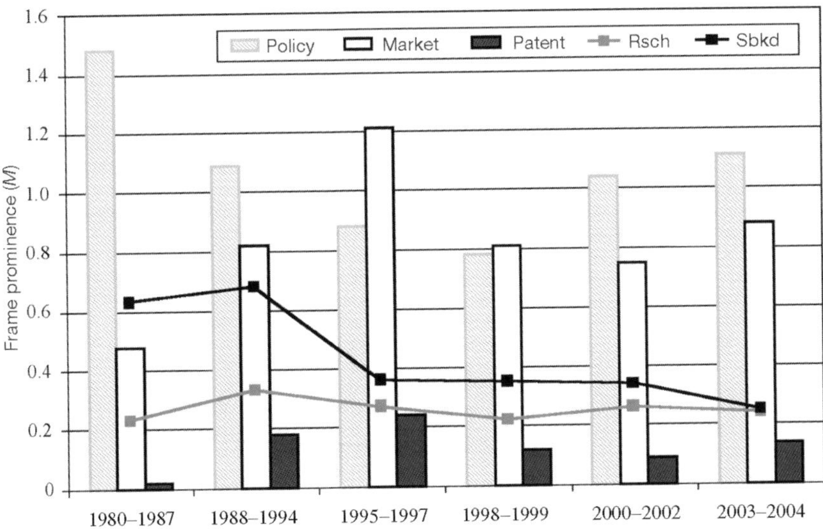


**Fig. 8.3.** Technical frames across stages of development. Bars and lines indicate mean prominence of each frame across policy stages of development, with each article scored for each frame as: 2 = dominant or lead frame of story; 1 = frame present but not dominant; and 0 = frame not present. Content analysis includes a population of articles, so all changes are significant.

release of a new scientific study or paper, the traditional package for science in the press (peaking at 0.35).

So, in examining the pattern of frames across time, interpretations have focused consistently and heavily on the technical details of policy background and economic developments. It is likely that these dominant technical interpretations have helped dampen wider social excitement and American concern over plant biotechnology, and defused the dramatic and narrative appeal of the issue to political and general/assignment journalists. Indeed, as Fig. 8.4 indicates, with the exception of scientific uncertainty, dramatic frames have not been nearly as prominent in coverage as technical interpretations.

When we examine the two key dramatic frames of strategy/conflict and ethics/morality, we see that these frames are almost completely absent from media coverage in the earliest stage of policy development ($M$ = 0.22 and 0.02, respectively). The strategy/conflict frame rose in prominence over the following years, with this emphasis still limited but peaking across the years 1998–2004, as social protest emerged in 1998–1999 ($M$ = 0.56), the StarLink affair surfaced in 2000–2002 ($M$ = 0.47) and the Bush White House paid brief attention to the issue in 2003–2004 ($M$ = 0.63).

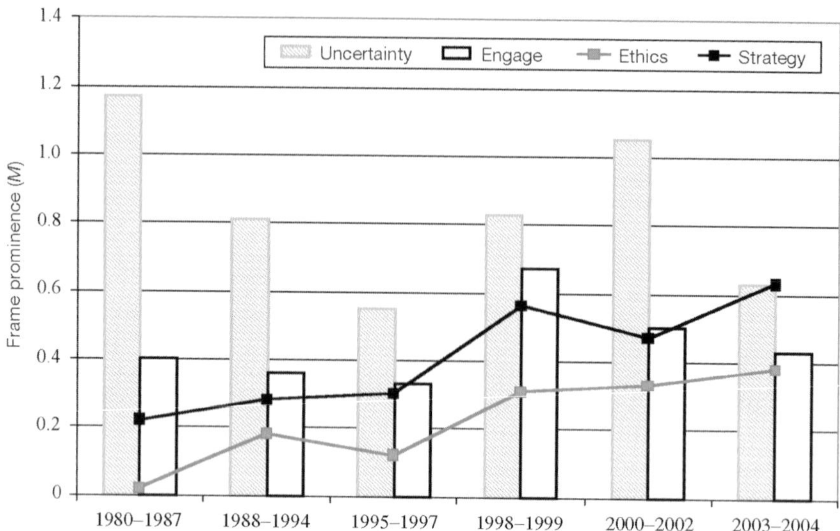


**Fig. 8.4.** Dramatic frames across stages of development. Bars and lines indicate mean prominence of each frame across policy stages of development, with each article scored for each frame as: 2 = dominant or lead frame of story; 1 = frame present but not dominant; and 0 = frame not present. Content analysis includes a population of articles, so all changes are significant.

The ethics/morality frame also rises in prominence, but was still not a major emphasis in later years, peaking in 2003–2004 ($M = 0.38$). Though not as prominent as technical interpretations, an emphasis on the public's acceptance of plant biotechnology has been relatively consistent across coverage, peaking in 1998–1999 with protest in the USA and Europe ($M = 0.67$).

Scientific uncertainty is the single dramatic frame that rivals technical frames in prominence, with its emphasis greatest during 1980–1987 ($M = 1.17$) and 2000–2002 ($M = 1.05$), the two stages featuring Downsian peaks in coverage and instances where critics were able to question – but with extremely limited success – the legitimacy of the US regulatory process. In between the rise and decline of these two peaks in the emphasis on uncertainty, notice that this frame is least prominent in 1995–1997, the key years where the biotech. industry achieved rapid and unprecedented market growth, and a period where media attention was minimal.

In sum, across the stages of development, the dramatic interpretations of strategy/conflict and ethics/morality remained relatively weak in emphasis, and were eclipsed by a greater media focus on technical frames of policy background and economic development. Strategy/

conflict and ethics/morality did rise in prominence as media attention to the issue picked up, yet the emphasis on these dramatic interpretations was still limited.

Does the relatively weak prominence of these dramatic frames in coverage provide another clue to why plant biotechnology has never climbed to the top of the US media agenda? To answer this question, it helps to place the relative emphasis on technical *versus* dramatic frames in context. Consider findings from a study of the stem cell controversy using a similar framing typology (Nisbet *et al.*, 2003).

Unlike plant biotechnology, stem cell research achieved celebrity status as an issue of widespread debate and concern in 2001 (or at least during the 8 months before the 11 September attacks). Stem cell research rose from relative obscurity in 2000 to the top of the media agenda during the summer of 2001, with nearly 500 articles appearing across the *New York Times* and *Washington Post*. Coverage by papers featured an almost exclusive and heavy emphasis on political strategy/conflict ($M$ = 1.02) and morality/ethics ($M$ = 0.62), with little or no emphasis on technical interpretations including market/economic development ($M$ = 0.12), scientific background ($M$ = 0.37) and the release of new scientific studies ($M$ = 0.30), though there was a comparatively strong emphasis on policy background ($M$ = 0.70).

## Risks, benefits and media evaluations

If news coverage has featured a heavy focus on technical frames, especially those interpretations emphasizing policy background and economic development, to what extent have the press also reflected the narrow emphasis in US policymaking on only those risks related to short-term human health and environmental risks? And to what extent has coverage of risks been balanced by a press emphasis on the benefits of plant biotechnology?

In Tables 8.1 and 8.2, we summarize the proportion of articles that mention various kinds of risks and benefits, categorized by their relation to human health, the environment/agriculture and social/economic dimensions. For human health risks, roughly one-quarter of articles (25.2%) have mentioned uncertainty because of a lack of safety testing. Approximately one of five articles have mentioned possible economic and social risks related to uncertain market/consumer demand (22.2%) or the threat of reduced consumer choice because of a lack of product labelling (21.2%). In terms of environmental risks, the most frequently mentioned risks were the possibility of environmental contamination (18.7%) and the transference to non-GM crops (15.7%).

**Table 8.1.** Types of risk mentioned in news coverage.

| Type of risk | Articles appearing (%) |
|---|---|
| Human health | |
| Uncertainty because of lack of safety testing | 25.2 |
| Food allergies | 14.5 |
| Contamination of non-GM food | 13.4 |
| Food toxicity | 6.7 |
| Increased antibiotic resistance | 1.5 |
| Environmental and agricultural | |
| Environmental contamination (general) | 18.7 |
| Gene transfer to non-GMO crops | 15.7 |
| Harm to non-pest insects or animals | 10.1 |
| Uncertainty because of lack of field trials or testing | 10.0 |
| Gene transfer to weeds (superweeds) | 7.0 |
| Increased resistance among pest insects | 4.2 |
| Loss of biodiversity | 3.4 |
| Undermining of sustainable practices | 1.4 |
| Release of new plant viruses | 0.4 |
| Excessive dependence on a single crop | 0.1 |
| Social and economic | |
| Uncertain market/consumer demand | 22.2 |
| Reduced consumer choice/rights (i.e. lack of labelling, notification, ability to buy non-GMO) | 21.2 |
| Undermining traditional life/farming practices | 11.1 |
| Financial losses/increased costs | 10.1 |
| Increased farmer dependence on seed companies (includes Terminator gene debate) | 7.5 |

In terms of mention of benefits related to human health, evidence shows that the press has not conveyed to the public the potential consumer advantages of plant biotechnology. Less than one of 10 articles mentioned either improved nutritional value (9.3%) or enhanced taste and food quality (8.6%). The heaviest emphasis in press coverage has been on agricultural benefits related to disease resistance (19.3%) and reduced use of pesticides (14.8%). In terms of social/economic benefits, roughly one of ten articles has mentioned reduced costs/increased profits (12.2%) or the alleviation of world hunger or improved world nutrition (10.9%).

In order to evaluate how the prominence of each category of risk/benefit has changed over time, in Table 8.3, for each article, we tallied up the total number of mentions of each kind of risk and benefit across health, environmental/agricultural and socio-economic dimensions and then computed a mean score across articles for each stage of policy development.

**Table 8.2.** Types of benefit mentioned in news coverage.

| Type of benefit | Articles appearing (%) |
|---|---|
| Human health | |
| Improved nutritional/vitamin content (i.e. Vitamin A in Rice) | 9.3 |
| Enhanced food quality (taste, texture, size, freshness) | 8.6 |
| Food vaccines/pharmaceuticals (i.e. bananas with pharmaceuticals) | 4.2 |
| Other medical treatments | 2.6 |
| Decrease in farmer pesticide exposure | 1.6 |
| Environmental and agricultural | |
| Increased disease resistance | 19.3 |
| Reduced use of pesticides | 14.8 |
| Increased pest resistance | 11.9 |
| Increased ability to farm in infertile, rugged areas | 10.3 |
| Increased frost resistance | 7.9 |
| Increased drought resistance | 3.7 |
| Increased heavy metal tolerance | 2.9 |
| Increased resistance to herbicide | 2.3 |
| Reduced use of water | 1.8 |
| Promotion of sustainable practices (general, unspecified) | 1.9 |
| Increased use of herbicides | 0.4 |
| Use as alternative fuel source (i.e. ethanol) | 0.3 |
| Use in building and construction materials | 0.3 |
| Social and economic | |
| Cost savings/increased profits | 12.2 |
| Alleviation of world hunger/improved world nutrition | 10.9 |
| Increased agricultural yield/acreage | 9.0 |
| Increased agricultural imports/exports | 4.0 |

In the first stage, as the Inter-Agency Framework was eventually adopted, the few articles devoted to the issue reflected the nature of the policy debate, focusing almost exclusively on enviro/agricultural risks and benefits, with little or no emphasis on dimensions related to human health or socio-economic considerations. With the introduction of the first food products between 1988 and 1994, the media began to focus on human health risks and benefits. Gradually across stages, attention to socio-economic risks and benefits increased, with a peak for risks in 1998–1999 and 2000–2002, as the media emphasized the potential for consumer backlash and reduced consumer choice because of a lack of labelling.

The emphasis on socio-economic benefits peaked in 2003–2004, as the Bush administration emphasized the potential for plant biotechnology for solving problems related to food security and nutrition. Of note, when media attention first began to rise significantly in 1999, there was a heavier emphasis on the various dimensions of risks than benefits, a pattern that continued through 2002. These results, however, should not be interpreted as journalists engaging in alarmist coverage. On the

**Table 8.3.** Risks, benefits and evaluation across stages of development.

| | Stage of policy development | | | | | |
|---|---|---|---|---|---|---|
| | Managing uncertainty, participation 1980–1987 (%) (*n*, 86) | Market development, precedent 1988–1994 (%) (*n*, 99) | Rise of Monsanto 1995–1997 (%) (*n*, 33) | Trade conflict, protest 1998–1999 (%) (*n*, 139) | Food contamination 2000–2002 (%) (*n*, 305) | Diplomacy and review 2003–2004 (%) (*n*, 105) |
| Risk | | | | | | |
| Health | 0.17 | 0.58 | 0.51 | 0.44 | 0.89 | 0.35 |
| Enviro/Ag | 1.13 | 0.51 | 0.39 | 0.86 | 0.66 | 0.42 |
| Socio-econ | 0.07 | 0.51 | 0.72 | 0.91 | 0.80 | 0.70 |
| Benefit | | | | | | |
| Health | 0.08 | 0.61 | 0.33 | 0.21 | 0.23 | 0.11 |
| Enviro/Ag | 1.27 | 0.88 | 0.45 | 0.73 | 0.73 | 0.61 |
| Socio-econ | 0.24 | 0.21 | 0.51 | 0.35 | 0.35 | 0.63 |
| Evaluation | | | | | | |
| Tone | 0.14 | 0.18 | 0.24 | −0.30 | −0.15 | −0.25 |

contrary, the figures for risk prominence show that, even at the height of attention to risks in 1998–99 and 2000–02, press coverage featured on average less than one mention of each category of risk per article.

We also rated the overall tone of coverage relative to plant biotechnology and calculated a mean across stages of development. In this case, we see that media coverage trended to be more positive towards plant biotechnology in the first three stages of development, but turned more negative with the emerging social protest in 1998–1999, and remained so through 2003–2004. Still, as we will see in the next section, much of this negativity in coverage derives not from news coverage, but from the opinion pages.

Two additional considerations deserve mention. First, our coding did not indicate whether risk was mentioned and then discredited, or characterized alternatively as credible. Secondly, as we noted earlier, though risks may be mentioned, in many cases coverage missed an opportunity to contextualize risks within a wider debate over the ecological issues linked to plant biotechnology, and whether or not regulators were addressing these questions.

## The shift across news beats and attention from opinion pages

As a final part of our analysis, we examined the relationship between media attention, media frames and the type of journalist assigned coverage, as well as the amount of attention to the topic from opinion pages. According to Table 8.4, news articles have dominated coverage but, across the 1990s, the proportion of letters-to-the-editor increases, peaking in 1998–1999 at more than one-quarter of coverage (26.6%), as various interest groups lobbied to get their messages directly across on opinion pages. This increased attention from the opinion pages in the late 1990s helps explain an increase in the total volume of coverage devoted to plant biotechnology, and the increase in attention to the topic during that period.

In terms of the type of journalist assigned coverage, the first two stages are dominated by science writers (37.6 and 23.4%, respectively) but, by the mid-1990s with the rise of Monsanto, business reporters emerged as the major specialist correspondents, accounting for more than one-quarter of articles in 1995–1997 (27.3%) and 1998–1999 (25.2%). Business writers increased in prominence in 2000–2002 (32.1%) and 2003–2004 (40.0%). Of importance, neither political/ general assignment reporters nor foreign correspondents accounted for substantial amounts of coverage, even in the peak period of coverage 2000–2002 (12.1 and 11.1%, respectively).

**Table 8.4.** Types of articles and journalists across stages of development.[a]

| | Stage of policy development | | | | | |
|---|---|---|---|---|---|---|
| | Managing uncertainty, participation 1980–1987 (%) (*n*, 86) | Market development, precedent 1988–1994 (%) (*n*, 99) | Rise of Monsanto 1995–1997 (%) (*n*, 33) | Trade conflict, protest 1998–1999 (%) (*n*, 139) | Food contamination 2000–2002 (%) (*n*, 305) | Diplomacy and review 2003–2004 (%) (*n*, 105) |
| Format | | | | | | |
| News | 95.3 | 84 | 78.8 | 61.9 | 79.3 | 81.0 |
| Op-ed | – | – | 3.0 | 7.2 | 2.0 | 4.8 |
| Editorial | 2.4 | 1.1 | 3.0 | 2.9 | 2.6 | 5.7 |
| Columnist | – | 1.1 | – | – | 1.0 | 1.0 |
| Letter | 2.4 | 9.6 | 15.2 | 26.6 | 13.1 | 6.7 |
| Magazine | – | 1.1 | – | 0.7 | 0.3 | 1.0 |
| Review | – | – | – | 0.7 | 0.7 | – |
| Journalist | | | | | | |
| Science | 37.6 | 23.4 | 15.2 | 15.1 | 9.1 | 3.8 |
| Business | 14.1 | 10.6 | 27.3 | 25.2 | 32.1 | 40.0 |
| Political | 1.2 | 11.7 | – | 1.4 | 12.1 | 1.9 |
| Foreign | – | 3.2 | 6.1 | 5.8 | 11.1 | 11.4 |
| Style | – | 2.1 | 3.0 | 3.6 | 1.3 | – |
| News wire | 7.1 | 7.4 | 12.1 | 1.4 | 6.6 | 5.7 |

[a] Percentages for type of journalist reflect proportion of total articles contributed, which includes opinion articles. Not all authors could be categorized by journalist type.

When we took a closer look at the nature of coverage submitted by each author, the relative dominance in coverage by specialist journalists explains the prominence of technical frames focused on policy background and market development, as these were the preferred interpretations of business reporters ($M$ = 1.00 and 1.13, respectively) and, to a lesser, extent science writers ($M$ = 1.09 and 0.49, respectively). The relative absence of political/general assignment reporters and foreign correspondents in coverage explains the weak emphasis in coverage of the strategy/conflict frame, as these reporters were most likely to feature this interpretation ($M$ = 0.51 and 0.85, respectively).

The low prominence of the strategy frame is probaly a threefold result because: (i) events related to StarLink did not lead editors and political journalists to interpret the issue as worthy of political coverage; (ii) few political journalists were assigned to the story; and (iii) the total potential volume of coverage remained constrained within science and business pages, where the strategy frame was unlikely to be applied by specialist correspondents.

Specialist journalists, in comparison to political and foreign correspondents, also differed substantially in their tone of coverage. Science writers and business reporters tended to be fairly neutral in coverage ($M$ = 0.01 for both), whereas political and foreign journalists tended to slant slightly more negative in their evaluations ($M$ = –0.17 and –0.10, respectively).

When we took a closer look at the preferred frames on the opinion pages, we find that much of the renewed emphasis on uncertainty in 1998–1999 derived from the increase in the number of letters-to-the-editor printed at the two newspapers ($M$ = 0.99), as various strategic actors used the opinion pages as a way to cast doubt on US regulatory policy. More so than news articles, these letters also emphasized the ethics/morality interpretation ($M$ = 0.40), though guest opinion-editorials had the heaviest emphasis on this angle ($M$ = 0.88). Perhaps more significantly, much of the increased negative tone in overall coverage of plant biotechnology that emerged in 1998–1999 flows from the upsurge in the number of letters to the editor appearing, as published letters tended to be strongly negative ($M$ = –0.43).

In sum, from an analysis of the type of journalists assigned coverage and the amount of attention from the opinion pages, we can see that the modest status of plant biotechnology on the overall media agenda – even in its peak years of coverage – is attributable in part to the fact that the issue never attracted heavy attention from political and general assignment reporters. Instead, the issue was covered mostly by business reporters and science writers who tended to interpret the issue in consistently technical ways.

It helps to place these journalist trends in context by comparing our results to similar features in stem cell coverage in 2001. During this period of peak attention for stem cell research, 25% of coverage was contributed by political/general assignment reporters, compared to just 12.1% of coverage for plant biotechnology during its peak in attention. And for stem cell research in 2001, political reporters interpreted the issue almost exclusively through the lens of political strategy ($M$ = 1.6). Moreover, 34% of coverage appeared on the opinion pages in 2001, compared to just 20.7% of coverage for plant biotechnology during its peak, and the emphasis on opinion pages focused heavily on morality/ethics ($M$ = 0.85) and strategy/conflict ($M$ = 1.1) (Nisbet, 2004).

## Conclusion

Previous research has attempted to understand why plant biotechnology has experienced limited political conflict in the USA, especially in comparison to that in the UK and several European countries. Applying the principles outlined in the model and the findings in our study, one major reason is that plant biotechnology proponents have been very successful at limiting the scope of participation surrounding the issue, as early policy decisions framed the issue in advantageous technical terms, establishing a virtual 'policy monopoly' within the administrative policy arenas of the FDA, the EPA, the USDA and various scientific advisory boards, with little significant attention from Congress or the Presidency.

Though increased media attention to plant biotechnology and more dramatic definitions of the issue have surfaced in recent years, challenging the *status quo* in regulation, the ability of biotech. proponents in early policy decisions to define the debate around short-term environmental and health risks have led to lasting and powerful feedback effects (Sheingate, 2005).

As we note, the early success of biotech. proponents is in part attributable to minimal media coverage, which made the 1986 Inter-Agency Framework and the precedent-setting early 1990s market approvals essentially 'non-decisions' for the wider public. This is in contrast to the UK and Europe where, from the beginning, Jasanoff (2005) and others have noted that there was a much wider scope of participation in policy decisions. The early inclusion of environmental, consumer and labour groups, and the comparatively stronger emphasis on transparency and public accountability, led to a very different European regulatory regime that took into account social and economic factors as well as the possibility of unknown future technical risks.

Despite attempts to shift debate towards more dramatic frames by various opposition groups, media discourse in the USA around plant biotechnology has remained predominantly technical. Because the issue has remained within administrative arenas, and because it has remained defined in technical and scientific terms, it is likely that journalists have been unable to place plant biotechnology into a larger narrative structure, giving greater meaning to passing events, thereby facilitating an increase in coverage of the issue.

Cycles of attention to plant biotechnology have appeared, but they remain small-scale perturbations rather than escalating into the large-scale news dramas that have surrounded media celebrity issues like stem cell research. Indeed, given the limited carrying capacity of the news media, competition with celebrity issues such as the stem cell debate, Presidential elections and, after 2001, terrorism and war, may all have significantly constrained attention to plant biotech., just when the conditions in terms of focusing events and drama might have otherwise propelled the issue into the wider media spotlight.

In addition, even though beginning in the mid-1990s, *in absolute terms*, the news media have featured an increased focus on risks and have drifted towards more negativity in overall coverage, the *relative prominence of risks* in coverage still remains relatively low. Most of the negativity in overall coverage derives not from the news articles authored by journalists, but rather in the rise in opinion articles and letters published at editorial pages.

A number of factors are likely to shape the trajectory of plant biotechnology as a political issue in the USA. Perhaps the strongest influence shielding the technology from wider public controversy is the decades-long bi-partisan support across Congress and the government for the technology, with the administrations of Presidents Ronald Reagan, George H.W. Bush, Bill Clinton and George W. Bush all strong proponents of the technology. Each Presidential administration has narrowly defined plant biotechnology in terms of economic growth and international competitiveness or, at times, when pressed by opposition from Europe, in moral terms emphasizing social progress and an end to world hunger.

In contrast, opposition in Germany to agricultural biotechnology, for example, originally derived from the active opposition of the Green Party. Similarly, in the UK during the early 1990s, initial opposition to agricultural biotechnology arose from the Green Alliance, an environmentalist branch of the Liberal Party (Jasanoff, 2005). Moreover, while the issue in the USA has been a priority for Greenpeace, Friends of the Earth and the Union of Concerned Scientists, plant biotechnology has never been a central issue for the larger, more influential environmental organizations such as the Sierra Club, League of Conservation Voters or Nature Conservancy.

Compounding this trend, in recent years, the ability of environmental groups to make progress – even on core issues such as climate change – has ebbed. Though far from the 'death of environmentalism' (*Grist Magazine*, 2005), any social movement is subject to limits on the number of issues they can devote resources to, and environmental groups have struggled with an identity crisis as many of their resources are taxed by the competing anti-war and progressive movements. All of these factors combine to shield plant biotechnology from debate within political contexts such as Congress or the White House.

There are two emerging trends, however, that might eventually weaken the ability of biotechnology proponents to control the scope of participation in policymaking about plant biotechnology. First, critics have added narrative fidelity to their framing efforts by connecting plant biotechnology to other contemporary issues. In her recent book, scientist and ecologist Jane Goodall (2005) links plant biotechnology to parallel controversies confronting the American food system, including childhood obesity, the survival of traditional farmers, organics and animal welfare.

If and when plant biotechnology becomes a topic of widespread attention and concern in the USA, it will probably be because it resonates and is framed in combination with these other food system issues. Secondly, evolving trends in international trade increasingly leave the USA as an outlier in its regulation and definition of the risks associated with plant biotechnology. And, as we reviewed in this chapter, while opponents have not had much success in changing the US policy regime, they have had success in shaping the actions and fortunes of industry members. It may be that changes in US regulation of plant biotechnology come about not through the domestic internal pressures channelled through the press/policy connection, but rather through the external pressures of international trade and market forces.

## References


Baumgartner, F. and Jones, B. (1993) *Agendas and Instability in American Politics.* Chicago University Press, Chicago, Illinois.

Becker, G.S. (2005) *Agricultural Biotechnology: Overview and Selected Issues.* Congressional Research Service Report for Congress, IB10109 (available at: http://www.ncseonline.org/NLE/CRS/abstract.cfm?NLEid=54715).

Bennett, L.W. (2001) *News: The Politics of Illusion.* Addison Wesley Longman, New York.

Bosso, C.J. (1987) *Pesticides and Politics: the Life Cycle of a Public Issue.* University of Pittsburgh Press, Pittsburgh, Pennsylvania.

Capella, J. and Jamieson, K.H. (1997) *Spiral of Cynicism: the Press and the Public Good.* Oxford University Press, New York.

Cobb, R.W. and Elder, C.D. (1983) *Participation in American Politics: the Dynamics of Agenda-Building*, 2nd edn. Allyn and Bacon, Boston, Massachusetts.

Cook, T.E. (1996) Afterword: Political values and production values. *Political Communication* 13, 469–481.

Cook, T.E. (1998) *Governing with the News: the News Media as a Political Institution*. University of Chicago Press, Chicago, Illinois.

Durant, J., Bauer, M. and Gaskell, G. (eds) (1998) *Biotechnology in the Public Sphere: a European Source Book*. Science Museum, London.

Downs, A. (1972) Up and down with ecology: The issue attention cycle. *The Public Interest* 28, 38–51.

Edelman, M. (1964) *The Symbolic Uses of Politics*. University of Illinois Press, Urbana.

Ferree, M.M., Gamson, W.A., Gerhards, J. and Rucht, D. (2002) *Shaping Abortion Discourse: Democracy and the Public Sphere in Germany and the USA*. Cambridge University Press, New York.

Fishman, M. (1978) Crime waves as ideology. *Social Problems* 25, 531–543.

Fishman, M. (1980) *Manufacturing the News*. University of Texas Press, Austin, Texas.

Gamson, W.A. and Modigliani, A. (1989) Media discourse and public opinion on nuclear power: a constructionist approach. *American Journal of Sociology* 95, 1–37.

Goodall, J. (2005) *Harvest for Life: a Guide to Mindful Eating*. Warner Books, New York.

*Grist Magazine* (2005) *Don't Fear the Reapers: a special series on the alleged 'death of environmentalism'* (available at http://www.grist.org/news/maindish/2005/01/13/doe-intro/).

Hallin, D.C. (1994) *We Keep America on Top of the World*. Routledge, New York.

Hilgartner, S. and Bosk, C.L. (1988) The rise and fall of social problems: a public arenas model. *American Journal of Sociology* 94, 53–78.

Jasanoff, S. (1987) Contested boundaries in policy-relevant science. *Social Studies of Science* 17 (2), 195–230.

Jasanoff, S. (2005) *Designs on Nature: Science and Democracy in Europe and the USA*. Princeton University Press, Princeton, New Jersey.

Kepplinger, H.M. (1992) Artificial horizons: how the press presented and how the population received technology in Germany from 1965–1986. In: Rothman, S. (ed.) *The Mass Media in Liberal Democratic Societies*. Peragon, New York, pp. 147–176.

Kepplinger, H.M. (1995) Individual and institutional impacts upon press coverage of sciences: the case of nuclear power and genetic engineering in Germany. In: Bauer, M. (ed.) *Resistance to New Technology: Nuclear Power, Information Technology, and Biotechnology*. Cambridge University Press, New York, pp. 357–377.

Krippendorf, K. (1980) *Content Analysis: an Introduction to its Methodology*. Sage Publications, Newbury Park, California.

Lawrence, R. (2000) Game-framing the issues: tracking the strategy frame in public policy news. *Political Communication* 17, 93–114.

Leiss, W. (2000) *The Trouble with Science: Understanding Risk Controversies*. Environment Canada Policy Research Seminar Series (available at: http://www.ec.gc.ca/seminar/WL_e.html).

Lyson, T.A. (2001) How do agricultural scientists view advanced biotechnologies? *Chemical Innovation* 31, 50–53.

Marks, L.A., Kalaitzandonakes, N., Allison K. and Zakharova, L. (2003) Media coverage of agrobiotechnology: did the butterfly have an effect? *Journal of Agribusiness* 21 (1), 1–20.

Maynard-Moody, S. (1992) The fetal research dispute. In: Nelkin, D. (ed.) *Controversy: Politics of Technical Decisions*, 3rd edn. Sage, Newbury Park, California, pp. 3–25.

Maynard-Moody, S. (1995) Managing controversies over science: the case of fetal research. *Journal of Public Administration Research and Theory* 5, 5–8.

McComas, K. and Shanahan, J.E. (1999) Telling stories about global climate change: measuring the impact of narratives on issue cycles. *Communication Research* 26 (1), 30–57.

McInerney, C., Nucci, M. and Bird, N. (2004) The flow of scientific knowledge from lab to the lay public: the case of genetically modified food. *Science Communication* 26 (1), 44–74.

Mooney, C. (2005) *The Republican War on Science.* Basic Books, New York.

Mooney, C. and Nisbet, M.C. (2005) When coverage of evolution shifts to the political and opinion pages, the scientific context falls away, unraveling Darwin. *Columbia Journalism Review*, Sep/Oct, 31–39.

Nelkin, D. (1975) The political impact of technical expertise. *Social Studies of Science* 5, 34–54.

Nelkin, D. (1992) Science, technology, and political conflict: analyzing the issues. In: Nelkin, D. (ed.) *Controversy: Politics of Technical Decisions.* Sage, Newbury Park, California, pp. vii–ix.

Nelkin, D. (1995) *Selling Science: How the Press Covers Science and Technology.* W.H. Freeman and Co., New York.

Nestle, M. (2004) *Safe Food: Bacteria, Biotechnology, and Bioterrorism.* University of California Press, Berkeley, California.

Nisbet, M.C. (2004) The stem cell controversy: towards a model of mediated issue development. Paper presented at the *Annual Conference of the International Communication Association*, New Orleans, Louisiana.

Nisbet, M.C. and Lewenstein, B.V. (2002) Biotechnology and the American media: the policy process and the elite press, 1970 to 1999. *Science Communication* 4, 359–391.

Nisbet, M.C., Brossard, D. and Kroepsch, A. (2003) Framing science: The stem cell controversy in an age of press/politics. *Harvard International Journal of Press/Politics* 8 (2), 36–70.

Pan, Z. and Kosicki, G.M. (2001) Framing as a strategic action in public deliberation. In: Reese, S.D., Gandy, O. and Grant, A. (eds) *Farming Public Life: Perspectives on Media and our Understanding of the Social World.* Lawrence Eribaum, Mahwah, New York, pp. 35–66.

Patterson, T.E. (1994) *Out of Order.* Vintage Books, New York.

Patterson, T.E. (2001) *Doing Well and Doing Good: How Soft News and Critical Journalism are Shrinking the News Audience and Weakening Democracy – and what News Outlets can do about it.* The Joan Shorenstein Center for Press, Politics, and Public Policy at Harvard University, Cambridge, Massachusetts.

Priest, S.H. (2001) *A Grain of Truth: the Media, the Public, and Biotechnology.* Rowman and Littlefield, Lanham, Maryland.

Priest, S.H. and Gillespie, A.W. (2000) Seeds of discontent: expert opinion, mass media messages, and the public image of agricultural biotechnology. *Science and Engineering Ethics* 6 (4), 529–539.

Reese, S.D. (2001) Prologue – framing public life: a bridging model for media research. In: Reese, S.D., Gandy, O.H. Jr. and Grant, A.E. (eds) *Framing Public Life: Perspectives on Media and our Understanding of the Social World.* Lawrence Earlbaum Associates, Mahweh, New Jersey, pp. 7–32.

Rochefort, D.A. and Cobb, R.W. (1994) Problem definition: an emerging perspective. In: Rochefort, D.A. and Cobb, R.W. (eds) *The Politics of Problem Definition: Shaping the Policy Agenda.* University of Kansas Press, Lawrence, Kansas, pp. 1–32.

Rodemeyer, M. and Jones, S.A. (2002) *When media, science, and public policy collide: The case of food and biotechnology.* Proceedings from a workshop sponsored by the Pew Initiative on Food and Biotechnology and the Joan Shorenstein Center on the Press, Politics, and Public Policy of the Kennedy School of Government at Harvard University, Cambridge, Massachusetts (available at: http://pewagbiotech.org/events/1121/proceedings.pdf).

Scheufele, D.A. (1999) Framing as a theory of media effects. *Journal of Communication* 29, 103–123.

Schnattschneider, E.E. (1960) *The Semi-Sovereign People.* Holt, Reinhart and Winston, New York.

Segarra, A.E. (2000) *Biosafety Protocol for Genetically Modified Organisms: Overview.* Congressional Research Service Report for Congress, RL20594 (available at: http://www.ncseonline.org/nle/crsreports/agriculture/ag-93.cfm).

Shanahan, J., Scheufele, D.A. and Lee, E. (2001) The polls – trends: attitudes about biotechnology and genetically modified organisms. *Public Opinion Quarterly* 65 (2), 267–281.

Sheingate, A. (2006) Promotion *versus* precaution: the evolution of biotechnology policy in the USA. *British Journal of Political Science* 36, 243–268.

Shoemaker, P.J., Tankard, J.W. and Lasorsa, D.L. (2003) *How to Build Social Science Theories.* Sage Publishing, Newbury Park, California.

Taylor, M.R. and Tick, J.S. (2001) *The StarLink Case: Issues for the Future.* Report of the Pew Initiative on Food and Biotechnology (available at: http://pewagbiotech.org/resources/issuebriefs/starlink/).

Weart, S.R. (1988) *Nuclear Fear: a History of Images.* Harvard University Press, Cambridge, Massachusetts.

9

# Opinion Climates, Spirals of Silence and Biotechnology: Public Opinion as a Heuristic for Scientific Decision-making

D.A. Scheufele

*School of Journalism and Mass Communication and Department of Life Sciences Communication, University of Wisconsin, Madison, Wisconsin, USA; e-mail: scheufele@wisc.edu*

## Introduction

Levels of awareness of agricultural biotechnology as an issue on the public agenda are low, in spite of a significant amount of scientific discourse on the ethical and societal implications of this technology (Shanahan *et al.*, 2001). Similarly, levels of scientific literacy and specific knowledge about agricultural biotechnology are far from where they ideally should be. This is not particularly surprising. In fact, research in communication, political science and related disciplines has consistently shown disturbingly low levels of information among the general public about current events, politics (Delli Carpini and Keeter, 1996), and scientific issues (Miller, 1998).

The fact that citizens are uninformed about an issue such as agricultural biotechnology, however, does not mean that they will not form opinions or judgements about the issue or its policy implications, such as labelling, federal funding for research, etc. Rather, people's opinions are influenced by a range of factors *other* than information, such as ideological predispositions, the way mass media present the issue and – most importantly – their perceptions of what the climate of opinion in the country looks like.

This type of decision-making has often been labelled 'low information rationality' (Popkin, 1994), implying that it makes rational sense for citizens to collect only limited amounts of information for any given decision. In fact, given the complexity of many scientific issues, it may make perfect sense for citizens to weigh the potential risks to themselves against the efforts necessary to

 *The Media, the Public and Agricultural Biotechnology* (ed. D. Brossard *et al.*)

understand an issue, such as agricultural biotechnology, in all its complexity. As a result, they may decide to rely on heuristic shortcuts – such as the opinions of others, rather than on scientific information.

This chapter provides an overview of opinions of others as one of the key shortcuts that people use when making decisions about agricultural biotechnology. It compares the competing influences of information and heuristic cues about public opinion. The research presented in this chapter relies on a theoretical model of opinion formation in societies called the 'spiral of silence' (Noelle-Neumann, 1993). Before discussing the larger theoretical model and its importance for public opinion formation about agricultural biotechnology, it is important to examine citizens' decision-making about new technologies and the risks associated with them more carefully.

## Low-information Rationality: how People Make Sense of Biotechnology

Public opinion research has struggled for decades with the question of how to systematically measure opinions of a public that is largely uninformed about scientific facts, as well as their implications for public policy (for an overview, see Page and Shapiro, 1992). Research from social psychology (Fiske and Taylor, 1991) and political science (Popkin, 1994) suggests that people are 'cognitive misers' or 'satisficers', who will only collect as much or as little information about a given issue as they think necessary to make a decision. And, in most cases, this means that they will make decisions with little, or at least insufficient, amounts of information (Kahneman, 2003).

In contrast to traditional scientific literacy models, however, which are mostly concerned with informational deficits among the general public (Bauer and Schoon, 1993; Miller, 1998), the cognitive miser model assumes that making decisions based on little or no information is not just part of human nature but may in fact make rational sense. When explaining the idea of 'low-information rationality' outlined earlier, Popkin (1994) argues that citizens are public consumers and invest efforts in information-seeking only if they see a reasonable pay-off.

For issues such as agricultural biotechnology, where developing an in-depth understanding would require *significant* efforts on the part of ordinary citizens, the pay-offs in terms of being able to make informed policy judgements may simply not be enough. As a result, it makes perfect sense for citizens to rely on shortcuts such as opinions of others when forming their own opinions and trying to make sense of different policy positions.

## Spirals of Silence and Opinion Climates about Biotechnology

In fact, in their work on public attitudes toward agricultural biotechnology, Scheufele *et al.* (2001) suggested that many of the opinion dynamics surrounding agricultural biotechnology can be explained by a theoretical model called the 'spiral of silence' model. Developed by Noelle-Neumann in the early 1970s, the spiral of silence theory is one of the most prominent theoretical models of opinion formation and consensus-building about controversial issues. It assumes that people are constantly aware of the opinions of people around them and adjust their behaviours (and potentially their opinions) to majority trends with the fear of being on the losing side of a public debate.

In particular, Noelle-Neumann assumes that individuals have a 'quasi-statistical sense' that allows them to gauge the opinion climate in a society, i.e. the proportions of people who favour or oppose a given issue. This quasi-statistical sense may be accurate, but very often it is not, i.e. people are wrong in their assessments of what everyone else thinks. This point is largely irrelevant for the spiral of silence theory, however, since it is the *perception* of opinion distributions rather than the *real* opinion climate that shapes people's willingness to express their opinions in public (Scheufele and Moy, 2000).

In addition to the quasi-statistical sense, Noelle-Neumann's theory introduces a second key concept: fear of isolation. This concept is based on the assumption that social collectives threaten individuals who deviate from social norms and majority views with isolation or even ostracism. As a result, individuals are constantly fearful of isolating themselves with unpopular views or behaviour that violates social norms.

Based on these assumptions, the spiral of silence predicts that groups who see themselves in a minority, or as losing ground, are less vocal and less willing to express their opinions in public. This, in turn, will influence the visibility of majority and minority groups and the minority group will appear weaker and weaker over time, simply because its members will be more and more reluctant to express their opinions in public. Ultimately, the reluctance of members of the perceived minority to express their opinions will establish the majority opinion as the predominant view or even as a social norm.

### The spiral of silence as a *process*: the public opinion dynamics surrounding agricultural biotechnology

The most critical component of the spiral of silence is also the one that has been overlooked most in previous research on the theory: its dynamic character. The spiral of silence is a process that works over time. As people who perceive themselves to be in the minority fall silent, perceptions of opinion climates shift over time and the majority opinion is established as the predominant one or even as a social norm.

Figure 9.1 illustrates this spiralling process over time using agricultural biotechnology as a hypothetical example. During the early stages of the issue cycle, awareness of the issue and support or opposition to the new technology among the general public tend to be low. In part as a result, it is during this stage of the debate when many of the long-term opinion dynamics are established. At this stage, anti-biotech organizations are beginning to aggressively vocalize their views, even though they may be at odds with the rest of the population, which is still fairly neutral toward agricultural biotechnology. These interest groups assert their influence through lobbying, dissemination of talking points among their supporters or other activities, which are all aimed at influencing media coverage.

Ultimately, all these groups try to take advantage of the ability of mass media to do one of two things. First, mass media provide important cues about what we think the climate of opinion looks like.

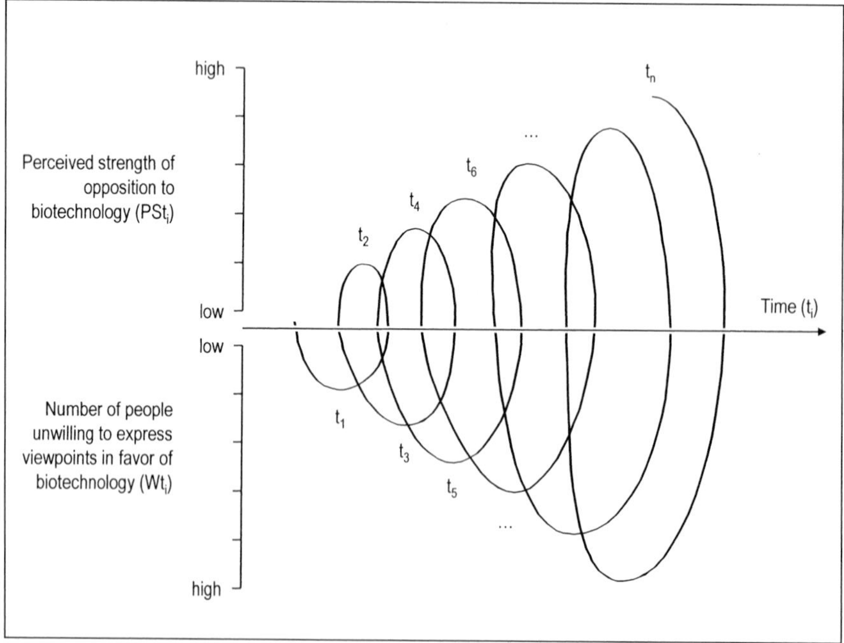


**Fig. 9.1.** The spiral of silence: on biotechnology as a dynamic process (based on ideas first outlined in Scheufele and Moy, 2000).

Which viewpoints are held by a majority of the population and which ones are held by a minority or are losing ground?

Second, mass media provide supporters or opponents of certain viewpoints with arguments to defend their position. Noelle-Neumann (1993) calls this the 'articulation function of mass media'. This includes frames such as 'Tony the Frankentiger' that will provide easily digestible storylines for journalists when they cover these new technologies. But it also includes providing arguments in favour or against certain policy options connected to agricultural biotechnology that will be used by supporters or opponents in public discourse.

But why do these initial strategies of interest groups matter? They matter because they are directly aimed at influencing perceptions among the public about the public early on. In other words, if silent minorities are successful in shaping public perceptions of which viewpoints are in the minority or losing ground, they will also influence the long-term spiralling processes outlined in Noelle-Neumann's theory. Figure 9.1 illustrates these processes nicely.

As more and more people think that opposition to agricultural biotechnology is a viewpoint shared by a majority of people, we will also see an increasing tendency among proponents of agricultural biotechnology to fall silent in public discussions about the issue. They will be less likely to write letters to the editor or to their congressman, they will be less likely to speak out at public meetings or dinner parties and they will be less likely to express their viewpoints in any type of public setting.

As people with minority viewpoints fall silent over time, of course, perceptions of the majority opinion gaining ground increase. This creates a mutually reinforcing spiral, where the reluctance of the minority group to speak out leads to perceptual biases in favour of the majority group which, in turn, further discourages the minority group from speaking out. The reluctance of proponents to express their views publicly will therefore play directly into the strategy of interest groups such as Greenpeace, who try to portray their view as the majority view. The more proponents of agricultural biotechnology that fall silent, the more prominent the position of Greenpeace, for instance, will appear to the general public, exacerbating the spiralling process and ultimately promoting an anti-biotech. stance as the accepted viewpoint in society.

While Fig. 9.1 provides a hypothetical overview of the spiral of silence in the case of agricultural biotechnology, Scheufele *et al.* (2001) tested some of these relationships empirically. In an experimental design they explored the factors influencing subjects' willingness to express their opinions on genetically modified organisms in various hypothetical and real settings. Their findings provided strong evidence in favour of a spiral of silence process. Respondents who were fearful of isolating themselves with unpopular viewpoints and respondents who saw themselves in the minority with their views on agricultural biotechnology were also significantly less likely to express their opinions in public. One interesting side note to the findings of Scheufele *et al.* was that more informed respondents were more likely to express opinions, regardless of their perceptions of a potentially hostile opinion climate.

## Key actors in the spiral of silence process for biotechnology

As this brief overview shows, the spiral of silence combines social-level explanations of the opinion dynamics surrounding agricultural biotechnology with individual-level analyses of citizens' opinions and predispositions. In this sense, the theory is one of the few true macro-social theories of public opinion, i.e. it links macro-, meso- and micro-levels of analysis.

As a micro-theory, the spiral of silence examines opinion expression, controlling for people's predispositions – such as fear of isolation, and also demographic variables that have been shown to influence people's willingness to publicly express opinions on issues, such as agricultural biotechnology. In particular, previous research has shown that younger respondents and male respondents were more likely to express their views in public, regardless of fear of isolation or perceptions of the dominant climate of opinion (for an overview, see Scheufele and Moy, 2000).

Figure 9.2 lists some of these individual-level controls in the context of agricultural biotechnology. It also shows the interplay between a person's own opinions and his or her perceptions of the opinion climate on agricultural biotechnology. If the two are incongruent, the person is less likely to express his or her views. This public expression of opinion is what moves the spiral of silence to a more macroscopic level of analysis. If more and more members of the perceived minority fall silent, as outlined earlier, public perceptions of the opinion climate and the societal level begin to shift. In other words, a person's individual reluctance to express his or her opinion, simply based on perceptions of what everyone else thinks, has important implications at the social level. These social, macro-level perceptions, of course, in turn influence individual perceptions and people's willingness to express opinions.

In addition to linking macro- and micro-levels of analysis, the spiral of silence theory also implements evidence from meso-levels of analysis. In particular, social groups are also directly relevant as

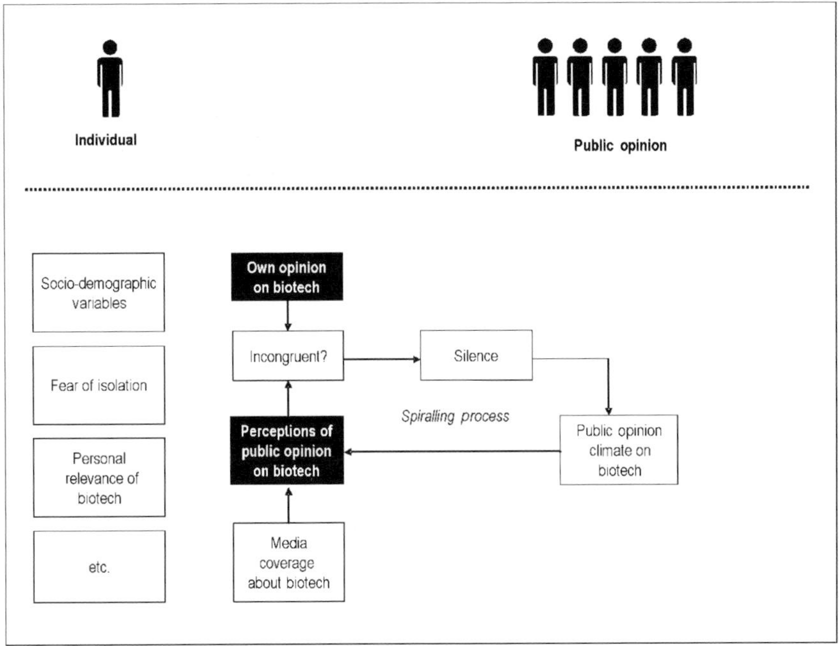


**Fig. 9.2.** The spiral of silence: linking individual and mass opinion on biotechnology (based on ideas first outlined in Donsbach, 1987).

explanatory mechanisms for two phenomena related to the spiral of silence. First, more recent research showed that reference groups can provide important social cues when people try to gauge the social climate of opinion. In particular, this research showed that, when trying to correctly assess the opinion distribution at the societal level, individuals often turn to experiences in their more immediate social circles or references group as a proxy. In other words, they project from the opinion distributions they encounter in their reference groups to the climate of opinion at the societal level (Scheufele *et al.*, 2001).

The second role that reference groups can play in the process of the spiral of silence is to provide a protective environment for people who resist or choose to counter hostile opinion climates. Noelle-Neumann (1984) refers to these group as 'avant-gardes' and 'hard cores.' Hard cores are groups who stick with a minority position even as the spiral of silence grows stronger and stronger against their own position. Avant-gardes may base their resistance to hostile opinion climates on strong ideological belief systems, on a strong concern about the issue being discussed, or – most important for our discussion here – on reference groups that reinforce their existing beliefs.

While hard cores hold on to existing issue stances, even as the opinion climate turns against them, avant-gardes promote *new*, unpopular viewpoints that go against existing social norms of predominant opinion climates. As a result, reference groups probably play an important role for avant-gardes – similar to hard cores – by creating a protective social environment that provides the necessary social support for members of the avant-garde who speak out against an existing majority opinion.

## Lessons for Biotechnology: when Spirals of Silence can Occur

The processes outlined in Figs 9.1 and 9.2 are, of course, idealized versions of what can take place. And, as some of the examples in this volume show, the opinion dynamics surrounding agricultural biotechnology have differed dramatically from country to country. Some of these differences can be explained by factors that have been identified as contributory conditions to the spiral of silence process. In other words, the strength of the spiralling effect depends on at least three additional variables.

### The Nature of the issue

The first variable is the nature of the issue that is being discussed. Previous research suggests that the spiral of silence works especially well for issues with a moral component, or value-laden issues 'by which the individual isolates or may isolate himself in public' (Noelle-Neumann, 1993, p. 231).

The recent debate about embryonic stem cell research in the USA is a good example of an issue where religious and moral concerns are intertwined with more rational, scientific arguments in public discourse. Public debate of this issue is therefore morally charged and it is impossible to *objectively* answer the question of whether the USA should proceed with this new technology and provide federal funds to do so. As a result, opinion climates provide critical cues for citizens when they have to decide whether they want to express their own views in public or not.

Similarly, agricultural biotechnology may have a different degree of moral loading in difference cultures. In Germany, for instance, agricultural biotechnology and genetically modified organisms may be directly at odds with the moral consensus of society. After all, Germany saw the transformation of the Green Party from an obscure splinter party in the 1980s to a member of the coalition government between the Social Democrats and the Greens (at least until the 2004 elections). It was also that particular government which endorsed concrete plans to gradually close down the country's 19 remaining nuclear power stations.

The same moral concerns that underlie public attitudes toward agricultural biotechnology in Europe, however, may not apply in Third World countries or in the USA, where some of the benefits of agricultural biotechnology or GMOs may be more directly tangible and where the electorate may not be as tuned in to environmental concerns.

## Media coverage

The second factor that is important when interpreting the process of the spiral of silence is the level of media attention to the issue of agricultural biotechnology. Given that few people have direct experience with agricultural biotechnology or even direct exposure to scientists and researchers working in the field of agricultural biotechnology, media play a crucial role in providing people with the information necessary to make decisions about policy options and the potential risks and benefits connected to agricultural biotechnology. In the spiral of silence context, however, mass media also play a second important role: they allow citizens to gauge the climate of opinion around them.

In many cases this is not necessarily an undesirable outcome and, in fact, mass media can play an important function in building societal consensus. If media coverage accurately reflects the real distribution of opinions among the public, it can help establish this opinion as the predominant one and avoid long, drawn-out discussions dominated by a vocal minority. The spiral of silence theory explicitly refers to this positive aspect of the spiral of silence. As a broad majority starts to support a certain viewpoint, the remaining minority tends to fall silent, establishing the predominant view as the norm or at least as the predominant viewpoint (Noelle-Neumann, 1993).

This is critical in democratic societies, where public debate about controversial issues is largely unmoderated and could theoretically go on forever unless there are elections, legislation or court decisions that cut it short. And as, examples like Roe *vs* Wade or the federal funding guidelines for stem cell research show, even legislation and court decisions often do not resolve some of these debates.

The role of media in promoting societal consensus, however, also has its downsides. One of these downsides is often referred to as the 'dual climate of opinion'. A dual climate of opinion exists when the majority of the population has a specific stance on an issue, but perceptions of which group is winning or losing the debate are just the opposite, and media are often the explanatory factor for these perceptual biases.

This could happen, for example, when a majority of the population supports agricultural biotechnology. A dual climate of opinion exists if there is also a prevalent perception among most citizens that a majority really opposes agricultural biotechnology. And, as outlined earlier, one explanation for this perceptual bias is mass media. If media portray agricultural biotechnology and the societal discourse surrounding it in a way that suggests that the majority of the citizens oppose research on and funding for agricultural biotechnology, we have a dual climate of opinion where collective perceptions deviate from collective preferences.

The implications for the spiral of silence are obvious. As a result of a dual climate of opinion we may see a spiral of silence against the real opinion distribution, i.e. against the majority who support agricultural biotechnology and are in favour of the opposing minority, simply because mass media inaccurately portray the opinion climate which, in turn, influences people's willingness to express their opinions and accelerates a spiralling process of the opinion climate and opinion expression against agricultural biotechnology.

In order to understand fully the idea of dual climates of opinion, it is necessary briefly to examine Noelle-Neumann's understanding of how media can shape public opinion. According to Noelle-Neumann, dual climates of opinion can develop since media coverage of controversial issues tends to be consonant and cumulative (Noelle-Neumann, 1993).

Consonance refers to the tendency of different media outlets to portray controversial issues in a consistent fashion. The idea of consonance is consistent with concepts such as inter-media agenda setting (McCombs, 2004) or news waves (Fishman, 1978), which both suggest that journalists' choices about what to cover and how to spin a story are often influenced by peer or elite media. According to Noelle-Neumann, consonant coverage of an issue therefore probably strengthens media effects since it undermines the ability of audience members to selectively expose themselves only to media messages that are consistent with their own views.

More importantly, however, Noelle-Neumann assumes that media effects are cumulative, i.e. they work *over time*. As a result, dual climates of opinion can develop when a cumulative stream of consonant media messages creates public perceptions of the opinion climate that deviate from the real opinion distribution in the population.

## Cross-cultural differences

The third factor that can help predict the occurrence of spirals of silence is cross-cultural differences. In fact, Scheufele and Moy (2000) suggested that many of the inconsistencies among different studies of the spiral of silence could be explained by taking into account intercultural differences, such as differences with respect to conflict styles and norms of opinion expression. Indeed, the theory has been tested in a number of countries, including Germany, Japan, Korea and the USA.

Across studies and countries, support for the spiral of silence hypothesis varied. Some researchers have argued that these differences are a function of how Noelle-Neumann measured people's willingness to express their opinions publicly. Specifically, these researchers criticized Noelle-Neumann's (1993) use of what she called the 'train test' for being too narrow and too culturally specific.

In order to simulate a public situation and test mothers' willingness to speak out about their views on spanking their children, for example, Noelle-Neumann asked a representative sample of mothers the following question: 'Suppose you are faced with a 5-hour train ride, and there is a woman sitting in your compartment who thinks ... that spanking is part of bringing up children/that spanking is basically wrong. Would you like to talk to this woman so as to get to know her point of view better, or wouldn't you think that worth your while?' (Noelle-Neumann, 1993, pp. 17–18). The questionnaire was based on a split-ballot design, and women who opposed spanking were asked to have a discussion with a proponent of spanking and vice versa.

This train test has been criticized for being potentially culturally biased, given that long train rides and conversations with strangers in train compartments may not be realistic enough to be used in surveys in the USA or other countries. Noelle-Neumann, however, suggested a whole range of other, more indirect measures. These include displaying campaign buttons and bumper stickers, participating in public meetings or other forms of public participation (Noelle-Neumann, 1993).

Subsequent studies have often used measures of people's willingness to speak out that were less culturally specific. In their study of the spiral of silence in the context of agricultural biotechnology, Scheufele *et al.* (2001) suggested a wording that takes the public element of discussion into account and also addresses the issue of speaking out in the face of a hostile environment:

> Imagine you're at a party where you don't know most people. You're talking to a group of people when somebody brings up the issue of

> genetic engineering. From the discussion you can tell that most people in the group do not support your point of view. In this kind of situation, some people would express their opinions and some would not ... How likely is it that you would express your own opinion in a situation like this?
>
> (Scheufele *et al.*, 2001, pp. 321–322)

Research suggests that, beyond the more methodological problem of finding appropriate indicators for concepts in a given culture, there is a substantial difference in personality traits for people living in different cultures. The concepts of 'culture' and differences between cultures, of course, are difficult to grasp conceptually and even more difficult to operationalize. Ting-Toomey (1988), however, proposed a conceptualization that is very useful in understanding why spirals of silence on issues, such as agricultural biotechnology, may be more pronounced in some cultures than in others. She suggests that the concept of individualism is a key variable in differentiating social behaviour, in general, and communicatory behaviour, specifically, across cultures.

Specifically, Ting-Toomey (1988) distinguishes between 'individualistic, low-context cultures and collectivistic, high-context cultures' (p. 213). Countries like Australia, Germany or the USA can be considered individualistic cultures, while Asian countries exemplify collectivistic cultures. In individualistic cultures, the consistency between private self-image and public self-image is of utmost importance, and expressing one's own viewpoints is a virtue in itself. What other people think is of only marginal importance. It seems, however, that the idea of individualistic is limited to certain cultures. In contrast, the 'self' collectivistic culture is situationally based and depends heavily on the social environment of the time the social interaction takes place.

Hui and Triandis (1986) summarize what can be called the 'collectivist personality': 'Collectivists are more likely to pay more attention to the influencing agent than are individualists. As a result, collectivists are more conforming than individualists ... It may be safe to say that the former are more willing to go along with the group, to avoid being rejected' (Hui and Triandis, 1986, p. 230).

This distinction between individualism and collectivism, therefore, is highly relevant for future spiral of silence research. If it is, indeed, possible to identify personality characteristics common to citizens of a given culture, these characteristics might prove to be important long-term predictors of people's willingness to speak out beyond more temporally bound perceptions of opinion climates.

More recent research has begun to explore these differences in greater detail. Willnat and his colleagues examined the spiral of silence model in Singapore, using two morally charged issues: interracial marriage and equal rights for homosexuals (Willnat *et al.*, 2002). In addition to the opinion climate and fear of isolation, they also tested a host of variables that might help explain culture-specific variations of people's willingness to speak out. These variables included communication apprehension, fear of authority and social interdependence. Interdependence is especially interesting as a potential attenuating force on people's willingness to speak out. As Willnat *et al.* point out: 'people with an interdependent self-concept ... value fitting in, and regard speaking out in such circumstances as a threat to group harmony and hence inappropriate' (Willnat *et al.*, 2002, p. 394).

These cross-cultural differences, of course, have immense implications for how interest groups and policymakers in different countries think about the opinion dynamics surrounding issues such as agricultural biotechnology. In light of some of the findings on the spiral of silence in the political arena (Scheufele and Moy, 2000), the contrast between the vehement opposition to agricultural biotechnology in Europe and the much more positive and subdued reactions in the USA may very well be a function of very different processes of opinion formation. Furthermore, interest groups and mass media have probably played very different roles in the two different cultural contexts with respect to influencing opinion perceptions and perpetuating opinion spirals.

## References


Bauer, M.W. and Schoon, I. (1993) Mapping variety in public understanding of science. *Public Understanding of Science* 2 (2), 141–155.

Delli Carpini, M.X. and Keeter, S. (1996) *What Americans know about politics and why it matters.* Yale University Press, New Haven, Connecticut.

Donsbach, W. (1987) Die theorie der schweigespirale [The theory of the spiral of silence]. In: Schenk, M. (ed.) *Medienwirkungsforschung.* Mohr, Tübingen, Germany, pp. 324–334.

Fishman, M. (1978) Crime waves as ideology. *Social Problems* 25, 531–543.

Fiske, S.T. and Taylor, S.E. (1991) *Social Cognition,* 2nd edn. McGraw-Hill, New York.

Hui, C.H. and Triandis, H.C. (1986) Individualism and collectivism: a study of cross-cultural reachers. *Journal of Cross-cultural Psychology* 17, 225–248.

Kahneman, D. (2003) Maps of bounded rationality: a perspective on intuitive judgement and choice. In: Frängsmyr, T. (ed.) *Les Prix Nobel: the Nobel Prizes 2002.* Nobel Foundation, Stockholm, pp. 449–489.

McCombs, M.E. (2004) *Setting the Agenda: the Mass Media and Public Opinion.* Blackwell, Malden, Massachusetts.

Miller, J.D. (1998) The measurement of civic scientific literacy. *Public Understanding of Science* 7 (3), 203–223.

Noelle-Neumann, E. (1973) Return to the concept of powerful mass media. *Studies in Broadcasting* 9, 67–112.

Noelle-Neumann, E. (1984) *The Spiral of Silence: Public Opinion, our Social Skin.* University of Chicago Press, Chicago, Illinois.

Noelle-Neumann, E. (1993) *The Spiral of Silence: Public Opinion, our Social Skin,* 2nd edn. University of Chicago Press, Chicago, Illinois.

Page, B.I. and Shapiro, R.Y. (1992) *The Rational Public.* University of Chicago Press, Chicago, Illinois.

Popkin, S.L. (1994) *The Reasoning Voter: Communication and Persuasion in Presidential Campaigns,* 2nd edn. University of Chicago Press, Chicago, Illinois.

Scheufele, D.A. and Moy, P. (2000) Twenty-five years of the spiral of silence: a conceptual review and empirical outlook. *International Journal of Public Opinion Research* 12 (1), 3–28.

Scheufele, D.A., Shanahan, J. and Lee, E. (2001) Real talk: manipulating the dependent variable in spiral of silence research. *Communication Research* 28 (3), 304–324.

Shanahan, J., Scheufele, D.A. and Lee, E. (2001) The polls-trends – attitudes about agricultural biotechnology and genetically modified organisms. *Public Opinion Quarterly* 65 (2), 267–281.

Ting-Toomey, S. (1988) Intercultural conflict styles: a face-negotiation theory. In: Kim, Y.Y. and Gudykunst, W.B. (eds) *Theories in Intercultural Communication.* Sage, Newbury Park, California, pp. 213–235.

Willnat, L., Lee, W.P. and Detenber, B.H. (2002) Individual-level predictors of public outspokenness: a test of the spiral of silence theory in Singapore. *International Journal of Public Opinion Research* 14 (4), 391–412.

10

# The Hostile Media Effect and Opinions about Agricultural Biotechnology

A.C. Gunther* and J.L. Liebhart**

*Department of Life Sciences Communication, University of Wisconsin-Madison, Madison, Wisconsin, USA;*
*e-mail: *agunther@wisc.edu; **jlliebha@wisc.edu*

## Introduction

Although the worldwide cultivation of genetically modified (GM) crops has burgeoned in recent years, in early 2004 in California's Mendocino County, local organic food producers, along with an alliance of local and national activists, spearheaded an initiative to ban the raising of any GM crops or animals. Despite the fact that no genetically modified organisms (GMOs) were actually raised in the county and that a national biotechnology trade organization had contributed more than US$0.5 million to counter the initiative, Mendocino voters became the first in the USA to adopt such a ban (Lau, 2004). In addition, supervisors in neighbouring Trinity County quickly followed suit without a public vote (Lee, 2004a).

Invigorated by success, anti-GM activists rallied to put similar initiatives on the ballots in five more counties before November, 2005, while local farm bureaux and growers took charge of the opposition (Lucas, 2004; Doyle, 2005). When these subsequent measures were defeated in all but urban Marin County, opponents claimed a mandate for biotechnology, but supporters, citing record levels of public awareness, promised new initiatives within and beyond California (Lee, 2004b; APSLW, 2005c). Since the Mendocino initiative, similar measures have been initiated in Hawaii and Vermont (AP, 2005b), and more than a dozen states have outlawed local bans of GMOs (AP, 2005a).

While the American public remains relatively ambivalent and uninformed about biotechnology (Pew Initiative on Food and Biotechnology, 2004, 2005), a smaller, highly involved group of individuals has been working ardently to influence the development and adoption of GMOs. Within this debate, and for highly controversial issues in general, partisans – those who, due to their livelihood or worldview, care deeply about a specific issue – are also those most willing to devote the necessary resources to sway others and events toward their side.

The pursuit of partisan interests similarly appears to be playing a vital role in the ongoing global debate over biotechnology. For example, in recent years, various anti-GM activists have uncovered illegal GM strains within commercial rice samples in China (Barboza, 2005; Brown, 2005a), influenced support for UN legislation by the Canadian government (Mick, 2005) and destroyed GM maize plots in France (AP, 2005d). Meanwhile, Brazilian farmers had planted smuggled GM seeds long before the technology was legal (Benson, 2005) and, while one biotech. company bribed Indonesian officials to skirt regulations (Birchall, 2005), another demanded government protection from activists in the UK before testing new crops (Brown, 2004b; Lean, 2004).

For contentious issues such as biotechnology, partisans often serve as principal agents – the outlaws and watchdogs, newsmakers and publicists. And, because highly involved individuals are typically quite concerned about public sentiments regarding the issue, they are often those most vocal regarding any inaccuracies, omissions or imbalances in pertinent media coverage.

Nevertheless, despite the fact that partisans may have access to a significant amount of personal knowledge about the issue when evaluating relevant mass media content, their assessments are not necessarily objective. In fact, partisans – those arguably most significant to a controversy – are subject to a well-documented, systematic bias when making such judgements, and those involved in the biotechnology debate appear to be no exception. This bias, known as the hostile media effect, occurs when partisans on both sides of an issue see media coverage of that issue as slanted against their own point of view. And, because such a 'contrast bias' might add a counterproductive element to the debate, it seems crucial that we explore the underpinnings and ramifications of this effect.

Although it is a relatively recent subject of empirical research interest, the hostile media effect has now been documented across a broad array of topics, including: (i) the Israeli–Palestinian conflict (Vallone *et al.*, 1985; Perloff, 1989; Giner-Sorolla and Chaiken, 1994); (ii) labour strikes (Christen *et al.*, 2002); (iii) primate research (Gunther and Chia, 2001; Gunther *et al.*, 2001); (iv) sports rivalries (Arpan and Raney, 2003); (v) physician-assisted suicide (Gunther and Christen, 2002); (vi) the 1992 election (Dalton *et al.*, 1998); and (vii) a range of social alliances (Gunther, 1992).

In this chapter, we will briefly describe the evidence documenting and characterizing this effect. However, we will focus primarily on a recent series of field experiments, examining this bias in the context of controversies surrounding genetically modified food (Gunther and Schmitt, 2004; Schmitt *et al.*, 2004; Gunther and Liebhart, 2006).

## Methods of Research

### *Recruiting participants*

We conducted these field experiments concerning the issue of biotechnology by recruiting partisans from special interest groups – groups whose members we expected to have strong opinions and high levels of involvement in GM food issues. For example, we recruited anti-GM food partisans from members of organic food cooperatives and GM food supporters from meetings of biotech. researchers working at universities or in industry.

To verify partisan involvement, we screened these participants via a series of questions about the strength of their support of, or opposition to, the development of GM food products. Because partisan position is an important variable in this research, and we considered the manipulation of this variable somewhat implausible, we recruited people who, based on group membership, were likely to have deeply felt opinions about the issue. However, because participants were self-selected into partisan camps, we also faced the difficulty that other distinctions between groups – such as dissimilarities in education level or income – might present rival explanations for any observed differences on dependent variables.

Analysis strategies can be used to control for known differences. However, the possibility of unanticipated third-variable explanations must be still acknowledged in quasi-experimental designs such as those described here and in most experimental tests of the hostile media effect.

### *Composing stimuli*

The primary goal of our experiments was to measure partisans' reactions to mass media coverage of GM food issues. Hence, we created stimulus material compiled from news articles about GM foods in a variety of sources, such as *USA Today*, the *New York Times* and the *Boston Globe*. The stimulus content concerned topics such as vitamin A-enhanced, or 'golden', rice, GM salmon designed for rapid growth and pigs genetically modified to produce more environmentally friendly low-phosphorus manure. It was important that the overall content be as neutral as possible. Hence, a small group of disinterested 'judges' rated early drafts of the information for slant and balance, and the text was adjusted accordingly.

### *Manipulating stimulus formats*

Because we wished to discover whether partisans preferentially judged information presented in mass media as biased, we chose to format the same content in two different ways. Across several experiments, we presented stimuli as either news articles attributed to media sources such as *USA Today* or as student essays written for a college composition class. For news articles, the text was prepared in news-story format with appropriate headlines, bylines and other journalistically graphic details. Student essays were presented as the typed, double-spaced, full-page-width documents one would expect a college student to produce for a class assignment. For all conditions, headings and text were identical.

### *Measuring bias perceptions*

To compare reactions to these stimuli, we developed an item index that included measures of respondents' perceptions of stimulus bias, percentage of favourable and unfavourable information and author bias. Respondents answered the first and last items using a nine- or ten-point scale and the other two items with percentages.

### *Measuring information processing*

A second goal was to assess the ways in which partisans processed GM food-related information. Various processing mechanisms have been hypothesized, and we have examined three: (i) selective recall, in which partisans on both sides selectively attend to or consider, and thus remember, a disproportionate amount of the disagreeable material in the stimulus; (ii) selective categorization, in which opposing partisans may pay attention to and recall identical content but still evaluate it differently – as relatively hostile to their own side; and (iii) different standards, in which rival partisans see content and valence in the same way but consider the disagreeable portions irrelevant or invalid. A graphic illustration of these three mechanisms can be found in Fig. 10.1 (from Schmitt *et al.*, 2004).

To examine the roles of these mechanisms, we tested recall by asking respondents to write down facts or arguments they remembered from the stimulus content. We then had both respondents and independent coders categorize these listings as supporting, neutral

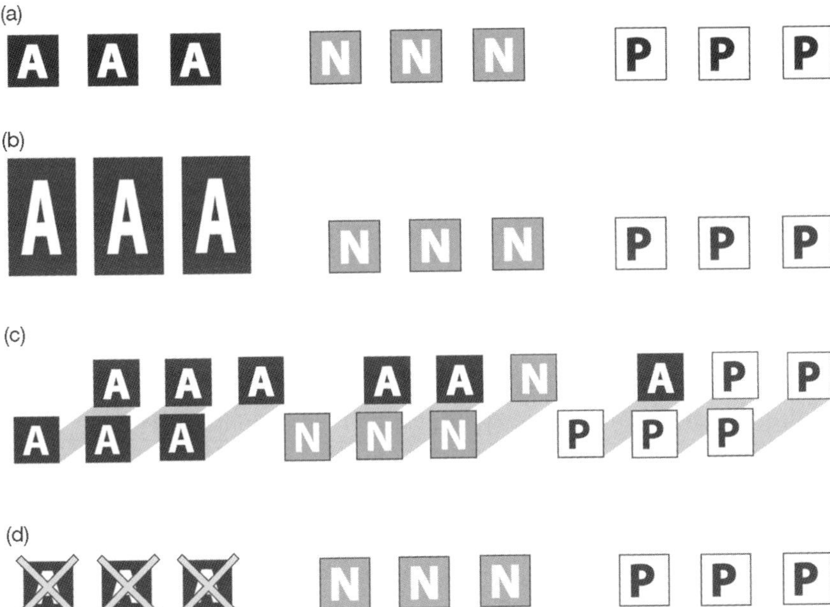


**Fig. 10.1.** Illustration of the distinctions among three potential mechanisms for the hostile media effect in pro-partisans. (a) Information distribution in what a neutral observer would see as a balanced news story, with equal proportions of anti- (A), neutral (N) and pro- (P) content; (b) Selective recall: pro-partisans disproportionately remember more unfavourable content; (c) Selective categorization: pro-partisans evaluate relatively more content as unfavourable; (d) Different standards: pro-partisans consider anti-content as being invalid or irrelevant. From Schmitt *et al.* (2004), copyright 2004 by Sage Publications, Inc. Reprinted by Permission of Sage Publications, Inc.

toward or opposed to GM food. We evaluated selective categorization by asking participants to rate specific excerpts taken from the stimulus as favourable, unfavourable or neutral. We assessed the different-standards hypothesis by asking respondents whether these same excerpts were accurate or pertinent to the debate, and subsequently controlling responses for perceived valence (selective categorization). We used these measures, in conjunction with our stimulus manipulations, to decipher the specific mechanisms likely to play a role in the hostile media effect.

## Overview of the Hostile Media Perception

The hostile media perception described in this chapter won't be news to journalists. To newspaper and broadcast news editors it is a commonplace experience – groups from different sides of an issue each complaining that the news coverage is flagrantly biased against their own point of view. In fact, news organizations often gleefully publicize such negative feedback when it comes from both sides.

Presumably, no explanation is necessary when, to borrow an example from Vallone *et al.* (1985), news magazine editors publish back-to-back letters referring to a profile of a political candidate: one letter writer calls it a poorly disguised endorsement, while the other derides the same article as a 'slick hatchet job'. But social science has only turned its attention to this phenomenon during the past two decades. And the controversy over GM food is not the first issue used to test this idea empirically.

In search of a situation that would elicit strong reactions from distinct groups of partisans, a graduate student and two social psychologists from Stanford University, California, USA, designed the first test of this hypothesis in the early 1980s (Vallone *et al.*, 1985). As a stimulus they used broadcast coverage of the 1982 'Beirut massacre' in Lebanon, and as partisan subjects they recruited groups of pro-Israeli and pro-Arab students to evaluate the news. As expected, each group said the broadcast news reports were unfair and biased in favour of the other side.

The authors also speculated about what their subjects were actually thinking as they made these judgements and used a colour analogy to describe two types of potential mechanisms. The first, which included different standards, was an 'evaluative' judgement in which partisans on two sides of an issue saw the truth as either black or white, and viewing media information as typically 'grey', both considered it inaccurate or misleading.

In contrast, the second type of mechanism, which included both selective recall and categorization, was 'perceptual' in nature, meaning that opposing partisans actually disagreed about the shade of the information before them. Using data, these authors argued that both processes occur – partisans on different sides not only evaluated the validity of the same content differently but they also actually perceived a different reality when they watched.

Two more experimental studies followed, both also using the Middle East conflict as the stimulus material and pro-Arab or pro-Israeli students as partisan participants. Both studies replicated earlier findings while also analyzing processing mechanisms in greater depth (Giner-Sorolla and Chaiken, 1994) and highlighting an interesting angle: that partisans believe neutral audiences will be influenced in undesirable ways by the 'hostile' content (Perloff, 1989).

After these initial studies, evidence for the generalizeability of the effect slowly accumulated. A study of the 1997 UPS strike showed that antagonists on both sides of the picket lines – UPS drivers *versus* managers – interpreted news reports of the strike as giving comfort to their adversaries (Christen *et al.*, 2002). Another quasi-experiment, in which participants were asked to evaluate genuine news articles about lab. research using primates, showed that animal rights activists and primate researchers were significantly divided in their perceptions of the stories – each finding relatively more hostility toward their own position (Gunther *et al.*, 2001). Other research showed that sports fans – perhaps one of the most visible manifestations of partisanship – also saw news coverage as biased against their own team, an effect moderated by the perceived affiliation of the source of the story (Arpan and Raney, 2003).

During this time, survey data provided additional evidence for the hostile media perception for a variety of political, scientific and social issues (Gunther, 1988, 1992; Dalton *et al.*, 1998; Gunther and Chia, 2001). Some results also hinted that partisans' perceptions of the slant of media coverage might have important consequences regarding perceived public opinion (Gunther and Chia, 2001).

In light of previous findings, our research team – noting the increasing public debate over biotechnologies and genetic engineering – turned its attention to GM food issues. As noted above, we sought groups of people who supported organic agricultural practices and were strongly opposed to genetically modified food products (and also research), and industry people who believed in the beneficial potential of GM food and strongly supported development of these technologies. Assuming that these groups and the controversy that polarized them would provide good tests of the hypothesis, we developed a series of experiments to explore the theoretical underpinnings of the hostile media perception.

## A media effect?

Up to this point, an unanswered question was whether the hostile media effect would also appear for non-mediated information. A related phenomenon in the social psychology literature – known as biased assimilation (Lord *et al.*, 1979) – added some zest to this question because it described a seemingly opposite outcome: that partisans would interpret balanced or neutral information as supportive of, rather than opposed to, their pre-existing opinions. Indeed, Vallone *et al.* (1985) had presented the contrast effect embodied in the hostile media effect as a curious exception to the assimilation bias, and thus raised a tantalizing question that has, nevertheless, received little attention in subsequent years.

In an attempt to answer this question, Gunther and Schmitt (2004) designed their experiment (described briefly above) so that the same information about GM food was alternatively presented as a news article or as a college student's composition, and subsequently randomized partisans on both sides of the issue to one of these conditions. In the media condition, respondents displayed the expected hostile media response; however, those reading the same information in student-essay format showed a dramatically different pattern.

On three of the six bias measures there was no difference, the effect disappeared; on the other three they reported that the content actually *favoured* their own position. Whether or not an assimilation bias occurred for those in the essay condition was not completely clear.[1] However, findings did give compelling support to the notion that the contrast bias is not universally invoked: it appears to be particular to mass media information.

Results of the manipulation in this experiment suggest a predictable pattern, but they do not explain why mass media should elicit the contrast bias while other information contexts do not. The authors speculated that an explanation might lie in one particularly distinctive characteristic of mass media – its broad reach. This sense of reach seemed to be an implicit element in earlier research (Vallone *et al.*, 1985; Perloff, 1989), demonstrating that partisans who saw media content as unfairly biased also felt that this unfavourable content would influence neutral audience members in a similarly unfavourable direction. This finding led the authors to ask: does a mass media context direct the attention of individuals highly involved in an issue to the potential influence of the information on others? If so, this awareness might put partisans in a defensive processing mode, causing them to see the content as disagreeable.

As a first test of this idea, Gunther and Schmitt (2004) asked respondents questions about persuasiveness, rather than bias *per se*. Did the weight of evidence in the content favour one side or the other? Was it more persuasive in support of, or against, the development of GM food? And further, respondents were asked to consider their answers in two different scenarios (in counterbalanced order): one when thinking about the influence of the information on their own opinions, the other when thinking about influence on other, neutral, readers.

1 When demographic controls were incorporated into the analysis, evidence of biased assimilation in the essay condition was no longer apparent, however the hostile media effect was still observed in the article but not the essay condition. These results suggested that demographic differences between groups did not appear to explain contrast effects, but that they might offer an alternate explanation for assimilation.

**Table 10.1.** Mean persuasiveness estimates as a function of reach and audience.[a]

| | News article | | | Student essay | | | | | |
|---|---|---|---|---|---|---|---|---|---|
| | Own opinion | Neutral reader's opinion | *t* | Own opinion | Neutral reader's opinion | *t* | Audience[b] effect *F* | Reach effect *F* | Reach X audience[b] interaction *F* |
| Arguments stronger on one side of issue | −0.2 | −0.6 | 1.98* | 0.7 | 0.2 | 2.65** | 10.7** | 6.4* | 0.6 |
| | (2.4) | (1.9) | | (2.2) | (2.0) | | | | |
| Evidence leans more toward one side | −0.1 | −0.4 | 1.98* | 0.8 | 0.3 | 3.00** | 12.2*** | 8.8** | 0.3 |
| | (1.8) | (1.9) | | (1.8) | (1.9) | | | | |
| Persuasiveness in general | −0.2 | −0.2 | 0.31 | 0.7 | 0.4 | 1.76* | 2.2 | 5.8* | 1.1 |
| | (2.1) | (1.9) | | (2.0) | (1.6) | | | | |
| Persuasiveness of 'golden rice' segment | 0.7 | 0.1 | 2.60** | 1.2 | 0.8 | 2.31** | 12.2** | 2.5 | 0.3 |
| | (2.7) | (2.7) | | (2.7) | (2.7) | | | | |
| Persuasiveness of GM salmon segment | −0.1 | −0.7 | 2.61** | 0.1 | −0.2 | 1.77* | 10.1** | 0.7 | 1.1 |
| | (2.6) | (2.5) | | (2.5) | (2.1) | | | | |

[a] Standard deviations are in parentheses. Sample sizes in the article and essay conditions were 82 and 87, respectively. Responses from both groups of partisans were recomputed so that positive values indicate support for the partisan's position and negative values indicate opposition to the partisan's position.

[b] Audience effect refers to question blocks with a preface asking either about influence on subject's own opinion or on a neutral reader's opinion.

* $P < 0.05$; ** $P < 0.01$; *** $P < 0.001$ (*t*-tests are one-tailed).

Framing the persuasiveness questions to vary in terms of reach – the locus of influence of the information on oneself or others – produced differences with apparent parallels to the hostile media effect (see Table 1). When partisans considered the likely influence of media content on their own opinions, rather than on others, they described the information as significantly more persuasive in a favourable, or relatively favourable, direction.

This result represents the converse to Perloff's notion (1989) that judgements of hostile content lead to inferences about undesirable influences on others. Instead, prompting individuals highly involved in biotech. issues to think about a larger audience – something that mass media may do by default – may unfavourably skew their judgements regarding the slant of that content.

## Separating media components

These results suggested that reach was likely to be one explanation for the hostile media effect in the debate over GM food. However, the article *versus* essay manipulation that demonstrated the reliance of this effect on a media context varied not only with respect to reach but also with respect to whether a professional journalist or college student had authored the information. Because either reach or source or both cues could potentially account for the hostile media effect, we used the GM food issue in a second experiment, aimed at disentangling these factors (Gunther and Liebhart, 2006).

Procedures for recruitment and stimulus construction were similar to those used for the previous study. However, participants here were randomized to one of four stimulus conditions. Two conditions were the same as those used previously: a high-reach journalist source (news article by a journalist) and a low-reach student source (classroom essay by a college student). Two other conditions were then added to complete the 2 × 2 design: a high-reach student source (award-winning college essay published as an op-ed article) and a low-reach journalist source (college composition by a journalist on leave of absence).

Analyses for this study were complicated somewhat by an inherent difficulty in fashioning manipulation checks to separate these closely aligned media factors. However, overall results supported two independent effects. Pro-GM and anti-GM partisans responded to either a high-reach context or a journalist source by moving their judgements of the information in a hostile direction, and both cues together produced an additive effect.

## Other Issues

### Processing mechanisms

The data described above contribute to explanations about *why* the hostile media effect occurs, by suggesting that mass media shift the partisan audience into a different processing mode, perhaps a defensive mode that somehow renders the content more unfavourable. But this observation, in turn, raises the question of *how*. How might this processing operate? What mechanisms might occur in the defensive mode that shift perceptions of content away from assimilation and in the direction of contrast? As noted above, this question has received some attention from past research, and three distinct mechanisms have been proposed.

Results of some studies in other contexts supported a role for a different standards mechanism but contradicted a role for selective recall and were equivocal regarding selective categorization (Vallone *et al.*, 1985; Giner-Sorolla and Chaiken, 1994; Arpan and Raney, 2003). Nevertheless, the measures used in these studies were sometimes confounded or otherwise problematic. In addition, some authors reported finding evidence for a fourth potential mechanism, the influence of prior beliefs about media bias concerning the issue (Giner-Sorolla and Chaiken, 1994). However, this factor may be a consequence, rather than a cause, of the hostile media effect (Gunther *et al.*, 2001).

As a result, studies relevant to the debate over GM foods used measures for the aforementioned three mechanisms that controlled for one another. And, because manipulations in these experiments produced quite different outcomes – a robust hostile media effect in some conditions and none in others – we were also able to identify specific mechanisms that were, and were not, associated with a contrast bias response.

In the first of these experiments, in which a news article by a journalist was compared with an essay by a college student, recall was similar across groups and conditions, so this could not explain the observed hostile media responses in the former condition and not the latter (Schmitt *et al.*, 2004). In contrast, groups used different standards to judge the validity of information: each side considered pro-attitudinal information more valid. However, they did so to a similar extent in article and essay conditions, so this mechanism also could not offer an explanation.

Finally, partisan groups also categorized information differently from one another – as relatively hostile to their own sides – and this contrast was only apparent in the article condition. Formal tests of this relationship similarly supported a role for selective categorization in mediating the hostile media effect.

The second study concerning this same issue (Gunther and Liebhart, 2006) narrowed the role for this mechanism but, in doing so, raised one important remaining question. In this experiment, which crossed reach and source conditions, selective categorization appeared to account only for the influence of a high-reach source – the potential of that information to influence a large audience. So, although biotechnology partisans may respond to both reach and source media cues with contrast effects, and the former appear to trigger the categorization of more passages as hostile, we currently have no mechanism to explain how a journalist source prompts a comparable overall judgement of bias.

## Partisanship

Another noteworthy, and perhaps centrally important, concern is the reliance of the hostile media effect upon the individual's type and level of involvement with the issue. Nevertheless, despite 20 years of research, the exact limits for this phenomenon have never been firmly established.

Clear boundaries for the hostile media response have been apparent since Vallone and colleagues (1985) conducted the first study of this effect – a telephone survey in advance of the 1980 election – and found that two-thirds of participants saw absolutely no bias in news coverage of candidates.[2] After the election, when these authors failed to detect either the same strong sentiments or perceptions of a hostile media bias, they sought issues and coverage that could strike a deeper and more personal note. As a result, the first two successful demonstrations of the hostile media effect occurred with partisans from pro-Arab and pro-Israeli student groups asked to judge news coverage of the Middle East conflict (Vallone *et al.*, 1985; Perloff, 1989).

Since that time, studies have provided disparate or incomplete clues regarding the types of sample selection methods required for demonstrating this effect. For example, results of some surveys suggest that the hostile media effect may be observable on the basis of attitude extremity alone (Dalton *et al.*, 1998; Gunther and Chia, 2001). However, some experiments in which partisan samples were chosen on this basis have produced weak or unsuccessful results (Giner-Sorolla and Chaiken, 1994).

2 A similar prevalence for this effect has since been reported in a general student sample (D'Allessio, 2003).

Other survey research identified group membership as a robust predictor of hostile media perceptions for news coverage of various social groups (Gunther, 1992). Such results were also congruent with Perloff's proposal (1989) that ego-involvement, typically characterized by a combination of extreme issue attitudes and strong levels of identification with one's partisan group, was actually at the root of the hostile media effect. In fact, this combination proved to be a highly reliable means of demonstrating this effect for a variety of issues (Perloff, 1989; Gunther *et al.*, 2001; Christen *et al.*, 2002), including biotechnology (Gunther and Schmitt, 2004; Gunther and Liebhart, 2006).

As described by social judgement theory, ego-involved attitudes are those closely associated with a person's self-esteem or central values, aligning with core aspects of identity and triggering strong emotions. Because such attitudes are essentially self-defining, they also tend to be aggressively defended, resulting in the rejection of a relatively large amount of issue-relevant information (Salmon, 1986).

Perloff proposed that such contrast effects extend to judgements of relevant information presented in mass media and that distorted social perceptions – known to occur in partisans – subsequently led them to infer that the attitudes of neutral others would be swayed in a hostile direction. He additionally distinguished ego-involvement from other potential types of involvement, such as relevance, or the perception that the issue had important future consequences.

Both types of involvement can be associated with high levels of information-seeking and knowledge about an issue. However, in opposition to ego-involvement, high levels of relevance reflect a motivation to hold accurate attitudes about an issue, promoting a relatively unbiased approach to information-processing (Salmon, 1986).

Few attempts have been made to actually measure the type and level of issue involvement that prompts the hostile media effect; however, results of several studies similarly support the notion that ego-involvement is an important factor. For example, Vallone *et al.* (1985) reported that high levels of two closely related factors – self-reported knowledge and emotional involvement regarding the issue – moderated the hostile media effect.

Also, in an experiment in which students were asked to evaluate the slant of an article about one of several issues, topics comparatively more likely to inspire ego-involvement (e.g. the performance of George Bush *versus* the campus parking shortage) also produced a significantly stronger hostile media effect (D'Allessio, 2003).

Similarly, in a recent survey about the possible resettlement of the Gaza strip among Jewish residents of the area, only those involvement measures that expressed deeply felt emotions or values (e.g. a strong sense of belonging to one's settlement or belief in the right to settle all of Israel) predicted partisan perceptions of unfavourable media coverage.

Concerning the debate over GM food, we recently included an expanded battery of involvement question in an experiment with a sample chosen to reflect a broad range of partisanship on the issue (Gunther and Liebhart, 2005). This instrument suggests that two separate involvement factors may play distinct roles in the hostile media effect. The first, like ego-involvement, was associated with group identification and extreme attitudes and predicted the perception that information was hostile in response to certain media cues. In contrast, a second factor – perhaps more like relevance – was more closely related to issue importance and, instead, predicted a tendency to judge balanced media information as impartial.

Additional studies refining the conception and measurement of issue involvement for the hostile media effect in anti- and pro-GM partisans would probably contribute to our understanding of underlying processes for this outcome. Further research may also offer clues as to how different types of involvement might be created, strengthened or transformed, and thus provide essential weaponry for preventing or defusing this response.

## Asymmetry of effects

Another interesting issue, perhaps one closely intertwined with partisan involvement, is the possibility that partisans on one side of the debate might be more susceptible to contrast effects than their adversaries. Historically, researchers have reported that some groups representing a minority position on an issue appeared to be both more fervent with respect to certain features of issue involvement and more likely to be moved to action for a given level of partisanship (Schumann and Presser, 1981).

Other investigators have found that partisan groups attempting to buck rather than uphold the *status quo* tend to have more negative emotions regarding the debate and be viewed by all sides as relatively extreme (Keltner and Robinson, 1996). However, it is unclear how such reports might pertain to the hostile media effect – whether such differences might preferentially sensitize underdogs to the potential influence of mass media coverage on others or spur them to action in response to perceived media bias. Nevertheless, at least a few experiments have made note of such asymmetries.

For example, in a previous study in which pro- and anti-primate research partisans were asked to judge relevant news articles containing different inherent slants, both groups saw counter-attitudinal articles as opposing their own side. However, while primate researchers judged pro-attitudinal articles as favourable to their position, animal-rights activists considered news coverage to be neutral at best (Gunther *et al.*, 2001).

More relevant to the biotechnology issue, in another recent experiment (Gunther and Liebhart, 2006), both pro- and anti-GM partisans categorized information excerpts as more hostile to their own view when presented in a high- reach, as opposed to low-reach, format. However, this tendency appeared to be considerably stronger for the anti-GM group (see Fig. 10.2). These observations might be explainable by differences in certain aspects of partisanship, other uncontrolled differences between groups or the types of excerpts or articles chosen for the experiment. Nevertheless, the potential for response asymmetry across groups remains a fascinating prospect.

## Consequences

Perhaps the most important question regarding hostile media effect is what it actually does. That is, are hostile media responses important or

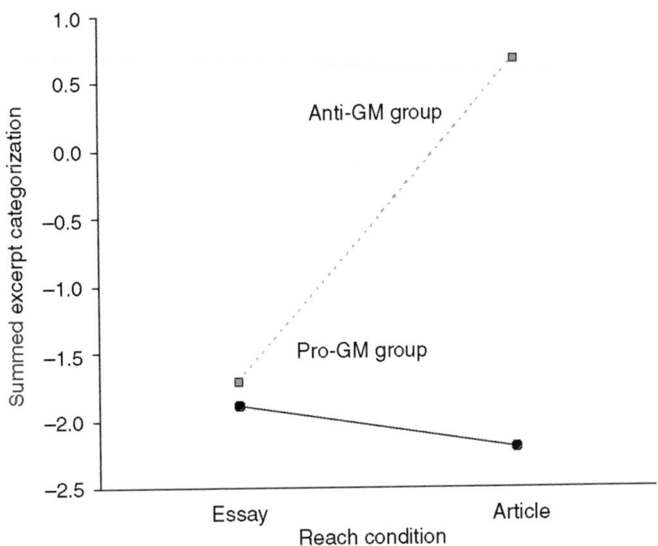


**Fig. 10.2.** Excerpt categorization as a function of reach condition and partisan group. Negative values indicate a perceived net excerpt bias opposing GM foods and positive values indicate a perceived net excerpt bias supporting GM foods.

just an intriguing peculiarity? Some answers to that question can be found in research, directly or indirectly relevant to consequences for this bias. One immediate implication is that partisans, like most people, are likely to extrapolate from their own sampling of media content. That is, because of the law-of-small-numbers bias suggesting that people will consider even a small sample as representative (Tversky and Kahneman, 1971), they will conclude that most media coverage is similar to that which they observe. Thus, partisans will assume that most other people are reading or viewing the same 'disagreeable or misleading' information that they are (Gunther *et al.*, 2001).

To the extent that partisans believe others are exposed to, and perhaps influenced by, such misleading content, they may feel public opinion is swayed in that direction. For example, data from a national probability sample showed both that people felt media coverage of the issue of using primates in lab. research was primarily unfavourable to such research, and that they had substantially overestimated public opposition to such research (Gunther and Chia, 2001).

Such cases of pluralistic ignorance – the collective misjudgement of public opinion – are also important because they, in turn, can result in a phenomenon sometimes called the spiral of silence (Noelle-Neumann, 1993; Scheufele, Chapter 9, this volume). This term describes the tendency to withdraw from public discussion and debate because of the perception that others do not share one's point of view. In line with these ideas, some researchers who have identified a 'hostile world phenomenon' in partisan groups also suggested that these perceptions may have arisen from hostile media responses (Shamir and Shikaki, 2002).

A third potential consequence of this effect is increased attitude polarization. In this regard, there is some evidence that, when individuals encounter information that they perceive as inconsistent with pre-existing attitudes, they may process that information in a counter-arguing mode. As a result, they become even more certain about or strongly committed to their own position on the issue (Tormala and Petty, 2004), perhaps also increasing their resistance to subsequent information.

The hostile media perception may also affect social-level beliefs and behaviours. For example, in a recent study of Jewish settlers in the Gaza Strip, surveyed shortly before the Israeli government compelled them to move out of occupied territories, Tsfati and Cohen found that the more settlers considered media coverage to be hostile toward their position, the more they were distrustful of democratic processes. Perhaps most importantly, this distrust in democracy was positively related to settlers' stated intentions to 'forcefully resist' resettlement (Tsfati and Cohen, 2005, p. 28).

Collectively, these data appear to support the inference that perceptions of hostile media content may not just be an interesting anomaly. Instead, they may contribute to a variety of important outcomes for highly involved individuals and for those impacted by their actions, including bolstering partisans' involvement levels and increasing their perceptions that others are unsupportive. These consequences could conceivably strengthen partisan resolve to mobilize others and instantiate their interests into law, such as has recently occurred in California.

However, for other individuals, such effects could instead reduce the likelihood of participating in sanctioned public forums. And from here, it is but a small step to suggest that the hostile media effect might ultimately help drive some of those most passionate about – and arguably most influential in – the debate over GM foods to engage in extreme actions, such as dodging environmental regulations in the Third World, digging up experimental plots of genetically engineered crops in Britain or engaging in other so-called 'eco-terrorist'[3] acts.

## References

Arpan, L.A. and Raney, A.A. (2003) An experimental investigation of news source and the hostile media effect. *Journalism and Mass Communication Quarterly* 80 (2), 265–281.

*Associated Press State and Local Wire* (2005a) Biotech bans in California counties spark push for state control (12 July).

*Associated Press State and Local Wire* (2005b) Early results against Sonoma ban on genetically modified crops (9 November).

*Associated Press State and Local Wire* (2005c) Voters reject Sonoma ban on genetically modified crops (9 November).

*Associated Press State and Local Wire* (2005d) Militant French farmer sentenced to prison for uprooting genetically modified crop (15 November).

Barboza, D. (2005) Illegal rice found again in China's food supply. *New York Times*, 14 June, p. C6.

Benson, T. (2005) Brazil passes law allowing crops with modified genes. *New York Times*, 4 March, p. C3.

Birchall, J. (2005) Monsanto agrees. Dollars 1.5 m penalty over crop bribe. *Financial Times*, 7 January, p. 4.

Brown, P. (2005a) Unlicensed GM rice may be in UK food chain: Greenpeace finds illegal strain in Chinese exports. *Guardian*, 14 April, p. 11.

3 The definition and use of this term alone appears capable of inspiring significant public debate, with some individuals contending that it has been wrongly used to label legal protests by environmentalists as extreme, but has not been applied to irresponsible behaviour toward the environment by governments or industry (Wikipedia, 2006).

Brown, P. (2005b) Despairing GM firms halt crop trials. *Guardian,* 15 April, p. 5.

Christen, C.T., Kannaovakun, P. and Gunther, A.C. (2002) Hostile media perceptions: partisan assessments of press and public during the 1997 UPS strike. *Political Communication* 19 (4), 423–436.

D'Alessio, D. (2003) An experimental examination of readers' perceptions of media bias. *Journalism and Mass Communication Quarterly* 80 (2), 282–294.

Dalton, R.M., Beck, P.A and Huckfeldt, R. (1998) Partisan cues and the media: information flows in the 1992 presidential election. *American Political Science Review* 92, 111–126.

Doyle, J. (2005) Campaign 2005: Sonoma County; measure to ban genetically altered crops causes split; farmers, ranchers, and grape growers have a lot at stake. *San Francisco Chronicle,* 26 October, p. B1.

Giner-Sorolla, R. and Chaiken, S. (1994) The causes of hostile media judgement. *Journal of Experimental Social Psychology* 30, 165–180.

Gunther, A.C. (1988) Attitude extremity and trust in media. *Journalism Quarterly* 65, 279–287.

Gunther, A.C. (1992) Biased press or biased public? Attitudes toward media coverage of social groups. *Public Opinion Quarterly* 56, 147–167.

Gunther, A.C. and Chia, S.C.Y. (2001) Predicting pluralistic ignorance: the hostile media effect and its consequences. *Journalism and Mass Communication Quarterly* 78 (4), 689–701.

Gunther, A.C. and Christen, C.T. (2002) Projection or persuasive press? Contrary effects of personal opinion and perceived news coverage on estimates of public opinion. *Journal of Communication* 52 (1), 177–195.

Gunther, A.C. and Liebhart, J.L. (2005) Broad reach or biased source? Decomposing the hostile media effect. Paper presented to the *Mass Communication Division of the International Communication Association,* New York, May.

Gunther, A.C. and Liebhart, J.L. (2006) Broad reach or biased source? Decomposing the hostile media effect. *Journal of Communication* 56 (3), 449–466.

Gunther, A.C. and Schmitt, K.M. (2004) Mapping boundaries of the hostile media effect. *Journal of Communication* 54, 55–70.

Gunther, A.C., Christen, C.T., Liebhart, J.L. and Chia, S.C.-Y. (2001) Congenial public, contrary press and biased estimates of the climate of opinion. *Public Opinion Quarterly* 65, 295–320.

Kahneman, D. (2003) A perspective on judgement and choice: mapping bounded reality. *American Psychologist* 58 (9), 697–720.

Keltner, D. and Robinson, R.J. (1996) Extremism, power, and the imagined basis for social conflict. *Current Directions in Psychological Science* 5, 101–105.

Lau, E. (2004) Anti-biotech measures approved by Mendocino. Measure H backers overcome a huge fund-raising disadvantage. *Sacramento Bee,* 3 March, p. 9.

Lean, G. (2004) GM firm finally gives up on planting in Britain. *Independent,* 21 November, p. 9.

Lee, M. (2004a) Trinity bans biotech farming. Rural county becomes only second in US to take a stand against biotechnology. *Sacramento Bee,* 4 August, p. D2.

Lee, M. (2004b) Biotech tide turns. Three defeats undercut ban momentum. *Sacramento Bee,* 4 November, p. D1.

Lord, C.G., Ross, L. and Lepper, M. (1979) Biased assimilation and attitude polarization: the effects of prior theories on subsequently considered evidence. *Journal of Personality and Social Psychology* 37, 2098–2109.

Lucas, G. (2004) Genetically altered crops: 2 counties reject ban; not Marin. *San Francisco Chronicle*, 3 November, p. B10.

Mick, H. (2005) Seeds of discontent: farming subsidies serve to ensure the rich get richer and the poor get thin; Canadian stance on 'terminator' seeds draws international protests. *Ottawa Citizen*, 12 February, p. B1.

Noelle-Neumann, E. (1993) *The Spiral of Silence: Public Opinion: Our Social Skin*, 2nd edn. University of Chicago Press, Chicago, Illinois.

Perloff, R.M. (1989) Ego-involvement and the third person effect of televised news coverage. *Communication Research* 16 (2), 236–262.

Pew Initiative on Food and Biotechnology (PIFB) (2004) Americans' opinions about genetically modified foods remain divided, but majority want a strong regulatory system (press release retrieved 14 January 2006 from http://pewagbiotech.org/newsroom/releases/112404.php3).

Pew Initiative on Food and Biotechnology (PIFB) (2005) Americans' knowledge of genetically modified foods remains low; majority are sceptical about animal cloning (press release retrieved 14 January 2006 from http://pewagbiotech.org/newsroom/releases/111505.php3).

Salmon, C.T. (1986) Perspectives on involvement in consumer and communication research. In: Dervin, B. and Voight, M. (eds) *Progress in the Communication Sciences: Vol. 7*. Ablex, Norwood, New Jersey, pp. 243–268.

Schmitt, K.M., Gunther, A.C. and Liebhart, J.L. (2004) Why partisans see mass media as biased. *Communication Research* 31 (6), 623–641.

Schumann, H. and Presser, S. (1981) *Questions and Answers in Attitude Surveys*. Academic Press, New York.

Shamir, J. and Shikaki, K. (2002) Self-serving perceptions of terrorism among Israelis and Palestinians. *Political Psychology* 23 (3), 537–557.

Tormala, Z.L. and Petty, R.E. (2004) Resistance to persuasion and attitude certainty: the moderating role of elaboration. *Personality and Social Psychology Bulletin* 30 (11), 1446–1457.

Tsfati, Y. and Cohen, J. (2005) Democratic consequences of hostile media perceptions – the case of Gaza settlers. *Harvard International Journal of Press-Politics* 10 (4), 28–51.

Tversky, A. and Kahneman, D. (1971) Belief in the Law of Small Numbers. *Psychological Bulletin* 76, 105–110.

Vallone, R.P., Ross, L. and Lepper, M.R. (1985) The hostile media phenomenon: biased perception and perceptions of media bias in coverage of the Beirut massacre. *Journal of Personality and Social Psychology* 49 (3), 577–585.

Wikepedia. Eco-terrorism article and discussion (retrieved 13 January 2006 from http://en.wikipedia.org/wiki/Eco-terrorism and http://en.wikipedia.org/wiki/Talk:Eco-terrorism).

11

# Risk Communication, Risk Beliefs and Democracy: the Case of Agricultural Biotechnology

S. Dunwoody[1*] and R.J. Griffin[2**]

[1]*School of Journalism and Mass Communication, University of Wisconsin-Madison, Madison, Wisconsin, USA;* [2]*College of Communication, Marquette University, Milwaukee, Wisconsin, USA;*
*e-mail:* *dunwoody@wisc.edu; ***robert.griffin@marquette.edu*

## Introduction

While much of the risk communication literature focuses on informational precursors to perceptions of risk at the individual level, some of the most intractable problems confronting us today are centred, instead, on the social allocation of risk. In that latter domain, the critical issue is not *whether* someone will incur a risk but *who* will incur it and how society will manage the resulting inequities – perceived or actual.

Far fewer scholars focus on the role of risk communication within this broader policy framework. And yet, almost every risk issue that confronts communities, countries and cultures is fraught with allocation questions. Among newer risks posed as of the present day are those by the use of stem cells to ameliorate health problems, by questions of whether to reinvest in nuclear power, by efforts to decipher the human genome in search of genetic causes of disease, by the rapid development of nanotechnology for just about any purpose one can imagine and by the subject of this book: the production of genetically modified foods for public consumption.

Certainly individual risk judgements are important to all of these issues, and scholars have made much progress in recent decades in helping us understand relationships among attributes of a risk, messages about that risk and a person's subsequent reaction to the risk. Less well understood is how to transform our relatively rich understanding of risk as a private issue into an understanding of risk as a public issue, that is, how to employ individual message effects in service to managing the allocation of risk in a democratic setting (for a summary of this argument, see Rich *et al.*, 1999).

We emphasize the 'democratic setting' component here because it allows us to privilege one important actor in this chapter: the audience. Frankly, a focus on audience is not optional, as the social management of controversial risks in recent years has succeeded or foundered on agencies' abilities to identify audiences, understand their questions and concerns with respect to a risk and then to manage that risk in ways that are responsive to those questions and concerns. Further, we will focus in this chapter on one specific audience concern – the stability of beliefs about a risk – and will address that through the lens of information-seeking and processing, important precursors to belief stability.

In the sections that follow, we start with what it means to adopt an audience-centred approach to risk communication, introduce concepts that have illuminated that approach in recent years, then, in service to illustrating these factors at work, discuss one model that has made an effort to incorporate some of those concepts in an exploration of information-seeking and processing and the stability of risk beliefs.

## The Centrality of Audience

Although an audience orientation to risk may seem to be a 'no-brainer' today, in the not too distant past it was a radical suggestion. Risk assessors and risk managers defined their responsibilities as calculating and then communicating the best possible estimate of the likelihood of coming to harm from, say, asbestos insulation in older buildings. If there was an ensuing goodness-of-fit issue – for example, if people seemed far more worried about that asbestos insulation than the level of risk suggested they should be – then the problem almost certainly rested with the flawed ability of people to understand the science.

It was easy for risk managers to shrug off this bad fit as long as audiences were powerless. But public reaction to such events as the accident at Three Mile Island and the Love Canal crisis in the late 1970s demonstrated that public opinion could be influential, indeed. Federal and state agencies responsible for risky technologies slowly began to build 'stakeholders' into their risk decision-making. Questions about how to manage a risk now included such important issues as: (i) who the most important subgroups of the public might be; (ii) what they knew about the risk and how they had acquired that knowledge; (iii) who they trusted to address a risk and who, in contrast, made them suspicious; and (iv) how best to involve them in deciding what to do about the risk.

An audience-centred approach received the imprimatur of The National Academy of Sciences in the late 1980s, when an academy study group, the Committee on Risk Perception and Communication, issued a report titled *Improving Risk Communication* (National Research Council, 1989). In it, the committee endorsed a definition of risk communication as 'a two-way street' (p. 10), with preference given to information that closely reflected 'the perspectives, technical capacity and concerns of the target audiences' (p. 11).

This initial sensitivity to audience is still a road down which agencies and scientists proceed with caution. Both authors have served on review and oversight committees of The National Academies (of Science, Engineering and Medicine), where risky technologies and processes figure prominently. Maintaining a focus on audience in such settings requires constant vigilance, as it is tempting for even Nobel Prize winners to default to a process of 'educating' publics about a risk rather than 'engaging' them in helping to devise ways to manage (allocate) the risk in specific social settings.

So what concepts matter to risk as a public issue? There are many, but here we will focus on those that figure significantly in the extent to which individuals engage in the kind of learning that encourages the construction of stable beliefs about a risk.

To some, a focus on the stability of beliefs, while ignoring the nature and valence of those beliefs, may seem nonsensical. After all, aren't most risk communication messages designed to promote particular beliefs, even particular behaviours? Risk managers tasked with convincing a public that genetically engineered soybean varieties are safe may not care how people develop beliefs, only that they arrive at the right beliefs.

In response, we return to the notion of audience-centred risk communication as a process not of understanding audiences so that they can be manipulated but of understanding audiences as principal actors in a democratic polity and as important partners in risk decision-making.

In that venue, belief stability is critical. Risk managers will rely primarily on public opinion surveys to capture public perceptions of and feelings about agricultural biotechnology. What they come to understand about public views, thus, will depend not just on their ability to ask skilfully worded questions of representative samples but also on the extent to which those questioned have developed beliefs about biotechnology that are firmly grounded: that is, beliefs that will not warp and shift with every passing biotech story in the media. In a setting where risk is a public issue, not a private one, risk managers need to value belief stability above all else, as it makes possible an understanding of how public knowledge and values are arrayed around an issue and makes feasible the process of developing policies responsive to that opinion variance.

One important precursor of stable beliefs is thoughtful, effortful information-gathering and processing. The rest of this chapter, then, focuses on factors that theorists believe influence information-seeking and processing among lay audiences in a risk setting.

## Information-seeking and Processing

Research on information-seeking and processing has a long and illustrious history in the social sciences. Primarily in service to better understanding the circumstances underlying the stability of beliefs, it was doubtless catalysed, in part, by the realization that people are 'economy-minded souls' who default to a variety of judgemental shortcuts when making decisions (Chaiken *et al.*, 1989). If, as theorists contend, the production of stable beliefs requires thoughtful and effortful information-gathering and processing (Chaiken and Eagly, 1983; Mackie, 1987), then what are we to make of a world of quick decisions rendered on the basis of one or two variables?

Scholars in this area divide seeking and processing into two main dimensions, often captured in the research literature as dual-processing models. Systematic information-seeking and processing is 'a comprehensive, analytic orientation ... in which perceivers access and scrutinize a great deal of information for its relevance to their judgement task' (Eagly and Chaiken, 1993, p. 326).

Heuristic processing, in contrast, is a cognitively truncated mode of decision-making that relies on a small number of cues. The former requires time, effort and an *a priori* sense that one can, indeed, accomplish the task while the latter occurs quickly, almost effortlessly. In heuristic mode, speed and efficiency are hallmarks; Gigerenzer identifies heuristic decision-makers as 'fast and frugal' processors (Gigerenzer and Goldstein, 1996; Gigerenzer and Selten, 2002).

Perhaps the dominant theory in this area is Eagly and Chaiken's Heuristic–Systematic Model of decision-making. Devised to better understand how individuals make accurate judgements in persuasive settings, the model argues that individuals employ both processing styles – they are not mutually exclusive – and that one can tap into real-world contingencies that lead to the dominance of one or the other. For example, systematic processing 'both demands and consumes cognitive capacity' (Eagly and Chaiken, 1993, p. 328), in contrast to heuristic processing. Thus, argue the two psychologists, when time and/or processing capacity is short, heuristic processing will dominate. On the other hand, the acknowledgement by an individual that he/she knows too little about an important arena may serve as a strong motivation to engage in more effortful, systematic processing.

The ubiquity of heuristic processing has led psychologists to unearth some of the most common heuristic cues. While these cues tend to vary by the informational situations in which a processing need occurs, a few of the more general ones include reliance on expertise, on likable individuals and something called 'the length–strength heuristic': an assumption that the more reasons a person has for a point of view, the more likely he/she is to be correct or that the more times a point of view is articulated, the more likely it is to be correct (Eagly and Chaiken, 1993, p. 334). Other judgemental heuristics commonly studied by researchers are:

**1.** The availability heuristic: estimates of the frequency of an event or the likelihood that it will occur may vary 'by the ease with which instances or associations come to mind' (Tversky and Kahneman, 1973, p. 208). Recent media accounts of agricultural biotechnology, thus, may influence risk perceptions disproportionately for the brief time that those stories are easily retrievable from memory (for a recent review, see Schwarz and Vaughn, 2002).
**2.** The representativeness heuristic: in making a judgement, individuals assess the degree of similarity between the salient features of an object and the class or category to which it might belong (Tversky and Kahneman, 1974). A risk that superficially resembles other risks, thus, may be grouped with them and assigned the same level of severity. Alternatively, a source of risk information may be judged more on the basis of his/her employer than on the attributes of his/her message. (For a recent review, see Kahneman and Frederick, 2002.)
**3.** The anchoring and adjustment heuristic: individuals may fasten initially onto information that readily comes to mind and then adjust their responses in a direction that seems appropriate (Tversky and Kahneman, 1974). Thus, initial perceptions of the risk of agricultural products produced through biotechnology may serve as longstanding anchors for future perceptions (for a recent review, see Epley and Gilovich, 2002).

Mass media scholars also have a stake in understanding information-seeking and processing strategies, particularly as they apply to individuals' use of news and public affairs information. Studies of audience interaction with news texts by McLeod and his students, for example, have isolated two dimensions much like the heuristic/systematic dimensions discussed above. In situations limited by time and capacity, explain Kosicki and McLeod (1990), individuals engage in 'selective scanning' of media messages, a process of ignoring the bulk of available messages and attending only to those of immediate interest or use.

More effortful processing of media messages, in contrast, involves attempts by audience members to make sense of a story and to embed the story information in existing interpretive frameworks. Eveland (2005) captures a variety of labels for this dimension with the phrase 'reflective forms of processing' and argues that such processing is associated with higher levels of knowledge about issues, as well as with more complex types of issue understanding.

A critical motivator of a more effortful information processing style is likely to be a perceived gap between what an individual already knows and what they believe they need to know in order to make an informed decision about something. According to the Heuristic–Systematic Model, it is this desire for sufficiency that motivates more intensive processing. The sufficiency principle, state Eagly and Chaiken, 'asserts that people will exert whatever effort is required to attain a "sufficient" degree of confidence that they have accomplished their processing goals' (Eagly and Chaiken, 1993, p. 330).

While what one does *not* know about a risk, thus, is an important determinant of information-processing mode, it is the *nature* of that missing information that may matter more than its volume. Information insufficiency is not just a measure of one's general ignorance about a risk but, rather, tries to capture a person's perceived deficit in the kinds of information that the individual deems important to his/her ability to deal reasonably with the risk. The concept, thus, calls on risk managers to care deeply about not just the size of the information gap but also the nature of information identified by a stakeholder that is required to close it. An example later in this chapter will explore the role of information (in)sufficiency in promoting more systematic information-seeking and processing.

## Risk Perception

Scientists routinely uncover what they diagnose as bad fits between the actual likelihood of coming to harm from risks and lay audiences' reactions to those risks. The scholar who has perhaps contributed most to our understanding of why those gaps occur is psychologist Paul Slovic, who, with a variety of colleagues, developed the 'psychometric paradigm' to better understand the multidimensional nature of risk perception.

The goodness of fit argument stems from a unidimensional view of risk perception: that one's reaction to a risk should be aligned with an estimate of the likelihood of coming to harm from that risk. Thus, if the health risk of ingesting bioengineered maize is estimated to be exceedingly low, then people eating maize products should be sanguine about their impacts. That they are sometimes not happy with the prospect of eating tacos made from genetically modified maize meal illustrates the kind of disjunction that led Slovic and colleagues to a series of studies that have unpacked multiple dimensions of risk perception.

The psychometric paradigm employs psychophysical scaling and multivariate analysis to create 'cognitive maps' of risk attitudes and perceptions (Slovic, 1987). Along the way, it illuminates the variety of factors that 'matter' to individual risk judgements. As Slovic puts it, his strategy asks people 'to characterize the "personality" of hazards' by rating them on selected characteristics (Slovic, 1992, p. 119). Likelihood of coming to harm is one of those characteristics, of course, but Slovic's work has shown that it is part of a suite of variables that will come into play at varying levels of intensity for different risks and different audiences.

Among the suite of factors that Slovic and colleagues uncovered are: (i) the extent to which encountering a risk is a voluntary or involuntary act; (ii) the severity of the health impacts and whether they could strike many people at once (catastrophic) or smaller numbers of people repeatedly over time (chronic); (iii) one's perceived level of knowledge about the risk; (iv) the extent to which a risk could harm future generations; (v) the possibility for harm to non-human aspects of the environment; and (vi) whether the risk appears to be something that can or cannot be controlled effectively.

Although their work led Slovic and colleagues to conclude that each risk produces a different 'personality profile' of characteristics, the related nature of many of the factors led these scholars to establish a smaller set of higher-order characteristics, most important among them being knowability and dread (Slovic, 1992). The former tracks the extent to which the individual and relevant others (scientists, for example) are perceived to know something about the risk. The latter captures a collective sense of one's perceived lack of control, a risk's catastrophic potential, the possibility that its consequences are not just harmful but fatal and a perception of the inequity of its distribution across people and geography.

As important as the cognitive – knowledge – dimension might be, reports Slovic, it is trumped routinely by the second dimension: dread. That has led Slovic and other risk perception researchers to a focus on emotional responses to risk.

## The Power of Affect

If individuals employ dual processes – heuristic and systematic – in order to make judgements, is it possible that we humans also utilize more than one system of reasoning? For a number of scholars, the answer is yes. Epstein and colleagues, for example, posit the existence of at least two conceptual systems: a rational system, which functions according to conventionally established rules of inference, and an experiential system, designed for rapid information assessment and quick action (Epstein *et al.*, 1992; Epstein, 1994). Similarly, Sloman (2002) characterizes two systems: a rule-based system, which employs abstract reasoning and symbol manipulation in a deliberative way, and an associative system, which makes inferences about the world based on personal experience and such entities as concrete concepts and stereotypes.

What makes these dual-reasoning models relevant to this portion of our chapter is that one system – Epstein's experiential and Sloman's associative – seems to dominate most reasoning situations and, further, appears to be acutely responsive to emotional appeals. Epstein, for example, found that his experiments supported the assumption that 'The experiential system is intimately associated with emotions, and the more a situation evokes emotions, the more it is processed in the mode of the experiential system' (Epstein *et al.*, 1992, p. 336).

Applying dual reasoning modes to risky situations, Slovic and colleagues posit two decision-making pathways: 'Risk as feelings refers to individuals' fast, instinctive and intuitive reactions to danger. Risk as analysis brings logic, reason and scientific deliberation to bear on risk management' (Slovic *et al.*, 2005, p. S35). The 'risk as feelings path', they argue, is susceptible to 'the affect heuristic, a process by which representations of objects and events in people's minds are tagged to varying degrees of affect' (Finucane *et al.*, 2000, p. 3).

While many social scientists over the years have acknowledged the ability of fear to serve as a powerful motivator (Witte and Allen, 2000; Dillard and Anderson, 2004), supporters of the heuristic affect refer to affect, in contrast, as 'a faint whisper of emotion' attached to concepts, images or concrete objects, which comes into play when an individual is called on to make a decision (Slovic *et al.*, 2005, p. S35).

Judgements about a risk, then, are susceptible not only to what people know about the risk but also – and occasionally primarily – to how they feel about it. If an individual has positive feelings about irradiated beef, for instance, he/she may be moved to judge the risks inherent in the product to be low and the possible benefits to be high, regardless of the nature of cognitions available to him/her; someone who associates irradiation with negative feelings, in contrast, may reverse that ratio, arguing that the risks outweigh the benefits.

Slovic and other affect researchers make the important point that, while emotional reactions often stem from what we learn about a risk, the reverse may be true. That is, an initial affective response to a risk may drive information-gathering and control the subsequent interpretation derived from that information rather than the other way around. Long posited by Moreland and Zajonc (1977, 1979), the ability of affect to occur via a mechanism that involves no prior cognitive process may underlie patterns seen in a number of risk communication studies in recent years.

For example, in one experiment where individuals encountered a news report of the presence of a (fictitious) parasite in the water supply, respondents seemed to interpret the news through the lens of an *a priori*, almost visceral, negative reaction to the parasite that became translated into a judgement that *any* level of risk posed by the organism was too high (Dunwoody *et al.*, 1992; Griffin *et al.*, 1995).

## Expectations of Self and Others

Another important group of mediators of information-seeking and processing are individuals' judgements about their own capacities to accomplish these acts, as well as their expectations of the attitudes and beliefs of others about what they (the individual) should be doing about a risk. Humans are social animals, and the very definition of risk at a practical level may be largely a social construct, dependent on time, place and the reactions of those around us. Two concepts important to this arena are self-efficacy, 'the extent to which an individual believes she can carry out the task ahead, and subjective norms, the effects on an individual of the perceived expectations of others.

Defined by Bandura as 'the conviction that one can successfully execute [a given] behaviour' (Bandura, 1977, p. 193), self-efficacy has long been recognized as an important mediator of decision-making and behaviours. When it comes to information-seeking and processing, one's perception of one's ability to find information and – if successful on that score – to actually understand the information may be critical. Even though you may be awash in scientific studies of the possible health impacts of a bioengineered product on lab animals, for instance, if you feel that you cannot make sense of what those reports say, or relate them in useful ways to your personal experience, then that perceived low self-efficacy may block you from examining information that might otherwise have been useful in closing an information gap.

Self-efficacy figures prominently in theories that attempt to predict beliefs about health. For example, in developing a model to better understand how individuals might react to fear appeals – messages that try to scare people into opting for safer behaviours in risky situations – Witte incorporated efficacy as a primary mediator of message-processing. If an individual feels that he/she can actually accomplish behaviours described as responsive to a threat, postulated Witte, he/she will act in self-protection.

On the other hand, if low self-efficacy leads him/her to believe that he/she is helpless in the face of a risk, he/she will act, instead, to control that fear rather than trying to ameliorate the danger posed by the risk. That is, he/she will engage in defensive behaviours such as downplaying the risk or denigrating the message in order to reduce the emotional reaction (Witte, 1994). The dramatic differences in these responses to a risk – one potentially lifesaving and the other possibly risk-increasing – illustrate the power of the self-efficacy concept.

Subjective norms, on the other hand, represent a person's perception of various social forces that might affect his/her intention to engage in a particular behaviour. Made salient by two powerful theories – the Theory of Reasoned Action and the Theory of Planned Behaviour – the norms attempt to measure the social pressure felt by an individual who is trying to decide what to do about a problem or an issue.

The Theory of Reasoned Action looks for causal antecedents of actions that people take that are voluntary. Initially articulated by Fishbein and Ajzen (Fishbein and Ajzen, 1975; Ajzen and Fishbein, 1980), the theory predicts that a voluntary behaviour will stem from a behavioural intention, which in turn will be determined by two sets of variables: a person's positive or negative evaluation of performing that behaviour and a perception of the extent to which relevant others believe he/she should perform the behaviour (subjective norms).

By way of illustration, determining the likelihood that someone will go to a hypnotist in order to try to stop smoking, according to the theory, will be a function of whether the individual has established a specific intention to visit the hypnotist, believes that hypnosis can, indeed, work for smokers who want to quit, and feels that friends and family believe he/she should make the appointment.

In a later formulation, Ajzen expanded the Fishbein–Ajzen model to accommodate behaviours that were less volitional. His Theory of Planned Behaviour (Ajzen, 1985, 1991) retained many of the earlier constructs but added the variable 'perceived behavioural control' – similar in many ways to the self-efficacy construct described above – to acknowledge that one's intention to perform a behaviour will be contingent on 'beliefs about the likelihood that one possesses the resources and opportunities thought necessary to execute the behaviour or attain the goal' (Eagly and Chaiken, 1993, p. 187).

In the example above, the person pondering using hypnosis to try to stop smoking may look like a foregone conclusion to the Theory of Reasoned Action, but not to the Theory of Planned Behaviour, which may tap into a belief about his/her own capacity – perhaps he/she worries about being one of those people who resists being hypnotized – that could lead the person to abandon the appointment.

Within those theories, subjective norms have emerged as powerful contributors to behavioural intention. That is, individuals seem to be heavily influenced by their perceptions of what others expect them to do. The power of those expectations may be even stronger in the realm of risk, where uncertainty tinges most of the 'facts' available to us. In an uncertain world, we may be particularly keen to develop a sense of what relevant others think to help us guide our own actions.

For example, Kahlor *et al.* {In press} found that subjective norms were strongly associated with a perceived information gap among individuals asked about their perceptions of risks to a nearby ecosystem. Since many of these individuals probably knew little about these risks, they seemed to rely, instead, on the extent to which they felt pressure from others to be concerned about the problem in order to develop their own sense of how much they needed to learn in order to make decisions about the ecosystem risks.

## Trust and Reliance

Another important mediator of information-seeking and processing is an individual's willingness to believe some information sources and channels more than others. People develop pronounced preferences for particular channels (i.e. newspapers, the Internet, individuals, television), as well as for particular sources (i.e. governmental officials, scientists, neighbours, family members). Differential reliance can be a function of many things; we take a look at three factors here: relevance, cost and recreancy.

### Relevance

As individuals become familiar with specific information channels, they develop a sense of the type and quality of information provided in them. When it is time to seek information about a particular topic, argues Chaffee (1986), individuals will seek out channels that they believe will be most likely to contain information 'relevant' to the topic at hand.

However, accessing channels likely to provide useful information may present real challenges. Take, for example, an individual who learns that his/her local dairy will no longer distinguish between products free of milk from BST-supplemented cows and those that contain milk from BST-supplemented herds. If this announcement propels that individual into information-seeking mode, he/she will encounter a series of difficult choices. His/her local mass media are unlikely to be 'covering' the topic when the issue is personally salient; mediated channels operate at scales that cannot be responsive to the time-sensitive needs of individual audience members. Friends and neighbours probably have little expertise on the topic, so the individual may be wary of relying on them. BST-related sites can be found on the World Wide Web (WWW), but their numbers will be legion and their information and advice will be contradictory, requiring time and effort to determine which sites are credible and possibly useful. Scientists would probably be regarded as highly relevant sources, but where can such information be found? Most channels available to the individual will feature information from scientists only sporadically at best.

### Cost

That 'time and attention' issue above illustrates a second contingency. Information channels vary dramatically in their cost to users. Sometimes that cost is literally in dollars; subscribing to some channels, such as prestige newspapers or trade newsletters, may cost hundreds of dollars, well beyond the budget of most Americans. In other cases, the cost is one of time and effort. Our hypothetical milk purchaser may simply not have the time to range across websites in search of a relevant subset. Or, upon learning that useful information may be available in a peer-reviewed journal article, he/she may not know how to find the publication.

Chaffee argues that channel selection, for most of us, involves an amalgam of relevance and cost considerations, fundamentally a cost *versus* benefit trade-off. Sometimes, a relevant channel will be too costly, or an easily accessible channel will contain too little useful information to be worth the effort. For example, when asked to select the ideal channel for learning about health matters, most Americans choose physicians.

But an assay of typical information-seeking patterns when we have a health issue indicates that physicians are rarely in the mix; instead, we tend to opt for mediated channels (television, WWW) or less expert interpersonal channels such as friends. Why? Querying physicians is simply too costly. Americans rarely drop in on the family doctor to pose a few questions. Although friends and family may offer advice that we suspect is not grounded in the relevant expertise, those sources are, at least, accessible. So, when we need informational help, low-cost channels will often prevail.

The cost issue may be why the public, while typically critical of mass media content, still relies fairly heavily on mediated channels for information about a range of topics. While they may judge the relevance of information in these channels to be less than ideal, they may none the less see the information as adequate for their purposes and the channels as accessible. Thus, media channels become the default, 'satisficing' choices for many people.

## Recreancy

The concept of trust is being treated in another chapter in this volume, so we will not develop it conceptually here. Instead, we introduce a societal level concept closely related to trust and developed by sociologist William Freudenburg, a concept that he calls 'recreancy'. The term, notes Freudenburg, refers to 'behaviours of persons and/or institutions that hold positions of trust, agency, responsibility or fiduciary or other forms of broadly expected obligations to the collectivity, but that behave in a manner that fails to fulfil the obligations or merit the trust' (Freudenburg, 1993, pp. 916–917).

The division of labour that characterizes modern society has made it impossible for individuals to be actively involved in the maintenance of all aspects of their own well-being, argues Freudenburg. While people may become adept at a few things, such as maintaining their computer or changing the oil in their car, most of the complex objects and processes in their life require expertise beyond what they possess. In such a world, one must rely on other members of society to do these jobs, and to do them competently. In other words, one has no choice but to trust other members of the culture to accomplish tasks important to one's daily wellbeing.

When an individual fails them – a mechanic makes a false claim that their car needs more expensive repairs than are warranted, a physician misdiagnoses a health problem or technicians monitoring a nuclear power plant fail to maintain safe operations and cause an accident – the only logical course open to individuals is to abandon trust in those persons or organizations.

Thus, individuals need to be aware not only of risks to their health from products and processes but also, according to Freudenburg, of the risk 'that socially consequential actors will fail to carry out their duties with the full degree of competence and responsibility that their fellow citizens need to expect' (Freudenburg, 1993, p. 927). This second domain introduces even more uncertainty into an already uncertain decision-making environment and can vastly complicate systematic information-seeking efforts.

## A Risk Information-seeking and Processing Model

To see a number of these variables in action, we offer some recent analyses conducted by a team of risk communication researchers of which we are a part. We gathered survey data over the course of 3 years in two metropolitan cities in the Midwest in order to explore factors important to systematic information-seeking and processing in service to understanding a variety of risks. Our model and data also will lead us to an exploration of the extent to which systematic seeking and processing can, indeed, contribute to more stable beliefs about these risks.

The risks chosen for this large study were three in number: (i) the risk of drinking contaminated water; (ii) the risk of eating contaminated fish from the Great Lakes; and (iii) the risks posed to the Great Lakes themselves from a variety of urban and agricultural processes. Our sample of respondents came from two cities bordering the Great Lakes: Milwaukee, Wisconsin, which sits on the shoreline of Lake Michigan, and Cleveland, Ohio, which is adjacent to Lake Erie. Both cities draw their drinking water from their respective lakes, and sport fishing is a major avocation in both places.

Figure 11.1 shows our model of risk information-seeking and processing. While other articles (see, for example, Griffin *et al.*, 1999) elaborate on aspects of the model, we will point out only a few of the key features relevant to this chapter.

The basic premise of the model is that information insufficiency – the gap between what people know and what they feel they need to know in order to make a decision or otherwise deal with a risk – will drive information-seeking and processing. That is, the larger the gap, the more likely an individual will invest in effortful seeking and analysis of additional information. Furthermore, the primary predictors of gap size will be a person's affective response to a risk (worry or anger, for example) and what the model terms 'informational subjective norms.' Put another way, individuals will judge their information insufficiency to be higher or lower by the level of worry they feel regarding the risk and relative to their sense that others think they should care about the risk and keep informed about it.

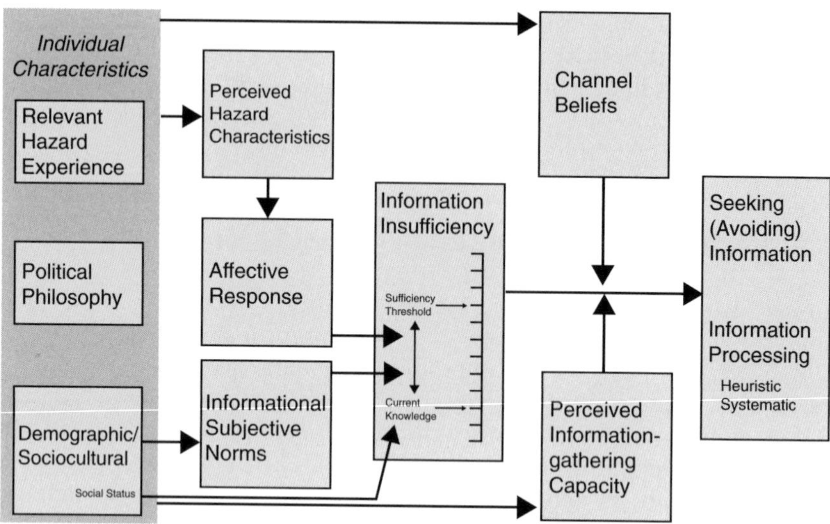


**Fig. 11.1.** Model of risk information-seeking and processing (adapted from Griffin *et al.*, 1999).

Finally, we expect that both self-efficacy (identified in the model as 'perceived information gathering capacity') and a person's beliefs about available information channels will mediate the relationship between information (in)sufficiency and information-seeking and processing. These factors can be seen as an extension of Chaffee's (1986) proposition that individuals perform an informal, optimizing analysis of information channels that then drives their seeking strategies.

For individuals who have a higher capacity to access and understand risk information, the seeking and processing of new information from various channels should be easier (less demanding in terms of psychological and physical 'costs'). Such capacity might result, for example, from formal education, economic resources or an existing, systematic knowledge of the risk.

The benefits – or lack thereof – that various channels might provide in terms of useful information and its presentation are probably central among the beliefs that individuals hold about the channels (channel beliefs). Following on from Ajzen's (1991) idea of behavioural beliefs (i.e. beliefs about the likelihood that a specific behaviour would produce a given outcome), channel beliefs could be seen as including the individual's perception that the behaviour of using a given channel for risk information might result in certain outcomes, some desirable to the individual and some not. The person seeking an optimal result presumably weighs these considerations, and might even develop a habitual use or dependence on channels that he or she has found to maximize benefits while minimizing access costs and information-processing costs.

This sort of cost–benefit approach, of course, depicts the consumers of risk information as cognitive misers. Given a certain capacity to seek and process information and a certain motivation to do so, a person will seek good enough channels to get good enough information to provide her/him with knowledge that, itself, will be good enough for dealing with the risk. While this rational-person assumption can be seductive, it ignores the jokers in the deck: the non-rational factors that can affect human behaviour, including communication behaviour. Thus, the model includes affective factors (for example, worry, anger or even positive emotions such as happiness or hope) and perceived social normative factors (informational subjective norms) that could influence information-seeking, avoidance and processing, directly or indirectly.

We have explored various pieces of this model (e.g. Kahlor *et al.*, 2003; Griffin *et al.*, 2004a, 2005) and, in a recent paper, tested the robustness of several components of the model (Griffin *et al.*, 2004b), concentrating on risk information-seeking and processing and their more immediate predictors: information insufficiency, capacity and channel beliefs. We will reflect briefly on some of the findings from that most recent paper.

First, a brief methodological note: the data reported in the paper reflect questions posed to 1123 adult residents of Milwaukee and Cleveland in the first wave of the 3-year study. Respondents were chosen via random digit dialling, and the combined response rate was 55.2% (61.3% in Milwaukee and 50% in Cleveland). Respondents who reported either eating Great Lakes fish or avoiding the fish for health reasons were asked questions about the risk of eating fish, which can contain a variety of chemicals, particularly polychlorinated biphenyls, or PCBs, a suspected cancer risk and associated with developmental problems in infants.

The rest of the sample was randomly assigned either questions about the risk of drinking tap water from the lake. (In 1993, the largest recorded outbreak of waterborne disease in the nation's history occurred when the parasite *Cryptosporidium* got into Milwaukee's drinking water.) or questions about risks to the Great Lakes themselves, an ecosystem that has been battered over the decades by industrial emissions and run-off from cities and farms.

We measured the information gap by first asking respondents to assess their current knowledge of the particular risk on a scale of 1 (know nothing) to 100 (knowing everything possible) and then, using the same scale, to estimate how much knowledge they would need to 'deal adequately' with the risk at hand. To determine the gap, we regressed the latter on the former.

Measures of risk information-seeking and avoidance and of heuristic and systematic seeking and processing were chosen from a slightly larger set of items via factor analysis and are displayed in Tables 11.1 and 11.2.

What did we find? Table 11.3 presents results of regression analyses for the three risks (termed fish, tap water and lake, respectively). There are several interesting patterns, but we'll comment on just a few:

1. Those who hold the channel belief that the media provide them with cues to the validity of messages are more likely to process the risk information they encounter systematically. Presumably, audiences see a benefit to these cues. Surprisingly, however, believing that the media provide distorted (e.g. biased or sensationalized) information has almost no relationship to seeking or processing information across these risks.
2. The measure of perceived information-gathering capacity did not perform well in this data set. However, a much improved measure used in a different data set reported in the same analysis (Griffin *et al.*, 2004b) did relate positively to the seeking of information about two natural environmental risks.
3. Most importantly, we saw robust patterns between a person's perceived information gap (*insufficiency*) and his information-seeking

**Table 11.1.** Factor analyses of information-seeking and -avoidance items by risk topic (Great Lakes Fish Advisory and Risk Communication Study).[a]

| | Fish topic factors (*n*, 634) | | Tap water topic factors (*n*, 252) | | Lake environment topic factors (*n*, 237) | |
|---|---|---|---|---|---|---|
| Item | Avoidance | Seeking | Avoidance | Seeking | Avoidance | Seeking |
| AV1[b] | **0.633** | −0.357 | **0.733** | −0.422 | **0.528** | −0.356 |
| AV2[c] | **0.629** | −0.173 | **0.522** | −0.170 | **0.667** | −0.136 |
| AV3[d] | **0.661** | −0.418 | **0.662** | −0.422 | **0.523** | −0.321 |
| SK1[e] | −0.273 | **0.770** | −0.238 | **0.759** | −0.204 | **0.642** |
| SK2[f] | −0.455 | **0.693** | −0.481 | **0.639** | −0.456 | **0.755** |
| *r* between factors | −0.43 | | −0.41 | | −0.40 | |
| Initial Eigenvalues | 2.340 | 1.063 | 2.292 | 1.064 | 2.115 | 1.078 |
| Rotation sums of squared loadings | 1.516 | 1.405 | 1.536 | 1.269 | 1.247 | 1.231 |
| Omega | 0.67 | 0.69 | 0.68 | 0.64 | 0.59 | 0.64 |

[a] Figures in bold are items with high enough factor loadings to be included in the scale.
[b] When [this topic] comes up, I'm likely to tune it out.
[c] Whenever [this topic] comes up, I go out of my way to avoid learning more about it.
[d] Gathering a lot of information on [this topic] is a waste of time.
[e] When it comes to [this topic], I'm likely to go out of my way to get more information.
[f] When [this topic] comes up, I try to learn more about it.

**Table 11.2.** Factor analyses of information-processing items by risk topic (Great Lakes Fish Advisory and Risk Communication Study).[a]

| | Fish topic factors (*n*, 634) | | Tap water topic factors (*n*, 252) | | Lake environment topic factors (*n*, 237) | |
|---|---|---|---|---|---|---|
| Item | Heuristic | Systematic | Heuristic | Systematic | Heuristic[b] | Systematic |
| HP1[c] | **0.567** | −0.324 | −0.320 | **0.555** | −0.418 | **0.343** |
| HP2[d] | **0.634** | −0.226 | −0.418 | **0.422** | −0.261 | **0.119** |
| HP3[e] | **0.500** | −0.154 | −0.159 | **0.466** | −0.174 | **0.469** |
| HP4[f] | **0.479** | −0.215 | −0.240 | **0.369** | −0.167 | **0.358** |
| SP1[g] | −0.341 | **0.677** | **0.734** | −0.275 | **0.556** | −0.428 |
| SP2[h] | −0.331 | **0.615** | **0.464** | −0.281 | **0.575** | −0.427 |
| SP3[i] | −0.072 | **0.386** | **0.355** | −0.210 | **0.671** | 0.092 |
| SP4[j] | −0.301 | **0.351** | **0.319** | −0.341 | **0.446** | −0.207 |
| *r* between factors | −0.40 | | −0.50 | | −0.35 | |
| Initial Eigenvalues | 2.468 | 1.269 | 2.243 | 1.117 | 2.341 | 1.175 |
| Rotation sums of squared loadings | 1.524 | 1.335 | 1.341 | 1.155 | 1.590 | 0.897 |
| Omega | 0.63 | 0.59 | 0.53 | 0.52 | 0.64 | 0.46 |

[a] Figures in bold are items with high enough factor loadings to be included in the scale.
[b] Third factor had been extracted, but was forced into the second.
[c] When I see or hear information about [this topic], I rarely spend much time thinking about it.
[d] There is far more information on [this topic] than I personally need.
[e] When I encounter information about [this topic], I focus on only a few key points.
[f] If I need to act on this matter, the advice of one expert is enough for me.
[g] After I encounter information on [this topic], I am likely to stop and think about it.
[h] If I need to act on this matter, the more viewpoints I get the better.
[i] After thinking about [this topic], I have a broader understanding.
[j] When I encounter information about [this topic], I read or listen to most of it, even though I may not agree with its perspective.

and processing behaviours. Specifically, across all three risks, the bigger the gap the more likely the individual reported engaging in more effortful seeking and processing. And the larger the gap, the less likely the person avoided paying attention to the risk information or embraced more heuristic strategies. With respect to these risks, at least, individuals who perceived a clear need to learn more were likely to invest more effort and time in doing so.

Are these patterns associated with the stability of risk beliefs? Initial evidence is promising. Our three-wave, two-city survey also gathered information, consistent with measures of the Theory of Planned Behaviour, on individuals' behavioural beliefs about the three risks, as well as their evaluations of those beliefs as good or bad. Analysis of data from the second wave found that deeper, more

**Table 11.3.** Regression of information-seeking, avoidance and processing on selected demographic/sociocultural variables and immediate predictors from model of risk information-seeking and processing (Standardized Regression Coefficients (betas)).

| | Information-seeking Topic | | | Information avoidance Topic | | | Systematic processing Topic | | | Heuristic processing Topic | | |
|---|---|---|---|---|---|---|---|---|---|---|---|---|
| | Fish | Tapwater | Lake | Fish | Tapwater | Lake | Fish | Tapwater | Lake | Fish | Tapwater | Lake |
| Demographic | | | | | | | | | | | | |
| Female gender | 0.06 | 0.10 | 0.13[a] | −0.08[a] | −0.04 | −0.15[b] | 0.05 | 0.03 | 0.10 | −0.08[a] | −0.02 | −0.19[b] |
| Age | 0.05 | −0.04 | 0.18[b] | 0.17[c] | 0.15[b] | 0.01 | −0.11[b] | −0.07 | 0.02 | 0.11[b] | 0.11 | 0.08 |
| Education | −0.03 | 0.10 | 0.02 | –0.16[c] | −0.18[b] | −0.24[c] | 0.08[b] | 0.12 | 0.17[b] | −0.14[c] | −0.18[b] | −0.22[c] |
| Income | −0.06 | −0.19[b] | 0.07 | 0.01 | 0.07 | −0.09 | −0.07 | −0.07 | 0.13[c] | 0.05 | 0.08 | −0.11 |
| Minority | 0.09[a] | 0.07 | 0.19[b] | 0.06 | 0.00 | −0.11 | 0.06 | 0.09 | 0.09 | 0.07[a] | −0.06 | −0.03 |
| $\Delta r^2$ | 0.03[b] | 0.07[b] | 0.08[c] | −0.06[c] | 0.08[c] | 0.12[c] | 0.04[c] | 0.04 | 0.09[c] | 0.05[c] | 0.06[b] | 0.13[c] |
| Channel beliefs: Media… | | | | | | | | | | | | |
| Media…distort | −0.05 | −0.01 | −0.04 | 0.05 | 0.10 | −0.02 | −0.03 | −0.03 | 0.04 | 0.09[a] | 0.09 | −0.06 |
| …Have validity cues | 0.01 | 0.15[b] | 0.15[a] | −0.09 | −0.24[c] | −0.08 | 0.17[c] | 0.20[c] | 0.15[b] | 0.00 | −0.09 | −0.05 |
| $\Delta r^2$ | 0.01 | 0.03[a] | 0.04[b] | 0.02[b] | 0.07[c] | 0.02 | 0.04[c] | 0.04[b] | 0.04[b] | 0.01 | 0.02 | 0.01 |
| Capacity | −0.15[c] | −0.22[c] | 0.01 | 0.17[c] | 0.07 | 0.06 | −0.19[c] | 0.02 | 0.02 | 0.19[c] | 0.02 | −0.04 |
| $\Delta r^2$ | 0.03[c] | 0.05[c] | 0.00 | 0.03[c] | 0.01 | 0.00 | 0.04[c] | 0.00 | 0.00 | 0.04[c] | 0.00 | 0.00 |
| Information sufficiency | | | | | | | | | | | | |
| Current knowledge | 0.14[c] | 0.12 | 0.28[c] | −0.14[c] | −0.04 | −0.20[c] | 0.13[c] | 0.07 | 0.16[b] | −0.11[b] | −0.13 | −0.14[c] |
| $\Delta r^2$ | −0.05[c] | 0.05c | 0.11[c] | 0.04[c] | 0.03[b] | 0.07[c] | 0.04[c] | 0.04[c] | 0.06[c] | 0.04[c] | 0.07[c] | 0.04[c] |
| Threshold (Insufficiency) | 0.33[c] | 0.25[c] | 0.22[c] | −0.26[c] | −0.33[c] | −0.24[c] | 0.24[c] | 0.34[c] | 0.28[c] | −0.31[c] | −0.34[c] | −0.24[c] |
| $\Delta r^2$ | 0.09[c] | 0.05[c] | 0.04[c] | 0.06[c] | 0.09[c] | 0.05[c] | 0.05[c] | 0.09[c] | 0.07[c] | 0.08[c] | 0.09[c] | 0.05[c] |
| Multiple $R$ | 0.46[c] | 0.50[c] | 0.53[c] | 0.46[c] | 0.53[c] | 0.51[c] | 0.44[c] | 0.46[c] | 0.50[c] | 0.47[c] | 0.49[c] | 0.48[c] |
| Adjusted $R^2$ | 0.19 | 0.22 | 0.25 | 0.20 | 0.25 | 0.23 | 0.18 | 0.18 | 0.21 | 0.21 | 0.21 | 0.20 |
| $n$ | 634 | 252 | 237 | 634 | 252 | 237 | 634 | 252 | 237 | 634 | 252 | 237 |

[a] $P \leq 0.05$; [b] $P \leq 0.01$; [c] $P \leq 0.001$.

systematic information-processing was indeed positively related to the number of strongly held behavioural beliefs actively considered by respondents, as well as the strength of respondents' evaluations of those beliefs (Griffin *et al.*, 2002).

## In Conclusion

What will it take to turn individuals into systematic processors of information about agricultural biotechnology? The risk communication literature examined in this chapter suggests that among the important factors are:

- An affective response to the topic. For some, this may be a sense of worry; for others, it may be a more positive affective response, perhaps about the benefit conferred on a product via biotechnology. Regardless, it will probably serve as a lens through which reading and learning about the topic will be filtered.
- A clear sense that respected others feel you should invest time in learning about the topic or risk.
- A perception on the part of the individual that he/she has the capacity to find and evaluate relevant and useful information.
- The perception that, although he/she may already know something about the topic, a person has much to learn before he/she can make a reasonable decision about whether to incorporate a bioengineered food into their life.

Once an individual reaches a point where he/she is ready to engage in more effortful processing of information, however, questions of where and how he/she can find useful information that meets his/her needs loom large. Society has been slow to construct an information environment that can provide possibly relevant information to individuals on an as-needed basis. The World Wide Web constitutes an important first step, but it will be years before the typical American can navigate its myriad paths carefully and critically.

Still, we need a world of systematic information processors and seekers who can build stable beliefs – of all shades and shapes – about agricultural biotechnology. Although few food products currently stem from biotech crops (Pollack, 2006), the next generation of products is on the horizon. As scientists, industry and policymakers work to place bioengineered products in our environment, they need to be able to interact with publics whose understandings of agricultural biotechnology are grounded in substantial information bases, understandings that can change but that will withstand the daily, almost haphazard variance in messages that pelt Americans in their daily life.

It is unlikely that those beliefs will align to form consensuses around policies and products. But democracies are not built with consensus in mind. The important goal, instead, is to develop socially acceptable strategies for managing technologies and allocating risk within these multivariate belief climates.

## References

Ajzen, I. (1985) From intentions to actions: a theory of planned behaviour. In: Kuhl, J. and Beckmann, J. (eds) *Action Control: from Cognition to Behaviour.* Springer-Verlag, New York, pp. 11–39.

Ajzen, I. (1991) The theory of planned behaviour. *Organizational Behaviour and Human Decision Processes* 50, 179–211.

Ajzen, I. and Fishbein, M. (1980) *Understanding Attitudes and Predicting Social Behavior.* Prentice-Hall, Englewood Cliffs, New Jersey.

Bandura, A. (1977) Self-efficacy: toward a unifying theory of behavioural change. *Psychological Review* 84, 191–215.

Chaffee, S.H. (1986) Mass media and interpersonal channels: competitive, convergent or complementary? In: Gumpert, G. and Cathcart, R. (eds) *Inter-Media,* 3rd edn. Oxford University Press, New York, pp. 62–80.

Chaiken, S. and Eagly, A.H. (1983) Communication modality as a determinant of persuasion: the role of communicator salience. *Journal of Personality and Social Psychology* 34, 605–614.

Chaiken, S., Liberman, A. and Eagly, A.H. (1989) Heuristic and systematic processing within and beyond the persuasion context. In: Uleman, J.S. and Bargh, J.A. (eds) *Unintended Thought.* Guilford Press, New York, pp. 212–252.

Dillard, J.P. and Anderson, J.W. (2004) The role of fear in persuasion. *Psychology and Marketing* 21 (11), 909–926.

Dunwoody, S., Neuwirth, K., Griffin, R.J. and Long, M. (1992) The impact of risk message content and construction on comments about risks embedded in 'letters to friends'. *Journal of Language and Social Psychology* 11 (1/2), 9–33.

Eagly, A.H. and Chaiken, S. (1993) *The Psychology of Attitudes.* Harcourt Brace Jovanovich, New York.

Epley, N. and Gilovich, T. (2002) Putting adjustment back in the anchoring and adjustment heuristic. In: Gilovich, T., Griffin, D. and Kahneman, D. (eds) *Heuristics and Biases: the Psychology of Intuitive Judgement.* Cambridge University Press, Cambridge, Massachusetts, pp. 139–149.

Epstein, S. (1994) Integration of the cognitive and the psychodynamic unconscious. *American Psychologist* 49 (8), 709–724.

Epstein, S., Lipson, A., Holstein, C. and Huh, E. (1992) Irrational reactions to negative outcomes: evidence for two conceptual systems. *Journal of Personality and Social Psychology* 62 (2), 328–339.

Eveland, W.P. Jr. (2005) Information processing strategies in mass communication research. In: Dunwoody, S., Becker, L.B., McLeod, D.M. and Kosicki, G.M. (eds) *The Evolution of Key Mass Communication Concepts.* Hampton Press, Cresskill, New Jersey, pp. 217–250.

Finucane, M.L., Alhakami, A., Slovic, P. and Johnson, S.M. (2000) The affect heuristic in judgements of risks and benefits. *Journal of Behavioural Decision Making* 13, 1–17.

Fishbein, M. and Ajzen, I. (1975) *Belief, Attitude, Intention, and Behaviour: an Introduction to Theory and Research.* Addison-Wesley, Reading, Massachusetts.

Freudenburg, W.R. (1993) Risk and recreancy: Weber, the division of labor, and the rationality of risk perceptions. *Social Forces* 71 (4), 909–932.

Gigerenzer, G. and Goldstein, D.G. (1996) Reasoning the fast and frugal way: Models of bounded rationality. *Psychological Review* 103 (4), 650–669.

Gigerenzer, G. and Selten, R. (eds) (2002) *Bounded Rationality: the Adaptive Toolbox.* The MIT Press, Cambridge, Massachusetts.

Griffin, R.J., Dunwoody, S. and Neuwirth, K. (1999) Proposed model of the relationship of risk information-seeking and processing to the development of preventive behaviours. *Environmental Research,* Section A 80, S230–S245.

Griffin, R.J., Neuwirth, K. and Dunwoody, S. (1995) Using the Theory of Reasoned Action to examine the impact of health risk messages. In: Burleson, B.R. (ed.) *Communication Yearbook 18.* Sage, Thousand Oaks, California, pp. 201–228.

Griffin, R.J., Neuwirth, K., Giese, J. and Dunwoody, S. (2002) Linking the heuristic–systematic model and depth of processing. *Communication Research* 29, 705–732.

Griffin, R.J., Neuwirth, K., Dunwoody, S. and Giese, J. (2004a) Information sufficiency and risk communication. *Media Psychology* 6, 23–61.

Griffin, R.J., Powell, M., Dunwoody, S., Neuwirth, K., Clark, D. and Novotny, V. (2004b) Testing the robustness of a risk information processing model. Paper presented at the *Association for Education in Journalism and Mass Communication Annual Conference,* Toronto, Canada.

Griffin, R.J., Yang, Z., Boerner, F., Bourassa, S., Darrah, T., Knurek, S., Ortiz, S. and Dunwoody, S. (2005) Applying an information-seeking and processing model to a study of communication about energy. Paper presented at the *Association for Education in Journalism and Mass Communication Annual Conference,* San Antonio, Texas.

Kahlor, L., Dunwoody, S., Griffin, R.J., Neuwirth, K. and Giese, J. (2003) Studying heuristic–systematic processing of risk communication. *Risk Analysis* 23, 355–368.

Kahlor, L., Dunwoody, S., Griffin, R.J. and Neuwirth, K. {In press} Seeking and processing information about impersonal risk. *Science Communication.*

Kahneman, D. and Frederick, S. (2002) Representativeness revisited: attribute substitution in intuitive judgement. In: Gilovich, T., Griffin, D. and Kahneman, D. (eds) *Heuristics and Biases: the Psychology of Intuitive Judgement.* Cambridge University Press, Cambridge, Massachusetts, pp. 49–81.

Kosicki, G.M. and McLeod, J.M. (1990) Learning from political news: effects of media images and information processing strategies. In: Kraus, S. (ed.) *Mass Communication and Political Information Processing.* Erlbaum, Hillsdale, New Jersey, pp. 69–83.

Mackie, D.M. (1987) Systematic and nonsystematic processing of majority and minority persuasive communications. *Journal of Personality and Social Psychology* 53, 41–52.

Moreland, E.L. and Zajonc, R.B. (1977) Is stimulus recognition a necessary condition for the occurrence of exposure effects? *Journal of Personality and Social Psychology* 35, 191–199.

Moreland, E.L. and Zajonc, R.B. (1979) Exposure effects may not depend on stimulus recognition. *Journal of Personality and Social Psychology* 37, 1085–1089.

National Research Council (1989) *Improving Risk Communication.* National Academy Press, Washington, DC.

Pollack, A. (2006) Biotech's sparse harvest. *The New York Times,* 14 February, pp. C1 and C7.

Rich, R.C., Griffin, R.J. and Friedman, S.M. (1999) Introduction: the challenge of risk communication in a democratic society. *Risk: Health, Safety and Environment* 10 (3), 189–196.

Schwarz, N. and Vaughn, L.A. (2002) The availability heuristic revisited: ease of recall and content of recall as distinct sources of information. In: Gilovich, T., Griffin, D. and Kahneman, D. (eds) *Heuristics and Biases: the Psychology of Intuitive Judgement.* Cambridge University Press, Cambridge, Massachusetts, pp. 103–119.

Sloman, S.A. (2002) Two systems of reasoning. In: Gilovich, T., Griffin, D. and Kahneman, D. (eds) *Heuristics and Biases: the Psychology of Intuitive Judgement.* Cambridge University Press, Cambridge, Massachusetts, pp. 379–396.

Slovic, P. (1987) Perception of risk. *Science* 236, 280–285.

Slovic, P. (1992) Perception of risk: reflections on the psychometric paradigm. In: Krimsky, S. and Golding, D. (eds) *Social Theories of Risk.* Praeger, Westport, Connecticut, pp. 117–152.

Slovic, P., Peters, E., Finucane, M.L. and MacGregor, D.G. (2005) Affect, risk, and decision making. *Health Psychology* 24 (4), S35–S40.

Tversky, A. and Kahneman, D. (1973) Availability: a heuristic for judging frequency and probability. *Cognitive Psychology* 5, 207–232.

Tversky, A. and Kahneman, D. (1974) Judgement under uncertainty: heuristics and biases. *Science* 185, 1124–1131.

Witte, K. (1994) Fear control and danger control: a test of the Extended Parallel Process Model. *Communication Monographs* 61 (2), 113–134.

Witte, K. and Allen, M. (2000) A meta-analysis of fear appeals: implications for effective public health campaigns. *Health Education and Behavior* 27 (5), 591–615.

# III Communicating about Agricultural Biotechnology: Practical Experiences in International Settings

12

# The GEO-PIE Project: Case study of Web-based Outreach at Cornell University, USA

T.C. Nesbitt[1]

*US Department of Agriculture, Animal and Plant Health Inspection Service, Biotechnology Regulatory Services, Riverdale, Maryland, USA; e-mail: thomas.c.nesbitt@aphis.usda.gov*

## Introduction

In 2000, genetically engineered crops had been grown in the USA for several years, and there was an accumulating body of evidence demonstrating that public opinion of the technology was mixed, at best (Shanahan *et al.*, 2001). Early results of survey research published in 2001 revealed that opinion of biotechnology among Cornell University faculty and extension staff was deeply divided (Wilkins *et al.*, 2001). Concurrently, there were a growing number of county- and state-level initiatives, particularly in New York and Vermont, for moratoria on outdoor plantings of genetically engineered crops and mandatory labelling of genetically engineered foods.

In direct response to the perceived needs to address public perception of agricultural biotechnology, and to support the informational needs of Cornell faculty and extension staff, researchers at Cornell University acquired funding to create the Genetically Engineered Organisms Public Issues Education (GEO-PIE) Project. Due in part to the nature of the project's funding, the primary audience of the project was Cornell Cooperative Extension (CCE) educators in the areas of agriculture and nutrition, and their local audiences.

In addition to summarizing the activities of the GEO-PIE Project, this chapter reviews several methods used to develop outreach strategy and to evaluate the impacts of the project's efforts. First, a

[1] Statements expressed in this chapter are the views of the author and should not be construed as representing the views of the federal government.

survey of New York State extension educators provided valuable data about the knowledge, opinions and information needs of the project's target audience. Second, analysis of automated logs of GEO-PIE website user activity gives an overview of user volume, frequency of visits and user behaviour while on the website. Finally, a 1-year web-based survey collected data on website user demographics, perception of website bias and opinion and knowledge about agricultural biotechnology.

## The GEO-PIE Project

The GEO-PIE Project was funded by two consecutive grants (2000–2002, 2002–2005) from the Smith-Lever Fund, a federal granting programme primarily supporting agricultural extension. The grants supported a single extension educator (the author), plus a limited budget for printing of educational materials and outreach-related travel. The Cornell faculty co-directors of the project were James Shanahan, Associate Professor in the Department of Communication and Jennifer Wilkins, Senior Extension Associate in the Division of Nutritional Sciences. A small group of faculty advisors and collaborators from several other departments at Cornell University and Cornell Cooperative Extension acted as an advisory committee to the project.

The Project was based upon a 'public issues education' (PIE) philosophy – that is, working to improve the public's understanding of all facets of an issue of importance to the public, without necessarily advocating any particular viewpoint (Dale and Hahn, 1994). The PIE approach is usually associated with extension-related outreach, and is advocated by many as a means of improving the relevance and utility of extension to a broader array of citizens (ECOP, 1992; Dale and Hahn, 1994; Patton and Paine, 2001). The goal of public issues education is to help the public understand the complexities of an issue in order to enhance critical thinking skills and to make well-informed decisions. To that end, the Project sought opportunities to work directly with county-level extension educators to strengthen public issues education capacity, so that they might better meet the informational needs of their local communities.

In a 2001 survey of New York State extension educators conducted by GEO-PIE (discussed in detail below), respondents strongly encouraged that outreach efforts remain neutral on the subject of agricultural biotechnology, and respondents' comments suggested that previous Cornell-associated outreach efforts were perceived as somewhat partisan (pro-biotech.). Further, the Wilkins *et al.* (2001) study showed that Cornell faculty and extension staff were nearly evenly split between technology proponents and those expressing some cautionary scepticism. Based upon these findings, GEO-PIE was developed with a conscious effort to be a neutral, non-partisan purveyor of up-to-date information on agricultural biotechnology.

Public opinion research showed that support for agricultural biotechnology was mixed, that knowledge and understanding of the technology were low and that there seemed to be strong support for consumer labelling of products derived from genetically engineered crops (Hallman, 1996; Hoban, 1998; Shanahan *et al.*, 2001). In general, literature opposing use of the technology stressed its unknown risks and unavoidable presence in foods (see, for example, Ticciati and Ticciati, 1998), often exaggerating its prevalence considerably.

On the other hand, promotional materials from industry and academia tended to focus on technical, scientific minutiae and, arguably, to over-generalize the technology's benefits (see, for example, SDCMA, 2000). Few resources had been developed that provided direct answers to questions of interest to the public. What exactly is out there? What exactly is known or unknown about the risks and benefits? With this background in mind, we deliberately chose to personalize the Project's message to target the citizen-consumer by posing the provocative, rhetorical question: Am I eating genetically engineered foods?

During its first 4 years, the GEO-PIE Project focused primarily on the development of a large, comprehensive website on the subject of agricultural biotechnology (http://www.geo-pie.cornell.edu). Aware of the public's expressed desire for consumer labelling, we organized the contents of the website to emphasize the prevalence of genetically engineered crops in the US food supply from a consumer's viewpoint. Sections were devoted to each crop for which genetically engineered varieties had been developed and commercialized in the USA. Each included discussion of technical details of the traits that had been engineered into them, some of the risks and benefits associated with the traits and the prevalence of the engineered varieties in the US agriculture and the marketplace.

Additional sections of the website were devoted to more general discussions of risks and benefits, US regulation of the technology and social and ethical issues. The site also included detailed discussion of some of the more familiar popularized stories and urban legends associated with genetic engineering, such as StarLink maize, monarch butterflies, fish-gene strawberries, the Percy Schmeiser drama and many more.

Responding to the preferences expressed by extension educators (see survey discussion below), we also developed a series of 11 fact sheets that condensed much of the content of the website for easy distribution (see Table 12.1).

Following the website's organization, the fact sheets individually described crops that had been genetically engineered and commercialized in the USA, the traits engineered into them and their distribution in agriculture and in the US food supply. Additional fact sheets focused on health-related issues, potential environmental impacts and the US regulatory system. We sent complete sets of the fact sheets to educators in all CCE offices and to 80 science and agriculture reporters at local newspapers throughout New York State. Printable PDF versions of the fact sheets were also posted on the GEO-PIE website, which were downloaded over 16,000 times during the project's first 4 years. The most downloaded fact sheets were one presenting a general overview of genetically engineered crops, and two fact sheets providing specific information about GE maize and GE tomatoes.

In collaboration with the New York State Experiment Station, the GEO-PIE Project also developed an educational brochure entitled *Genetically Engineered Foods 2001: a Consumer's Guide to What's in Store*, which was distributed to customers of the Wegmans Food Markets, a grocery chain in north-eastern USA.

Additionally, the GEO-PIE extension educator gave nearly 60 public lectures on agricultural biotechnology to a wide variety of local audiences around New York State (see Box 12.1). Most public talks were organized either by CCE educators or by local groups hosting seminars or workshops about biotechnology. The GEO-PIE Project

**Table 12.1.** GEO-PIE Project fact sheets[a] (with number of times downloaded, May 2003–August 2004).

| Fact sheet | Number downloaded |
|---|---|
| Genetically Engineered Foods in the Marketplace | 1722 |
| Genetically Engineered Foods: Maize | 1121 |
| Genetically Engineered Foods: StarLink Maize in Taco Shells | 644 |
| Genetically Engineered Foods: Soybeans | 848 |
| Genetically Engineered Foods: Canola | 533 |
| Genetically Engineered Foods: Cotton | 599 |
| Genetically Engineered Foods: Tomatoes | 1073 |
| Genetically Engineered Foods: Resistance to Plant Viruses | 882 |
| Genetically Engineered Foods: US Safety Regulation of Genetically Engineered Crops | 673 |
| Genetically Engineered Foods: Food Safety of Genetically Engineered Crops | 941 |
| Genetically Engineered Foods: Environmental Safety of Genetically Engineered Crops | 870 |

[a] Available online at: http://www.geo-pie.cornell.edu/educators/educators.html

**Box 12.1.** Sample of talks given by GEO-PIE extension educator (on the subject of genetically engineered organisms in US agriculture).

Agribusiness Economic Outlook Conference (Ithaca, New York).
Albany Area Extension Educator Workshop, Saratoga County Cooperative Extension (Clifton Park, New York)
Association of Cornell Cooperative Extension Educators Conferences (Ithaca, New York; Kerhonkson, New York).
Bethlehem Public Library, Albany County Cooperative Extension (Albany, New York).
Biotechnology Seminar: Is This 'Frankenfood'? Steuben County Cooperative Extension (Corning, New York).
Cornell Alumni Association Meetings (Charlotte, North Carolina; Greesnboro, North Carolina; Greenville, South Carolina; Ithaca, New York; Jacksonville, Florida; Leesburg, Florida; Minneapolis, Minnesota; Milwaukee, Wisconsin; New Windsor, New York; Omaha, Nebraska; Westbury, New York).
Cornell Breast Cancer and Environmental Risk Factors Workshop (Ithaca, New York).
Cornell Seed Conference, 62nd and 63rd Annual (Geneva, New York).
Day of Discovery, Southern Tier Biotechnology Conference, Owego County Cooperative Extension (Owego, New York).
Farmer to Farmer: Sharing Experiences Related to Agricultural Biotechnology (Manila, Philippines).
Hobart and William Smith Colleges (Geneva, New York).
LEAD-NY: Cornell Biotechnology Day (Ithaca, New York).
Long Beach Island Garden Club (Long Beach Island, New Jersey).
News from the Fronts, Genesee Dietetic Association (Greece, New York).
Nutritional Concerns Conference, 24th Annual (Albany, New York).
Organic Turfgrass Management Conference (Melville, New York).
Rochester Dietetic Association, Monroe County Cooperative Extension (Rochester, New York).
Shenendehowa Public Library, Saratoga County Cooperative Extension (Clifton Park, New York).
Sullivan West Central School (Jeffersonville, New York).
Unitarian Church of Ithaca (Ithaca, New York).
Vassar Brothers Institute (Poughkeepsie, New York).

advisory panel also referred many speaker requests to the project's extension educator, as word of the project began to spread. Talks also included guest lectures in Cornell University undergraduate and graduate courses in plant breeding and other agriculture-related fields, nutrition and communication.

## Survey of New York State extension educators

At the beginning of the GEO-PIE Project, in late 2000 and early 2001, we conducted a survey of New York State extension educators in order to assess their knowledge, opinions and informational needs related to agricultural biotechnology. A list of 127 survey participants was developed from directories of Cornell Cooperative Extension (CCE) educators, who were selected to include at least one nutrition educator and one agricultural educator from each of the 58 New York State counties. An invitation letter and printed survey were mailed to the members of this list. Participants were invited to respond either by returning the completed survey via postal mail or by completing an identical survey posted on the GEO-PIE website. The invitation to participate in the survey was also distributed through local extension educator 'listserves' and newsletters. The original survey instrument, with a tabulation of individual responses, is included in Appendix I of this chapter.

Seventy-four educators responded to the survey, most (82%) using the online version; 61% of respondents were women, 68% held an advanced degree (MS or higher) and the majority were educators in the fields of agriculture (45%) and nutrition (20%). Thirty-three respondents represented 25 individual New York State counties, 11 identified themselves as regional or multi-county educators, six were Cornell University campus-based educators and 24 did not identify their location.

The survey provided important information about the knowledge of New York State extension educators. Most educators overestimated (37%) or did not know (26%) the actual prevalence of genetically engineered crops in the US food supply. When provided with a list of 20 food crops, most educators could correctly identify the most common engineered crops, maize and soybeans. However, educators were less certain whether GE varieties of other crops existed in US markets: for each of the 20 crops listed, there was at least one educator who believed a GE version could be found in the marketplace.

In responding to knowledge questions, most educators answered most questions correctly (64% of educators answered six or more of eight questions correctly). Confidence in their knowledge, however, was much lower: of those providing correct answers, on average only 59% thought their answers were 'definitely' correct (as opposed to 'probably' correct). Not surprisingly, confidence was lower among those providing incorrect answers: on average, only 30% thought their incorrect answers were 'definitely' correct.

The survey also provided useful details about how the topic of genetic engineering related to the professional activities of extension educators. Most educators characterized their professional need for information about generic engineering as 'very high' (32%) or 'some need' (52%). Given a list of 14 topics associated with genetic engineering, educators overwhelmingly believed that it was important or very important to be able to answer their audience's questions on almost every topic. Ironically, one of the topics of least importance was 'how genetically engineered crops were made', a topic emphasized by many proponents of biotechnology. Educators identified printed fact sheets and web-based resources as preferred tools for communicating about GE foods and crops.

Nevertheless, despite the expressed need for information, educators did not appear to be receiving many questions on the subject from their audiences: only 21% had been asked questions on the subject more frequently than 'once every week or two', and 12% had never been asked about the subject at all. The setting in which educators were most likely to receive questions about genetic engineering was in casual conversations with friends, family and co-workers: 47% ranked this setting as first, second or third most likely).

There was a strong consensus among respondents that educational efforts about genetic engineering should not be promotional in nature. Although most (79%) definitely or somewhat agreed that 'education will increase consumer acceptance of GE food and crops', an even greater majority (83%) definitely or somewhat agreed that 'CCE educators should not advocate a position for or against GE foods or crops'. Only 37% agreed definitely or somewhat that 'the role of CCE educators is to restore confidence in the safety of GE foods and crops'; 77% definitely agreed and an additional 20% somewhat agreed that 'values and beliefs need to be acknowledged when planning educational programmes about GE'. The survey also included an open-ended question to allow additional comment, and many of those who provided comments stressed the importance of the GEO-PIE Project's neutrality.

The extension educator survey provided important insights that influenced the development of GEO-PIE Project outreach strategy. First, educators strongly preferred printed fact sheets as their primary educational tool, with web-based resources as a distant second. Secondly, educators overwhelmingly supported the neutrality of educational efforts regarding agricultural biotechnology, i.e. that outreach materials should be informational and not promotional. Although most did believe that education would probably increase consumer acceptance of genetically engineered foods and crops, they did not agree that it was their responsibility as educators to promote acceptance.

Finally, although educators felt that it was 'very important' that they be able to respond to questions on almost any specific issue related to genetic engineering, educators actually appeared to be receiving very few questions on the subject from their audiences. This suggested that educators may have been anticipating that public desire for information about genetic engineering would increase in the future.

## Analysis of website tracking data

Throughout the first 4 years of the GEO-PIE Project, we monitored and analysed web server logs generated by visits to the website. Each individual file distributed by the web server to a website visitor – a 'hit' – generates an entry in a computer log which records the file loaded, time of the event and some very basic information about where the file was sent. Using a commercial software package to analyse daily activity logs, we were able in generate a wealth of information about visitor activity, such as number of individual visitors, length of visit, pages viewed and the paths visitors travelled through the website. Additionally, the Project used 'cookies' (a marker file placed on a web visitor's computer) to determine how often visitors returned to the website.

From September 2000 to August 2004, the GEO-PIE Project website had over 150,000 unique visitors reading nearly 400,000 pages of HTML text (a total of 4.5 million 'hits'). At its peak, the website averaged over 300 unique visitors per day (679 unique visitors on the highest-activity day). Most visitors (73%) viewed only a single page, but 8% of visitors viewed five or more pages. Similarly, 50% of individual page views lasted less than 1 minute, while 38% of lasted greater than 10 minutes; 13% of visitors returned to the website after the first visit and 2% visited the website more than ten times. These patterns are consistent with typical website use: the majority of visitors enter the website, briefly view a single page and exit, while a smaller proportion of visitors stay longer, view more individual pages and return to the website at a later date.

The volume of GEO-PIE website traffic over time is shown in Fig. 12.1. The first observable trend is a dramatic increase in the number of visitors over time. This pattern is expected: over time, awareness of the website grows, more search engines index the site and more websites add their own links directly to GEO-PIE. Secondly, visitor traffic is strongly seasonal, with peaks in mid- to late autumn and mid- to late spring, and valleys during the summer and winter holidays. This pattern clearly follows the seasonal academic cycle, suggesting that many (if not most) of website visitors are students or educators. This inference is strongly supported by our survey of web users, discussed in the section below.

## Survey of website users

In order to assess the demographics and opinions of visitors to the GEO-PIE Project website, we conducted a 1-year online survey that

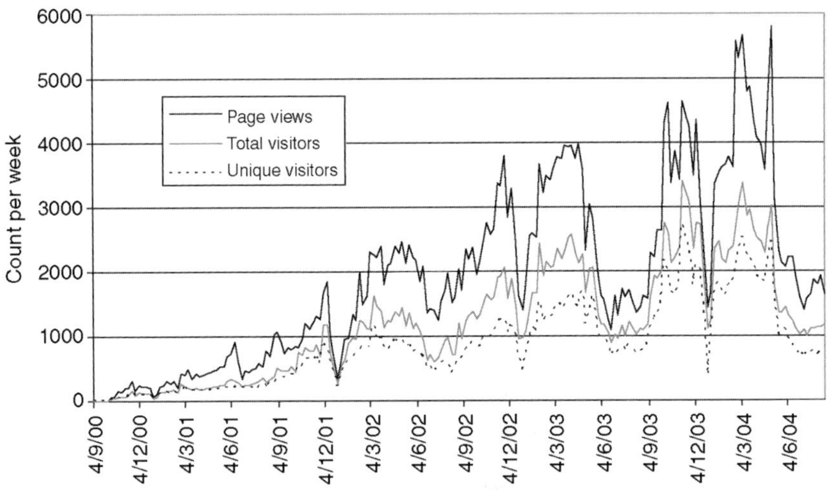


**Fig. 12.1.** Volume of weekly visitor activity to the GEO-PIE Project website (http://www.geo-pie.cornell.edu), September 2000 to August 2004. 'Page views' is the number of individual HTML pages viewed by visitors over a given week. 'Total visitors' is the total number of separate visits (counting repeat visitors) to the website over a given week. 'Unique visitors' is the total number of different visitors (not counting repeat visitors) over a given week. Thus, the difference between total visitors and unique visitors is an indication of the number of people returning to the site within the same week. Similarly, the difference between unique or total visitors and page views is an indication of the number of pages visitors view while at the website. Data in this figure have been filtered to remove visits from 'spiders' (automated programmes that scan websites for indexing purposes).

ran from July 2003 to July 2004. The survey invitation was designed so that a box would appear at the bottom of the left-margin navigation bar of the GEO-PIE Project homepage that stated: 'Tell us about you! Care to take our two-minute user survey?' (with buttons: 'yes', 'no' and 'later'). Visitors who clicked 'yes' were sent to the 11-question pop-up survey. Completion of the survey placed a file (a 'cookie') on the visitor's computer, which prevented the appearance of the invitation text and blocked access to the survey page on later visits to the site. While not an entirely perfect system – a knowledgeable user could easily delete the 'cookie' file – it reduced the likelihood that visitors would take the survey multiple times. The original survey instrument, with a tabulation of individual responses, is included in Appendix II of this chapter.

During the year-long run of the survey, 424 visitors responded. The respondents were 60% female and the majority (57%) was composed of students at high school or at the undergraduate level (78% of those describing themselves as students). The second largest demographic group was educators (12%), primarily of high school, undergraduate and graduate level students (23, 15 and 12% of those describing themselves as educators, respectively). Most visitors found the GEO-PIE website by using a search engine (53%), by following a link from another website (18%) or from an address provided by an educator (18%). A large majority of visitors were using the website to research a school project (73%).

Most website visitors found the website to be useful (74% responding 4 or 5 on a scale of 1–5), accurate (64% 4 or 5 on a scale of 1–5) and either well-balanced (34%) or biased slightly in favour of genetic engineering (29.1%). The majority of visitors either overestimated (65%) or did not know (17%) the actual prevalence of genetically engineered crops in the US food supply. Visitor opinion of genetic engineering varied widely and was slightly more positive than negative: 37% thought benefits outweighed risks, 20% thought risks outweighed benefits, 21% thought the risks and benefits were roughly equal and 23% were undecided.

Because respondents to the survey represent an extremely small fraction of the more than 72,000 unique visitors to the website during the survey period, it is difficult to judge how representative this sample is of all website visitors. However, the observation that temporal fluctuations in the total number of site visitors closely follow the annual academic cycle supports the conclusion that most visitors to the GEO-PIE Project website were likely to be high school and undergraduate students conducting research for class projects.

## Conclusions

The GEO-PIE Project consciously emphasized two central themes in developing its outreach strategy. First, we interpreted public opinion survey results indicating strong support for labelling of genetically engineered foods as an indication of both: (i) public desire for more information about the prevalence of genetic engineering in consumer products; and (ii) a need for more control to avoid genetic engineering if so desired.

Slovic (1987) and many others have suggested that increased public concern about a technology may be associated (among many other factors) with an inability to avoid perceived risks associated with that technology. Information originating from both proponents and opponents of the technology tended to avoid straightforward information about the precise distribution of GE-derived foods in consumer products. We chose instead to focus outreach efforts on the prevalence of genetic engineering in the food supply as a proxy for food labelling, and thereby to personalize our communication message.

Second, cursory surveys of publicly available information about genetic engineering have suggested that most information on the subject was highly partisan (both pro and con). Because of the long-established negative association between perceived information source bias and perceived source credibility (discussed in other chapters in this book), we chose to emphasize neutrality on the subject in order to boost credibility (even to the extent of frequently communicating the funding source of the project). Accordingly, in all communications, we stressed that the intent of the project was to provide reliable information about what exactly was known and unknown about both the risks and benefits of the technology, and to let consumers make informed decisions for themselves.

Yet, despite our observations that information provided by the GEO-PIE Project website, fact sheets and educator seminars was extremely well received, the future of the project remains uncertain. Unfortunately, no quantitative surveys measuring the project's impacts have been conducted. Since August of 2004, the content of the website has not been substantially updated, and is growing increasingly out of date (although surprisingly only one new crop – alfalfa – has been added to the list of 'US genetically engineered crops' since the project began in 2000).

Future funding prospects for the project are also unclear. Nevertheless, the GEO-PIE stands as an example of a successful, if short-lived, small-budget, outreach project with a unique approach that has reached a wide range of local, national and global audiences.

## Acknowledgements

The author wishes to acknowledge the important contributions of the project's co-directors James Shanahan and Jennifer Wilkins, in addition to the project collaborators and technical advisors: Margaret Smith Einarson, Jim Grace, Bruce Lewenstein, Steve Kresovich, David Pelletier, Anthony Shelton, Ed Staehr and Christina Stark.

## References

Dale, D.D. and Hahn, A.J. (eds) (1994) *Public Issues Education: Increasing Competence in Resolving Public Issues.* University of Wisconsin Cooperative Extension, Madison, Wisconsin.

Extension Committee on Organization and Policy (ECOP) (1992) *Public Issues Education: the Cooperative Extension System's Role in Addressing Public Issues*. Extension Service, USDA, Washington, DC.

Hallman, W.K. (1996) Public perceptions of biotechnology: another look. *Nature Biotechnology* 14, 35–38.

Hoban, T.J. (1998) Consumer acceptance of biotechnology: an international perspective. *Nature Biotechnology* 15, 232–234.

Patton, D.B. and Paine, T.W. (2001) Public issues education: exploring extension's role. *Journal of Extension* 39 (4) [online]. Available at: http://www.joe.org/joe/2001august/a2.html

San Diego Center for Molecular Agriculture (SDCMA) (2000) *Food from Genetically Modified Crops*. SDCMA, San Diego, California. Available at: http://www.sdcma.org/GMFoodsBrochure.pdf

Shanahan, J., Scheufele, D. and Lee, E. (2001) The polls-trends: attitudes about agricultural biotechnology and genetically modified organisms. *Public Opinion Quarterly* 65, 267–281.

Slovic, P. (1987) Perception of risk. *Science* 236, 280–285.

Ticciati, L. and Ticciati, R. (1998) *Genetically Engineered Foods: are they Safe? You Decide*. Keats Publishing, Chicago, Illinois.

Wilkins, J.L., Kraak, V., Pelletier, D.L., McCullum, C. and Uusitalo, U. (2001) Moving from debate to dialogue about genetically engineered foods and crops: insights from a land grant university. *Journal of Sustainable Agriculture* 18 (2/3), 167–201.

## Appendix I. Survey of New York State Extension Educators (December 2000–January 2001)

The following is the survey instrument used to assess the knowledge, opinions and informational needs of New York State's extension educators (see chapter text for more details). For each question in the survey, numerals denote the percentage of respondents supplying a given answer (*n*, 74). Numbers may not add up to 100% due to rounding errors or no response.

The following questions assess your informational needs about genetic engineering as it relates to your responsibilities as faculty/staff of Cornell Cooperative Extension.

| | Very high | Some need | Not much | No need |
|---|---|---|---|---|
| **1.** How would you characterize your current need for information about GE foods and crops? | 32.4 | 52.7 | 14.9 | 0.0 |

**2.** There are many topics related to GE foods and crops that may be of interest to our stakeholders. As an extension educator, how important do you feel it is that you are able to answer your stakeholders' questions about each of the following topics?

| | Very Important | Important | Not sure/ no opinion | Not important |
|---|---|---|---|---|
| Food and food products containing GE | 59.5 | 29.7 | 5.41 | 4.1 |

| | Very Important | Important | Not sure/ no opinion | Not important |
|---|---|---|---|---|
| Economic impacts for growers | 37.8 | 37.8 | 5.41 | 13.5 |
| Consumer choice and labelling | 48.7 | 39.2 | 9.46 | 2.7 |
| Antibiotic resistance | 39.2 | 27.8 | 8.10 | 9.5 |
| Allergenicity | 51.4 | 36.5 | 10.8 | 6.8 |
| Rate of pesticide use with GE crops | 37.8 | 46.0 | 2.7 | 5.4 |
| Movement of genes to wild plants | 43.2 | 37.8 | 8.1 | 8.1 |
| How GE plants are made | 24.3 | 51.4 | 8.1 | 10.8 |
| Potential environmental risks and benefits | 55.4 | 35.1 | 10.8 | 2.7 |
| Potential health risks and benefits | 66.2 | 29.7 | 2.7 | 1.4 |
| Regulation of GE foods | 46.0 | 43.2 | 1.4 | 5.4 |
| Safety testing | 54.1 | 33.8 | 4.1 | 4.1 |
| International implications | 25.7 | 40.5 | 6.8 | 13.5 |
| Ethical aspects of GE | 37.8 | 39.2 | 17.6 | 4.1 |

| | Very confident | Somewhat confident | Not very confident | Not at all confident |
|---|---|---|---|---|
| **3.** In general, how confident are you in your ability to answer questions about genetic engineering? | 16.2 | 52.7 | 27.0 | 2.7 |

| | Very difficult | Difficult | Not too difficult | Easy | Never sought |
|---|---|---|---|---|---|
| **4.** In general, when you have sought information about genetic engineering, how difficult has it been for you find the information you needed? | 5.4 | 21.6 | 46.0 | 12.2 | 12.2 |

**5.** How trustworthy do you believe information about genetic engineering is from each of the following sources?

| | Very trustworthy | Trustworthy | Not sure, depends | Untrust- worthy | Very untrust- worthy |
|---|---|---|---|---|---|
| Industry sales representative | 4.1 | 2.7 | 55.4 | 29.7 | 6.8 |
| Government official | 5.4 | 6.8 | 59.5 | 6.8 | 2.7 |
| Organic farmer | 8.1 | 8.1 | 52.7 | 21.6 | 8.1 |
| Member of activist group | 6.8 | 0.0 | 48.7 | 29.7 | 13.5 |
| County extension agent | 18.9 | 43.2 | 36.5 | 0.0 | 0.0 |
| Cornell research faculty | 27.0 | 54.1 | 16.2 | 1.4 | 0.0 |
| Cornell extension faculty | 29.7 | 50.0 | 18.9 | 0.0 | 0.0 |
| Website | 4.1 | 12.2 | 75.7 | 5.4 | 0.0 |
| Television | 4.1 | 6.8 | 52.7 | 33.8 | 4.1 |
| Radio | 4.1 | 6.8 | 51.4 | 35.2 | 4.1 |
| Newspaper | 5.4 | 6.8 | 56.8 | 27.0 | 2.7 |
| Popular magazine | 4.1 | 4.1 | 60.8 | 25.7 | 2.7 |
| Scientific journal | 28.4 | 43.2 | 20.3 | 5.4 | 0.0 |

**6.** How frequently in the last 3 months have you been asked questions related to genetic engineering?

| | |
|---|---|
| (i) Daily | 10.8 |
| (ii) A few times each week | 10.8 |
| (iii) Once every week or two | 17.5 |
| (iv) Only a few questions in the last 3 months | 45.9 |
| (v) Never; I have not yet been asked questions about GE foods | 12.2 |

**7.** From which stakeholder groups are you most likely to receive questions related to GE? (please rank your top three, with 1 = most likely).

| | Ranked 1 | Ranked 2 | Ranked 3 |
|---|---|---|---|
| (i) Consumer or interested citizen | 43.2 | 25.7 | 8.1 |
| (ii) Parents | 4.1 | 4.1 | 6.8 |
| (iii) Youth | 6.8 | 12.2 | 6.8 |
| (iv) Grower/producer | 21.6 | 16.2 | 4.1 |
| (v) Local business person | 2.7 | 4.1 | 2.7 |
| (vi) Public official | 1.4 | 4.1 | 0.0 |
| (vii) Other local extension agent | 1.4 | 6.8 | 17.6 |
| (viii) Limited resource audience | 8.1 | 13.5 | 21.6 |
| (ix) Other | 6.8 | 4.1 | 1.4 |
| (x) I have not yet been asked questions about GE foods | 8.1 | | |

**8.** In what settings do you receive questions about GE? (please rank your top three, with 1 = most frequent).

| | Ranked 1 | Ranked 2 | Ranked 3 |
|---|---|---|---|
| (i) Phone call | 13.5 | 10.8 | 9.5 |
| (ii) E-mail | 6.8 | 4.1 | 2.7 |
| (iii) Office visitor | 2.7 | 2.7 | 8.1 |
| (iv) Visit to client | 2.7 | 6.8 | 4.1 |
| (v) Formal presentation | 12.2 | 4.1 | 9.5 |
| (vi) Informal talks | 10.8 | 17.8 | 5.4 |
| (vii) Extension-organized group activity, workshop or class | 8.1 | 9.5 | 12.2 |
| (viii) Casual conversation with family, friends or co-workers | 21.6 | 18.9 | 6.8 |
| (ix) Other | 2.7 | 0.0 | 2.7 |
| (x) I have not yet been asked questions about GE foods | 9.5 | | |

**9.** In communicating information about GE foods and crops to the public, what educational tools would you prefer to use? (please rank your top three, with 1 = highest preference).

| | Ranked 1 | Ranked 2 | Ranked 3 |
|---|---|---|---|
| (i) Printed fact sheets | 51.4 | 10.8 | 10.8 |
| (ii) PowerPoint presentation | 16.2 | 6.8 | 14.1 |
| (iii) Web page | 4.1 | 18.9 | 13.5 |
| (iv) An e-mail listserv | 2.7 | 5.4 | 4.1 |
| (v) Group workshops | 9.5 | 16.2 | 12.2 |
| (vi) Instructional video | 1.4 | 10.8 | 13.5 |
| (vii) Other | 5.4 | 5.4 | 2.7 |

**10.** For each of the following statements please indicate how strongly you agree or disagree.

| | Definitely agree | Agree somewhat | Disagree somewhat | Definitely disagree |
|---|---|---|---|---|
| Values and beliefs need to be acknowledged when planning educational programmes about GE | 77.0 | 18.9 | 2.7 | 0.0 |
| Education will increase consumer acceptance of GE foods and crops | 28.4 | 40.5 | 21.6 | 5.4 |
| The role of CCE educators is to restore confidence in the safety of GE foods and crops | 16.2 | 23.0 | 16.2 | 33.8 |
| CCE educators should not advocate a position for or against GE foods and crops | 56.8 | 25.7 | 9.5 | 4.1 |

The following questions examine your personal knowledge about issues related to genetic engineering. We're asking this question of a wide range of Extension staff, to assess educational needs.

**11.** How many different genetically engineered plant species are currently present as either fresh produce or as a part of a processed food product in US food markets? (Correct answer is in bold).

| | |
|---|---|
| (i) 0 | 8.1 |
| **(ii) 1–10** | **28.4** |
| (iii) 11–50 | 16.2 |
| (iv) 51–100 | 2.7 |
| (v) > 100 | 17.8 |
| (vi) I don't know | 25.7 |

**12.** Which of the following plants have been genetically engineered and are currently present as either fresh produce or as a part of a processed food product in US food markets? Check each choice that applies (correct answers are in bold, although at the time of the survey it was unlikely that GE tomatoes were still marketed).

| | | |
|---|---|---|
| apple (16.9) | lettuce (7.0) | strawberry (17.6) |
| banana (5.4) | oats (9.5) | sweet potato (5.6) |
| **canola/rape seed (51.4)** | pea (5.6) | rice (29.7) |
| carrot (5.6) | peanut (12.2) | **tomato (59.5)** |
| **maize (71.6)** | pear (1.4) | wheat (25.7) |
| aubergine (4.1) | **potato (55.4)** | **yellow squash (19.7)** |
| grape (10.8) | **soybean (70.3)** | I don't know (19.2) |

**13.** Which of the following statements do you believe are true? (Correct answers are in parentheses: T, true; F, false).

| | Definitely true | Probably true | Probably false | Definitely false |
|---|---|---|---|---|
| Most plants we eat have been altered by cross-pollination and selection by humans (T) | 79.7 | 16.2 | 2.7 | 0.0 |
| Some plants we eat contain naturally occurring toxins (T) | 81.1 | 14.9 | 1.4 | 0.0 |

| | Definitely true | Probably true | Probably false | Definitely false |
|---|---|---|---|---|
| In nature, plants transmit their genes to unrelated kinds of plants through the process called pollination (F) | 23.0 | 10.9 | 33.8 | 29.7 |
| 'Bt' (as in 'Bt maize') is short for biotechnology (F) | 5.4 | 9.5 | 9.5 | 71.6 |
| Some tomatoes on the market today have been genetically engineered to contain fish antifreeze genes (F) | 9.5 | 24.3 | 33.8 | 24.3 |
| If a GE food has a different nutritional content than its non-GE counterpart, the FDA requires that it be labelled (T) | 20.3 | 27.0 | 32.4 | 13.5 |
| No permits are required before growing an experimental GE plant in an outdoor field (F) | 9.5 | 12.2 | 37.8 | 37.8 |
| Food products containing StarLink maize were recalled when it was determined that StarLink caused allergic reactions (F) | 14.9 | 24.3 | 18.9 | 36.5 |

**14.** Please give us a little information about yourself. Your answers in this section are optional, and all answers will be kept confidential and will not be connected with you in the final data.

What is your gender? Female (60.8) Male (35.1)

What is your age? Mean = 46.13 (SD = 7.61)

What is the highest level of education you have attained?

| | |
|---|---|
| Associates degree | 8.1 |
| BS/BA | 20.27 |
| MS (or similar advanced degree) | 46.0 |
| Advanced professional degree | 1.4 |
| PhD | 20.3 |

Which of the following most closely describes your specialty area within CCE?

| | |
|---|---|
| Agriculture and food systems sustainability | 44.6 |
| Children, youth and families | 6.8 |
| Community and economic vitality | 4.1 |
| Environmental and natural resources | 4.1 |
| Nutrition, health and safety | 20.3 |
| Other | 10.1 |

How many years have you worked for Cornell Cooperative Extension?

Mean = 7.09 years (SD = 7.29)

In what New York county do you work? (number of responses > 1 listed in parentheses)

Individual counties: Albany, Allegany (2), Columbia, Cortland (2), Delaware, Dutchess, Erie (3), Essex, Jefferson, Lewis, Oneida (2), Onondaga, Ontario, Orange, Saratoga, Schoharie, Schuyler, Steuben, Suffolk, Sullivan, Tompkins (3), Ulster, Washington, Westchester (2), Wyoming

Regional or multi-County (11)

Cornell University campus (6)

No response (24)

[a] Throughout this survey, 'GE' is used as an abbreviation for 'genetic engineering' or 'genetically engineered', and is meant to refer to the specific process of inserting new genes into an organism using modern biotechnological methods.

## Appendix II. Survey of visitors to the GEO-PIE Project (July 2003–July 2004)

The following is the online survey instrument used to assess demographics and perceptions of visitors to the GEO-PIE Project website (see chapter text for more details). For each question in the survey, numerals denote the percentage of respondents supplying a given answer (*n*, 424). Figures may not add up to 100% due to rounding errors or no response.

**1.** Age

| | |
|---|---|
| (i) < 10 | 0.0 |
| (ii) 10–19 | 31.6 |
| (iii) 20–29 | 25.9 |
| (iv) 30–39 | 12.0 |
| (v) 40–49 | 9.9 |
| (vi) 50–59 | 11.1 |
| (vii) 60–69 | 4.0 |
| (viii) 70 and Older | 1.9 |

**2.** Gender Male (42.7) Female (55.9)

**3 (a).** Which one of the following categories best describes your occupation?

| | |
|---|---|
| (i) Student | 56.8 |
| (ii) Homemaker | 0.9 |
| (iii) Retired | 2.1 |
| (iv) Self-employed | 2.8 |
| (v) Administrative | 1.9 |
| (vi) Sales/ Marketing | 1.7 |
| (vii) Trade/Labour | 1.4 |
| (viii) Educator | 12.3 |
| (ix) Medical | 2.1 |
| (x) Management | 2.1 |
| (xi) Professional | 9.2 |
| (xii) Other | 6.4 |

**3 (b).** If you selected 'Student' in Question 3 (a), please select your grade level. If you selected 'Educator', please select the grade level of your students.

| | Student | Educator |
|---|---|---|
| (i) Elementary | 2.1 | 3.9 |
| (ii) Middle School | 4.6 | 3.8 |
| (iii) High School | 39.8 | 23.1 |
| (iv) Undergraduate | 38.2 | 15.4 |
| (v) Graduate | 12.5 | 11.5 |
| (vi) Adult Education/Extension | 0.1 | 9.6 |
| (vii) Other Professional Level | 1.7 | 5.8 |

**4.** How did you find the GEO-PIE Project website?

| | |
|---|---|
| (i) Search engine | 52.7 |
| (ii) Link from another website | 18.3 |
| (iii) Address was on a printed publication | 1.7 |
| (iv) Address was provided by an educator | 17.8 |
| (v) Address was provided in a public talk | 0.0 |
| (vi) Website was recommend to me by a friend/colleague | 1.7 |
| (vii) Other | 4.1 |

**5.** Why are you visiting the GEO-PIE Project website?

| | |
|---|---|
| (i) Personal interest in the topic | 7.9 |
| (ii) Research for a school project | 72.6 |
| (ii) Research for a mass media story (newspaper, TV, radio, etc.) | 0.0 |
| (iv) Looking for materials/information to use for educational purposes | 12.9 |
| (v) Just wandered in | 1.7 |

**6.** How many different genetically engineered plant species do you believe are currently present as either fresh produce or as part of a processed food product in US food markets? (Correct answer is in bold).

| | |
|---|---|
| (i) 0 | 1.2 |
| (ii) 1–10 | 17.1 |
| **(iii) 11–50** | **29.9** |
| (iv) 51–100 | 11.6 |
| (v) > 100 | 23.7 |
| (vi) I don't know | 16.6 |

**7.** Which of the following best describes your opinion on the current use of genetic engineering in agriculture?

| | |
|---|---|
| (i) The benefits definitely outweigh the risks | 17.0 |
| (ii) The benefits mostly outweigh the risks | 19.9 |
| (iii) The benefits and risks are roughly equal | 20.8 |
| (iv) The risks mostly outweigh the benefits | 9.5 |
| (v) The risks definitely outweigh the benefits | 10.0 |
| (vi) I don't know/I haven't decided | 22.8 |

**8.** Based on what you have read on the GEO-PIE Project website, in your opinion how useful is the information that is provided (on a scale of 1 to 5, with 5 = very useful)?

| | |
|---|---|
| 1 Not useful | 2.5 |
| 2 | 2.9 |
| 3 | 12.9 |
| 4 | 38.6 |
| 5 Very useful | 36.1 |
| I have no opinion | 6.6 |

**9.** Based on what you have read on the GEO-PIE Project website, in your opinion how accurate is the information that is provided (on a scale of 1 to 5, with 5 = very accurate)?

| | |
|---|---|
| 1 Very inaccurate | 2.5 |
| 2 | 2.1 |
| 3 | 11.2 |
| 4 | 33.6 |
| 5 Very accurate | 30.3 |
| I have no opinion | 19.9 |

**10.** Based on what you have read on the GEO-PIE Project website, in your opinion how balanced is the information that is provided (on a scale of –2 to +2, with –2 = biased against, +2 = biased in favour, zero = balanced)?

| | |
|---|---|
| −2 Biased against genetic engineering | 2.9 |
| −1 | 7.5 |
| 0 Balanced | 34.0 |
| +1 | 29.1 |
| +2 Biased in favour of generic engineering | 7.1 |
| I have no opinion | 17.8 |

**11.** If you have any additional comments about the GEO-PIE Project website, please use the space below. We are interested in how you think we can improve our site.

13

# Governing Controversial Technologies: Consensus Conferences as a Communications Tool

J. Medlock,* R. Downey** and E. Einsiedel***

*Communication Studies, University of Calgary, Calgary, Alberta, Canada; e-mail: *jemedloc@ucalgary.ca; **rdowney@ucalgary.ca: ***einsiede@uclagary.ca*

## Introduction

Agricultural biotechnology represents a cluster of new technologies, such as GM food and plant molecular farming, with profound and far-reaching societal and ethical implications, and has been a recurring and contentious public policy issue across the globe. The controversy over GM food in Europe, for example, has led to outright public rejection of the technology across much of the continent. Front and centre in these debates is the notion that the successful integration of a technology into everyday life rests as much on social viability as on economic viability.

Implicit in the evolution of agricultural biotechnology policy has been a controversy over expertise – that the notion of expertise encompasses diverse forms of knowledge and is necessarily multi-disciplinary and should include input from various stakeholders as well as civil society (Liberatore, 2001). Governments are finding themselves operating in an environment of increasing public dissatisfaction, decreased trust and have responded with more vocal rhetoric in support of more participation on science and technology issues (UK House of Lords Science and Technology Committee, 2000; European White Paper on Governance, 2001; External Advisory Committee on Smart Regulations, 2004).

It seems that everyone – from policymakers to civil society groups to industry representatives – agrees that there should be more participation from citizens but, at the same time, there seems to be little reflection as to whether there is a common understanding of what it means in practice (Nielsen *et al.*, 2004).

Addressing this issue, Rowe and Frewer (2004) contend that: 'among the key questions that need to be answered regarding public participation are why it has caught the attention of policy institutions at the present time, and why public participation is perceived by institutions as potentially facilitating governance and institutional practices' (p. 513). A starting point to address these questions in the realm of agricultural biotechnology is to consider the changing nature of governance mechanisms themselves.

According to Boucher *et al.* (2002), 'Governance comprises the traditions, institutions and processes that determine how power is exercised, how citizens are given a voice and how decisions are made on issues of public concern' (p. i). Furthermore, they believe that: 'the question of how we may need to adapt our governance mechanisms to cope with biotechnology is worth posing because this has been described by some as "the ultimate horizontal", or cross-cutting policy issue' (p. iii), where 'horizontal' refers to more than cross-government coordination and also includes interests from the private sector, universities and civil society.

Within this shift from traditional top-down models of government to more cooperative, inclusive governance arrangements: 'How to involve citizens in policy making is at the core of discussions over modernizing governance and building a stronger civil society' (Phillips and Orsini, 2002, p. iii).

An alternate perspective on governing science and technology issues comes from the literature on Technology Assessment (TA). Classic or traditional TA, as embodied by the Office of Technology Assessment (OTA) in the USA, was premised on the idea that neutral, objective analysis could assist governments in predicting any adverse consequences of technology (essentially an 'early warning' function) (Van Den Ende *et al.*, 1998). Newer forms of TA developed in the 1980s have broadened this expert-driven, scientifically focused process by recognizing the social complexity of the entire technological design, development and implementation process (Schot, 1992).

Technologies are no longer black-boxed, but are seen as influencing and being influenced by the society in which they are embedded. In this broader conception, TA is defined as 'a scientific, interactive and *communicative* process, which aims to contribute to the formation of public and political opinion on societal aspects of science and technology' (Butschi *et al.*, 2004, p. 14 [emphasis added]). And, within this context, the inclusion of knowledge and input from various 'publics' (citizens, consumers and other stakeholders) in debates about technology development has become a key issue (Joss, 1998).

In response, experimentation with new modes of public participation is ongoing, encompassing a variety of approaches: from solicitation of opinions (e.g. focus groups and surveys) to appointments on advisory committees, from informal town hall meetings to more formal, learning-intensive consultations. Participatory Technology Assessment (PTA) encompasses a family of consultation methods with the common feature of being based on a model of discussion and deliberation (e.g. consensus conferences, citizens' juries, scenario workshops, etc.). Within these methods, citizens are given the time and resources to learn about new technologies and their potential implications, and the opportunity to talk about them with representatives from government, industry and interested groups from civil society.

In the realm of agricultural biotechnology, much experimentation has been done with deliberative consultations in a range of countries and contexts, and especially with the consensus conference model (discussed in detail in the next section). The spread of this technology assessment innovation can be variously viewed as an indicator of disaffection with current modes of decision-making (which often have relied on technocratic expertise), as a different forum for bringing together voices not normally in the same space, or as an outcome of the realization that useful public input on complex technological innovations requires opportunities for learning and interaction.

Examining PTA from the perspective of changing governance practices, Kluver *et al.* (2000) prescribe a dual role for modes of PTA. On the one hand, PTA aims to improve decision-making by making the process more informed and more egalitarian and, on the other, it provides a forum for social learning and deliberation that can aid in social acceptance of decisions. However, debates continue over how to measure the influence of PTA both on decision-making and on broader social debate, and how these models can be integrated into governance practices.

Kluver *et al.* (2004) remind us that, at its core: 'the process of TA in itself is a communication tool' (p. 97) and it is from this perspective that we position this chapter. Framing the consensus conference as a 'communication tool', we ask the question: *How can deliberative models contribute to the evolving governance arrangements surrounding agricultural biotechnology?*

This chapter is structured in the following way. First, we will discuss the evolution of participation techniques from one-way information campaigns through to interactive, two-way models. We will provide descriptions of various deliberative models (with a focus on the consensus conference) and examine these models from a communications perspective, using examples from agricultural biotechnology. Two aspects of deliberation will be highlighted: the process and the outcome.

In the next section of the paper, we will discuss the variety of 'communication events' that take place as part of the consensus conference process. By recognizing that policy development is an iterative process, and that deliberative consultations can be an important part of the policy process over time, it will be important to recognize communications within the conference process itself, but also to understand how communication flows from the conference participants to outside audiences (policymakers, media, public and other stakeholders) and vice versa, and what this process looks like over a longer time frame.

The levels of communications we will examine, starting from the micro-level and expanding from there, are:

- Communication at the citizen–citizen level: discussions between participants in the process as well as participants and those outside the process.
- Communication between citizens and experts: discussions that act as part of the demonopolization of expertise.
- Communication as a catalyst for societal dialogue: discussions that extend beyond the boundaries of the consensus conference event (i.e. the use of the media to extend the debate).
- Communication for the policymaking sector: the communications that happen between such consultation events and other stakeholders and decision-makers.

The final section of the chapter will outline why this kind of communication is important in the development of governance practices surrounding agricultural biotechnology (and of technology in general). We will situate the communication aspects of the consensus conference within the broader governance environment and discuss the implications for future decision-making on agriculatural biotechnology-related issues.

## Participation and the Concept of Deliberation

The case supporting the use of participation mechanisms has been made elsewhere (e.g. Fiorino, 1990; Durant, 1999; Kluver *et al.*, 2000), so we will only summarize here. The support centres around two main issues: satisfying democratic conditions and improving decision-making. Within the context of democratic procedures, the rights of citizens include input into decisions made by governments and, at the same time, it can be seen as an obligation of citizens to become involved to support a well-functioning democracy.

Participation is also claimed to improve government decision-making processes. If done appropriately, consultations and participation can foster legitimacy and increase public trust in institutions, as well as lead to better decisions being made. Critical public perspectives need not be seen as a negative for technology development, but rather as a way to encourage more sustainable technologies in the technological, environmental and social arenas.

Rowe and Frewer (2000) define public participation as: 'a group of procedures designed to consult, involve and inform the public to allow those affected by a decision to have an input into that decision', noting that 'input' is the key word in differentiating participation from other communication strategies. The methodologies used in participation procedures can vary, depending on whether the focus of the participation is to consult, to involve or to inform. The Danish Board of Technology distinguishes between the following types of participation and provides examples of each type:

- Informing: providing information to various publics (pamphlets, websites).
- Consulting: taking feedback from the public (surveys) or getting into dialogue between publics and decision-makers (citizen hearings).
- Involving: supporting articulation of views from the public (consensus conferences), giving influence to the public (mediation) or giving power to the public (direct democracy such as referenda) (Joly, 2005).

Different styles of participation are suitable for different contexts. In the realm of agricultural biotechnology, consultation procedures might be most useful for discerning policy directions and developing regulation, while information provision may be suitable at all stages of decision-making. More successful public participation programmes use a mix of techniques, complementing informational campaigns with more deliberative, involved processes.

Up until the early 1980s, participation procedures focused only on disseminating information to the public. Called the Public Understanding of Science (PUS) model of public involvement (*aka* the 'deficit model'), this mode emphasizes the lack of understanding of scientific issues on behalf of the public. The widespread belief was that if members of the public just received more information about science and technology issues, their new understanding would lead to support of new innovations. No dialogue or deliberation was incorporated into public participation procedures. This model has largely been cast aside as insufficient and, on its own, potentially harmful for technological development.

In the area of GM food, for example, the European PABE (Public Perceptions of Agricultural Biotechnology in Europe) project demonstrated that institutional behaviour that followed the deficit model was partially responsible for GM opposition (Joly, 2005) and, further, that lack of knowledge on the part of the public was not linked with scepticism and resistance to the technology. In contrast, the study showed that: 'Public concerns were not so much about the risks of this technology per se, but more about trusting the institutions that are responsible for its regulation' (Hails and Kinderlerer, 2003, p. 821).

Newer models of TA recognize the limitations of the PUS model and have complemented learning activities with more interactive, dialogical consultations. In these newer models, technology trajectories are seen as influenced and being influenced by the societal environment that surrounds them. This 'co-production' of technology and society means that decisions made at all stages of the development process influence the ultimate technological product.

Within this context, the perception of the 'public' has shifted: there is recognition of many publics with diverse interests, and technologies are now seen as open-ended and continually impacted by social and political factors and therefore as a legitimate matter for public debate (Mayer and Stirling, 2004). Furthermore, expertise has become a contentious issue. New, scientific knowledge is not always certain, and in fact the public has recognized that scientific knowledge is often 'partial, provisional and even on occasions deeply controversial' (Durant, 1999, p. 315).

This uncertainty has opened up space to allow public knowledge as another legitimate sort of expertise, one that can help complete the picture for decision-making. Rip (2005) uses the term 'reflexive co-evolution' to describe newer TA processes that attempt to open up new spaces for societal discussion and negotiation and include a place for publics.

## Deliberative Models of Consultation

Joly (2005) highlights the diversity of the participation toolbox and discusses the complementarity of various methods and how they might be mixed to create innovative forms of governance. Thus, while we are focusing in this chapter on deliberative models, it should be noted that they are only one type of tool in the toolbox, and will generally be integrated with other techniques. The deliberative model and the consensus conference illustrate a mode of participation mechanism which has features of dialogue and deliberation (see Box 13.1 for descriptions of other kinds of deliberative models). The rationale behind the consensus conference is to give lay citizens an opportunity to learn about a technological issue, to assess its

**Box 13.1.** Description of a selection of deliberative models.

**Consensus conference**

This process is typically used as a technology assessment tool, ideally in the early stages of a technology's development. It is designed to consider the broad features of a technology by identifying key issues that ought to be addressed. A group of 15–18 lay citizens are selected (either through media advertising or by invitation to a randomly selected pool). This panel is ideally balanced demographically (age, gender, socio-economic attributes). The process takes place over three weekends, the first two as preparatory weekends and the third as a public conference. The citizens' panel identifies the key issues and the types of experts they need to hear from on these issues. These expert presentations are made in a public meeting with an opportunity for lay panel and public questions. The citizen panel then deliberates overnight or over the next day to arrive at their consensus conclusions and recommendations.

**Citizens' jury**

A citizen jury is very similar to a consensus conference in that it involves a citizens' panel convened to assess a particular issue, typically in the form of a key policy question (the 'charge' to the jury). The jurors are given a forum in which to hear experts, discuss and debate the issue and arrive at conclusions and recommendations. The charge becomes the focus of the deliberation sessions and provides the framework for the jury's final report. This process usually takes place over 3–4 days.

**Deliberative polling**

Deliberative polling attempts to incorporate deliberation into the traditional opinion poll. The process uses a statistically representative sample of citizens and provides 2–3 days of discussion and debate. The outcomes of the process are in the form of polling individual opinions.

**Scenario workshops (Citizen foresight)**

A scenario workshop starts with a problem that needs a solution. The aim of the process is to create an action plan for dealing with an identified problem. Participants are presented with a set of possible future scenarios (which have been pre-formulated), describing alternative ways of solving the problem. Workshops typically last 1–2 days and go through the following three phases: (i) a criticism phase, where positive and negative aspects of each scenario are discussed; (ii) a vision phase, with participants crafting their vision or solutions to the problem; and (iii) a realization phase, where participants create an action plan which includes identifying potential barriers to realizing their vision and how these might be overcome. Generally, there are 25–30 participants from different groups of society: lay citizens are joined by policymakers, business representatives and technical experts.

**Future search conference**

This method, developed in the USA, is very similar to a scenario workshop in that participants are from various sectors of society and their focus is to build a common vision and action plan. However, unlike in a scenario workshop, the participants are not initially presented with scenarios to critique, but build their common vision from square one. Participants first review key events from the past for themselves, their community and the world, then they create a mind map of which factors are currently most important in influencing the issue under discussion; and, finally, they use the mind map to forecast a realistic and satisfactory future scenario. Future search conferences can have up to 64 participants who are split into eight groups and generally last 3 days.

**Study circle**

A study circle is a group of 8–12 people from different backgrounds and viewpoints who meet several times to talk about an issue. In a study circle everyone has an equal voice, and people try to understand each other's views. They do not have to agree with each other. The idea is to share concerns and look for ways to make things better. A facilitator helps the group focus on different views and makes sure the discussion goes well. The approach is usually employed to address neighbourhood or community issues.

**Internet dialogue**

This is a generic label for any form of interactive discussion (asynchronous or real time) over the Internet, a process increasingly used for direct public consultation. Participants could be self-selecting and unrepresentative in an open process or could be limited to those chosen specifically for the consultation through a restricted-access procedure.

possibilities through dialogue with experts and amongst themselves and to develop a position in a final report.

The consensus conference methodology is centered on a panel of 12–15 'ordinary citizens'. After the members of the citizen panel have been recruited (generally by random selection or through media advertisements), they undergo an intensive learning process that is structured through two preparatory weekends held before the conference itself. These weekend meetings perform the following functions:

- to provide an opportunity for the panellists to get to know each other;
- to get an overview of the consensus conference process;
- to learn the basics of the overarching issue; and
- to develop key questions on the issue to present to an expert panel.

One component that differentiates the consensus conference process from most other deliberative models is this first 'scoping' stage of the consultation. In other models such as the citizens' jury, citizen participants are given more specific direction on the policy issue or question under discussion (i.e. should we go ahead with GM food labelling, and if so, how? Or, should GM wheat be approved in Canada?).

In the consensus conference, the participants begin with a very general overarching policy area (i.e. GM food) and take on the added task of developing a list of key questions. In consensus conferences that have taken place on GM food, these have ranged from environmental considerations, to human health and safety, to public involvement in regulation to the economic implications.

Based on these key questions, an expert panel is selected and is invited to participate in the final conference (which is open to the public and the media). On the first day of the conference, the experts give presentations on the key issues defined by the citizens' panel. Time is available for cross-examination and rebuttal. Day 2 begins with questions from the audience as well as further queries from the panel. The rest of Day 2 (and sometimes Day 3) is devoted to writing the report. On Day 3 (or 4), the report is presented and the expert panel responds.

The management of such a process involves a project management team, along with an Advisory Committee. In this instance, the Committee generally consists of representatives from government, industry, civil society, consumers, the scientific community and the project director and manager.

The consensus conference process has been touted as suitable for the following situations: (i) when there is a technology question of current societal interest with significant implications for the future; (ii) when there is controversy surrounding such an issue (usually when there is a clash of social, political, ethical or economic values); (iii) when the issue is complex and involves unresolved questions; and (iv) when there are many (and competing) interests at stake (Mayer and Geurts, 1998).

Given these features of an issue to which the consensus conference might be applied, it is not surprising that this method has been most frequently used for examining agricultural biotechnology issues. The range of countries that have used this approach is quite broad (see Table 13.1) and provides some indication of the degree of controversy generated and a growing interest in exploring more meaningful ways for involving publics in these technological debates.

The process of deliberation is a cornerstone of consensus conferences. Deliberation has been variously described as 'a conversation whereby individuals speak and listen sequentially before making a collective decision (Gambetta, 1998, p. 19); participation from those affected, with a decision made on the basis of an exchange of arguments under conditions of impartiality and rationality (Elster, 1998); and deliberation as either a particular sort of discussion – one that involves the careful and serious weighing of reasons for and against some proposition – or to an interior process by which an individual weighs reasons for and against courses of action' (Fearon, 1998, in Abelson *et al.*, 2003).

**Table 13.1.** Citizen consensus conferences on biotechnology (for additions made to this list see http://www.loka.org/pages/worldpanels.html).

| Country | Year | Subject |
|---|---|---|
| Argentina | 2000 | GM food |
| Australia | 1999 | Gene technology in the food chain |
| Austria | 2003 | Genetic data |
| Canada | 1999 | GM food |
| Denmark | 1999 | GM food |
| | 1998 | Citizens' food policy |
| | 1992 | Transgenic animals |
| | 1987 | Gene technology in industry and agriculture |
| France | 1998 | GM food |
| Germany | 1994 | GM crops |
| Italy | 2004 | GM crop trials |
| Japan | 2000 | GM food |
| Netherlands | 1993 | Genetically modified animals |
| New Zealand | 1999 | Plant biotechnology |
| | 1999 | Biotechnological pest control |
| | 1996 | Plant biotechnology |
| Norway | 2000 | GM food |
| | 1996 | GM food |
| South Korea | 1998 | GM food |
| Switzerland | 1999 | GM food |
| UK | 1994 | Plant biotechnology |
| USA | 2002 | GM food |
| | 2003 | Future of food (New England) |

Leading from these definitions, deliberation can mean different things to different people – some see it as a process (of debate and discussion), while others relate it to the outcomes (reaching collective decisions). In the consensus conference model, the 'process' aspect involves aspects such as understanding who is involved in the deliberation (citizens, experts, policymakers, etc.), the kind of information participants have access to, how much time they have for consideration of information and diverse perspectives and the format of those discussions.

The 'outcome' aspect has to do with the content of the final report of the panel and how it is distributed, how the conference is promoted and covered in the media, and if and how the panellists themselves talk about the process and the recommendations once the conference is complete. Common features of the deliberative consultation process include (not an exhaustive list):

- Deliberation: there is time provided for 'careful consideration' of information, of diverse views and perspectives, and the dialogue is respectful.

- Inclusion: public dialogue and deliberation can be most effective when participation is inclusive of a broad range of perspectives and diverse participants. While many organized stakeholders often have opportunities to participate in policy discussions, lay citizens typically have been less involved.
- Access to appropriate information: in order for deliberation to proceed effectively, participants should have access to information on the subject. This means, where possible, the most current and accurate information. While this often means scientific information, it could also mean experiential expertise (e.g. the experience of farmers, patients, etc.). In addition, participants should gain an understanding of the areas of uncertainty on the subject.
- Diverse discussion modes: these processes are based primarily on social interactions face-to-face. However, they do not just rely on oral discussions: other modes of communication could include audio and visual material and to complement face-to-face interaction.
- Opportunity for reflection: the scope of this opportunity is dependent on the procedures chosen but, at the very least, deliberation and reflection should enable participants to examine and re-evaluate their positions in light of diverse views and new information.

## Communication Events in a Consensus Conference

We recognize that emerging governance practices are iterative in nature, and that consultations will punctuate the process over time. Thus, it will be important to recognize the 'communication events' that occur within the consensus conference process, but also to recognize the communication flows from the conference participants to outside audiences (policymakers, media, public and other stakeholders), and what this process looks like over a longer time frame. These 'communication events' include the following:

- Briefing packages: this is a one-way information flow from the organizers to the panel members. This is, however, merely a starting point as panellists can — and often do — search for and utilize their own information sources.
- Preparatory weekends: these meetings are based on discussion amongst panellists and a facilitator and will then, on request of the participants, bring in selected experts to provide more information.
- Expert presentations and debate: this part of the process provides

an opportunity for one-way presentations of information followed by two-way debating and questioning between panellists and experts and members of the public.

- Public conference: this provides an opportunity for the general public (beyond the panel members) to learn more about the issues and become involved with the debate. It also provides an opportunity for media to attend and disseminate the conference happenings to a wider audience.
- Final Report: this report is completed after extended deliberation by the panel members. It is circulated to the expert panel, the media and also to the relevant policymakers.

Each of the above elements represents a different communication process that emanates from a consensus conference. At the macro-level, consensus conferences are representative of what is truly dialogic between citizens and policy makers. Much has been made of the one-way form of communication to publics so prevalent of earlier policymaking and still present today. What can policymakers learn from publics, their views and preferences, their expectations and concerns? What are the conditions under which a technological application is more (or less) acceptable? What governance mechanisms are appropriate? What standards ought to be implemented (e.g. should GM food be labelled and if so, how?)? What criteria ought to be employed for assessing impacts?

From the various reports and recommendations of citizens' panels, it is also evident that, while risk and benefit or health and safety issues are important, citizens' concerns extend beyond these factors to include broader considerations such as environmental impacts, socio-economic impacts and ethical dimensions (Einsiedel *et al.*, 2001). These questions are all illustrative of the kinds of issues that deliberations between citizens and policymakers can cover.

As has been emphasized earlier, consensus conferences are simply one approach in the battery of public consultation mechanisms that can be utilized to enlarge discussion and debate about technological directions. The process is useful for technological issues that are socially relevant, controversial, complex and involve many competing interests. That the deployment of consensus conferences in a majority of international cases has been on the issue of GM food is an illustration of the link between public conflicts and the use of deliberative approaches (Seifert, 2006).

Even under these conditions of ideal use, consensus conferences can be a complex exercise, expensive and, like any other tool, can be utilized for reasons that have little to do with an interest in consulting the public. A recent consensus conference experience in Austria was characterized as a way of learning ‘a new communication technique in vogue’, but having little to do with policy decision-making (Seifert, 2006).

The cultural context in which public consultation mechanisms are deployed must also be considered. For example, variations in the application of consensus conferences have been documented and the use and interpretations of outcomes have also varied in different countries. Nielsen *et al.* (2004) compared consensus conferences on GM food across France, Norway and Denmark. They found each country had a prevailing view of democracy (procedural, communitarian and deliberative), and this philosophy impacted on how the consultation process was perceived.

In the context of France’s tradition of procedural democracy, there was a suspicion of the process as interfering with its representative democracy institutions. The utility of the consensus conference process was viewed as an attempt to bring public knowledge closer to experts. In Norway, the communitarian political tradition saw the consensus conference as a new forum in which to infuse decisions with the ideas, values and morals of the general community. Denmark’s long tradition of community deliberation with citizen debate and participation, long entrenched in society, made for decisions linked directly to parliament, adding legitimacy to its consensus conference practices.

For both Norway and Denmark, Nielsen *et al.* (2004) found that the political traditions of communitarianism and deliberative democracy made the process of achieving consensus an important consideration to provide political weight to the citizen reports. In France, on the other hand, less importance was attributed to arriving at consensus. In Austria where administrative elites are dominant and where decision-making behind closed doors remains the norm, the likelihood of deploying a consensus conference approach linked to political decisions remains dim (Seifert, 2006).

In Italy, a citizens’ panel on GMOs illustrated the importance of institutional sponsorship and underlying distrust of governing bodies:

> One of the indications emerging from the project was the need for an *independent body or agency* to perform and organize such participatory assessment procedures. Citizens, in fact, seem to be quite sceptical when such initiatives are organized by those same institutions which are set to take policy decisions, i.e. regional or national governments.
>
> (Public Participation and Governance of Innovation, 2004)

## What this Communication Means for Governance

Consensus conferences and other deliberative models of consultation have matured in an era of evolving governance structures. The emergence of the term 'governance' has shifted the concept of the exercise of power from traditional notions of steering and control by a dominant political institution (i.e. 'government') to a system of negotiations within a multi-actor and multi-institutional context (see Pierre, 2000 or Hajer and Wagenaar, 2003). Science and technology applications, in particular, are influencing the shift in governance practices:

> Scientific and technological change introduce novelty all the time and so lead to rearrangement of sectors and institutions. According to German sociologist Beck it is the development of technology and its (unintended side-) effects that undermine traditional ways of dealing with collective risks, the role and status of technical knowledge (technocracy) and traditional forms of governance. In a sense it is the development of technical knowledge that … adds to the disembedding of traditional institutions and leads to the development of new forms of governance.
>
> (Institute for Governance Studies, 2003, p. 10)

The evolution of GM food technologies in Europe is a case in point, demonstrating the deficiency of expert-based policy processes. Public concerns were not rooted in ignorance, but were outside the realm of prevailing scientific expertise, and thus a challenge to the prevailing technocratic approach to policy (Felt, 2005). Shifting governance practices are recognizing this deficiency: an EU White Paper on European Governance states: 'The advent of bio-technologies is highlighting the unprecedented moral and ethical issues thrown up by technology. This underlines the need for a wide range of disciplines and experience beyond the purely scientific' (Liberatore, 2001, p. 19).

Traditional, expert-oriented structures have not disappeared, but now exist alongside more *ad hoc*, diverse arrangements that involve a new range of political practices between state institutions and societal organizations as well as members of the public (Hajer and Wagenaar, 2003). The networked component of emerging governance structures introduces 'new sites, new actors and new themes' into decision-making processes (Hajer and Wagenaar, 2003, p. 3). Within the context of public consultation, deliberative models serve as 'access points' (Giddens, 1990) for citizens, an entry point to the network of actors and institutions involved in governance practices.

Joly (2005) positions consensus conferences and other deliberative models of consultation as 'hybrid' deliberations where knowledge may be co-produced by both scientists and citizens. Lay knowledge is presented as an alternative and valid source for decision-making. Einsiedel *et al.* (2001), when examining GM food consensus conferences in Canada, Denmark and Australia, characterize the process as an attempt to 'de-monopolize expertise' and to bridge the incommensurability between expertise and participation.

The final reports of consensus conference panels on agricultural biotechnology consistently advocate the inclusion of social and ethical issues alongside scientific issues when making policy decisions on new technological innovations. By promoting a deeper understanding of informed public views on a technology, consensus conferences can aid technology developers and policymakers in being better placed to understand whether the technology is socially viable, rather than focusing solely on technological or scientific viability.

However, once placed in the context of policy decision-making, the 'hybrid' nature of the process can raise tension about what kinds of information are deemed legitimate for policy input. Levidow and Marris (2001) contend that most efforts are simply overlaying stakeholder participation onto a dominant model of technocratic decision-making that still maintains a dichotomy between science and values, which allows scientific information to maintain objectivity and all other information to be defined as subjective.

So, while the 'deficit model' approach to the public may be disappearing in participation methodologies through the development of consensus conferences and the like that allow for expression of social concerns alongside scientific concerns, there are suggestions that it can linger in the realm of policy development. Decision-makers tend to retreat to science-based models which advocate a 'sound science' view and the necessity of public education to bring the public onside (Irwin and Michael, 2003).

Further tensions can arise because policymakers often expect clear solutions or preferences to surface from consultations. Deliberative consultations, and technology assessment processes in general, tend to *add* complexity to policy situations by providing a fuller picture of the problem under examination, for example by including different social perspectives and shedding light on areas of uncertainty (Hennen *et al.*, 2004). Here again, as above, there is a tension between the spirit of the consultation method (inclusive) and the policy context into which the consultation results are being introduced (finding one viable solution).

This tension can be examined by looking at whether the aim of the consultation (or 'social appraisal' as termed by Stirling, 2005) is to 'open up' or 'close down' wider policy discourses. If it is about closing down, then it is about providing policymakers with a 'prescriptive recommendation to inform decisions'. When opening up, as in models of deliberative consultation, the focus shifts dramatically. The advice becomes 'plural and conditional' and may pose alternative questions, consider new options and include new voices. This, of course, adds complexity to decision-making by multiplying the number of interests to be accountable to and the types of knowledge to investigate and incorporate. But, at the same time, it provides a broader base of learning from which to anticipate potential future impacts of a given technology and, more importantly, to explore what a preferred outcome might look like and how it might be achieved.

## Conclusion

Rip (2005) describes deliberative models and other forms of technology assessment processes as '*de facto* governance', where actors create patterns of interaction in emerging science and technology that enable and constrain further action. The focus of this chapter has been to understand the role of consensus conferences in shaping these patterns of interaction. Consensus conferences open communication channels between citizens and many of the actors involved in network governance practices (industry, academic, civil society and government representatives) that are not readily available elsewhere.

Some contend that this kind of public involvement process serves mostly as window-dressing for policy decisions, with technocratic approaches prevailing, but even getting publics in the same room as decision-makers and other stakeholders signifies a new locus for finding solutions. The consensus conference process has been used so consistently in the context of agricultural biotechnology issues that, at a minimum, it can be seen as a tool for enabling communication between the many interested parties. And further, by introducing social, environmental and ethical issues that have traditionally been left out of policy debates, it shifts the patterns of interaction between the various actors, offering opportunities for new technology trajectories.

## Acknowledgements

The authors wish to acknowledge financial support for this research from Genome Canada.

## References

Abelson, J., Forest, P.-G., Eyles, J., Smith, P., Martin, E. and Gauvin, F.-P. (2003) Deliberations about deliberative methods: issues in the design and evaluation of public participation processes. *Social Science and Medicine* 57, 239–251.

Boucher, L., Cashaback, D., Plumptre, T. and Simpson, A. (2002) *Linking In, Linking Out, Linking Up: Exploring the Governance Challenges of Biotechnology.* Report for the Institute On Governance, Ottawa, Ontario, Canada, February 2002.

Butschi, D., Carius, R., Decker, M., Gram, S., Grunwald, A., Machleidt, P., Steyaert, A. and van Est, R. (2004) The practice of TA: science, interaction and communication. In: Decker, M. and Ladikas, M. (eds) *Bridges between Science, Society and Policy: Technology Assessment – Methods and Impacts.* Springer-Verlag, New York.

Durant, J. (1999) Participatory technology assessment and the democratic model of the public understanding of science. *Science and Public Policy* 26 (5), 313–319.

Einsiedel, E.F., Jelsoe, E. and Breck, T. (2001) Publics at the technology table: the consensus conference in Denmark, Canada and Australia. *Public Understanding of Science* 10, 83–98.

Elster, J. (ed.) (1998) *Deliberative Democracy.* Cambridge University Press, Cambridge, UK.

External Advisory Committee on Smart Regulation (2004) *Smart Regulation: a Regulatory Strategy for Canada.* Report to the government of Canada (http://www.smartregulation.gc.ca/en/08/rpt_fnl.pdf, accessed 18 October 2004).

Fearon, J. (1998) Deliberation as discussion. In: Elster, J. (ed.) *Deliberative Democracy.* Cambridge University Press, Cambridge, UK.

Felt, U. (2005) Session 2: Science, technology and democracy: *Report to the European Commission on Science in Society – Forum 2005.* Brussels, 9–11 March 2005 (http://europa.eu.int/comm/research/conferences/2005/forum2005/docs/felt_report_en.pdf (accessed 10 October 2005).

Fiorino, D. (1990) Citizen participation and environmental risk: a survey of institutional mechanisms. *Science, Technology and Human Values* 15 (2), 226–243.

Gambetta, D. (1998) 'Claro!' An essay on discursive machismo. In: Elster, J. (ed.) *Deliberative Democracy.* Cambridge University Press, Cambridge, UK.

Giddens, A. (1990) *The Consequences of Modernity.* Polity, Cambridge, UK.

Hails, R. and Kinderlerer, J. (2003) The GM public debate: context and communication strategies. *Nature Reviews Genetics* (October), 819–824.

Hajer, M. and Wagenaar, H. (eds) (2003) *Deliberative Policy Analysis: Understanding Governance in the Network Society.* Cambridge University Press, Cambridge, UK.

Hennen, L., Bellucci, S., Berloznik, R., Cope, D., Cruz-Castro, L., Karapiperis, T., Ladikas, M., Kluver, L., Menendez, S., Staman, J., Stephan, S. and Szapiro, T. (2004) Towards a framework for assessing the impact of technology assessment. In: Decker, M. and Ladikas, M. (eds) *Bridges between Science, Society and Policy: Technology Assessment – Methods and Impacts.* Springer-Verlag, New York.

Institute of Governance Studies (2003) *Research Programme and Organization of the Institute*. Twente University, Netherlands, August (http://www.igs.utwente.nl/research-progs/def_instituutsplan_igs.pdf, accessed 10 October 2005).

Irwin, A. and Michael, M. (2003) *Science, Social Theory and Public Knowledge*. Open University Press, Maidenhead, UK.

Joly, P. (2005) Debates and participatory processes: lessons from the European experience. Presentation at the *Science in Society Forum*, OECD, Brussels, 9–11 March 2005.

Joss, S. (1998) Danish consensus conferences as a model of participatory technology assessment: an impact study of consensus conferences on Danish Parliament and Danish public debate. *Science and Public Policy* 25 (1), 2–22.

Kluver, L., Peissl, W., Torgersen, H., Gloede, F., Hennen, L., van Eijndhoven, J., van Est, R., Joss, S., Bellucci, S. and Butschi, D. (2000) *EUROPTA: Participatory Methods in Technology Assessment and Technology Decision-Making. Danish Board of Technology* (http://www.tekno.dk/pdf/projekter/europta_Report.pdf).

Kluver, L., Bellucci, S., Berloznik, R., Butschi, D., Carius, R., Cope, D., Cruz-Castro, L., Decker, M., Gram, S., Grunwald, A., Hennen, L., Karapiperis, T., Ladikas, M., Machleidt, P., Sanz-Menendez, L., Peeters, W., Staman, J., Stephan, S., Szapiro, T., Steyaert, S. and van Est, R. (2004) Technology assessment in Europe: conclusions and wider perspectives. In: Decker, M. and Ladikas, M. (eds) *Bridges between Science, Society and Policy: Technology Assessment – Methods and Impacts*. Springer-Verlag, New York.

Levidow, L. and Marris, C. (2001) Science and governance in Europe: lessons from the case of agricultural biotechnology. *Science and Public Policy* 28 (5), 345–360.

Liberatore, A. (2001) *White Paper on Governance Work Area 1: Broadening and Enriching the Public Debate on European Matters*. Report of the Working Group's 'Expertise and Establishing Scientific Reference Systems'. European Commission, Brussels, May.

Mayer, S. and Guerts, J. (1998) Consensus conferences as participatory policy pnalysis: a methodological contribution to the social management of technology. In: Wheale, P., von Schomberg, R. and Glasner, P. (eds) *The Social Management of Genetic Engineering*. Ashgate, Aldershot, UK, pp. 279–301.

Mayer, S. and Stirling, A. (2004) GM crops: good or bad? *European Molecular Biology Organization* 5 (11), 1021–1024.

Nielsen, A., Lassen, J. and Sandoe, P. (2004) *Involving the Public: Participatory Methods and Democratic Ideals*. Danish Centre for Bioethics and Risk Assessment, Copenhagen, December.

Pierre, J. (ed.) (2000) *Debating Governance*. Oxford University Press, Oxford, UK.

Phillips, S. and Orsini, M. (2002) *Mapping the Links: Citizen Involvement in Policy Processes*. Discussion Paper No. F21, Canadian Policy Research Networks, Ottawa, Ontario, Canada, April.

Public Participation and the Governance of Innovation (2004) Main results from the research project promoted by Lombardia Region, Irer, Bassetti Foundation and Observa (http://www.fondazionebassetti.org/06/argomenti/2004_11.htm #000324 (accessed 10 Oct 2005).

Rip, A. (2005) Technology assessment as part of the co-evolution of nanotechnology and society: the thrust of the TA Program in nanoned. Paper presented to the *Conference on Nanotechnology in Science, Economy and Society*, Marburg, Germany, January.

Rowe, G. and Frewer, L. (2000) Public participation methods: a framework for evaluation. *Science, Technology and Human Values* 25 (1), 3–29.

Rowe, G. and Frewer, L. (2004) Evaluating public-participation exercises: a research agenda. *Science Technology Human Values* 29 (4), 512–555.

Schot, J. (1992) Constructive technology assessment and technology dynamics: the case of clean technologies. *Science, Technology and Human Values* 17 (1), 36–56.

Seifert, F. (2006) Local steps in an international career: a Danish-style consensus conference in Austria. *Public Understanding of Science* 15, 73–88.

Stirling, A. (2005) Opening up or closing down? Analysis, participation and power in the social appraisal of technology. In: Leach, M., Scoones, I. and Wynne, B. (eds) *Science and Citizens: Globalization and the Challenge of Engagement.* Zed Books Ltd., London.

UK House of Lords Select Committee on Science and Technology (2000) *Science and Society: 3rd Report of Session 1999–2000* (http://www.parliament.the-stationery-office.co.uk/pa/ld199900/ldselect/ldsctech/38/3801.htm (accessed 14 March 2005).

Van Den Ende, J., Mulder, K., Knot, M., Moors, E. and Vergragt, P. (1998) Traditional and modern technology assessment: toward a toolkit. *Technology Forecasting and Social Change* 58, 5–21.

# 14

# The Bt Maize[1] Experience in the Philippines: a Multi-stakeholder Convergence

M.J. Navarro[1*], M. Escaler[1**], M.I. Ponce de Leon[1***] and S.P. Tababa[2****]

[1]*International Service for the Acquisition of Agri-biotech Applications (ISAAA), Metro Manila, Philippines:* [2]*SEAMEO SEARCA Biotechnology Information Center, Laguna, Philippines;*
*e-mail:* [*]*m.navarro@isaaa.org;* [**]*m.escaler@isaaa.org;*
[***]*illustria@gmail.com;* [***]*spt@agri.searca.org*

## Introduction

The Philippines has the distinct honour of being the first country in Asia to have a biotech. crop for food and feed approved for commercialization. Although the country has been involved in biotechnology research and development initiatives since the early 1980s, it was only in 2003 that the country was able to initiate commercial planting of Bt maize.

Next to rice, maize is the second most important crop in the Philippines, with yellow maize accounting for about 70% of livestock mixed feeds. Despite its importance, maize production cannot meet domestic demand, primarily due to the extensive damage by the Asian maize borer. Bt maize is expected to provide another option for farmers to increase their yield and help the country attain self-sufficiency in this grain.

The introduction of Bt maize in the Philippines was a story that spanned 7 years of rigid scientific study and evaluation, with the public and private sectors involved in various stages of the research and development (R & D) process. However, Bt maize also had its share of controversy, which resulted in a polarized and emotional debate between those for and against the technology.

Critics accused proponents of biotechnology of rushing the introduction of Bt maize into agriculture without adequately considering

[1] The Bt maize referred to in this paper is the event Mon 810.

 *The Media, the Public and Agricultural Biotechnology* (ed. D. Brossard *et al.*)

health, environmental, socio-economic and ethical risks. They felt that the public was not consulted and informed about the technology's potential consequences. On the other hand, proponents believed that such risks were exaggerated and were not based on scientific evidence.

The Bt maize story of the Philippines brought to light the noticeable divide between scientists and the public sphere. Although a growing number of scientists took a deliberate effort at communicating with the public, many of them preferred to work quietly in their laboratories and disclose their findings only to their peers. Communication with the public was thus limited and not a regular activity planned with other science communicators. Now, however, scientists are being asked to communicate not only to a diverse group of stakeholders but also to a more concerned, involved public which, with today's information revolution, can understand the merits and pitfalls of a technology.

The commercialization of Bt maize in the country is a unique case study demonstrating the crucial role of science communication in technology development and acceptance. In this chapter, we will relate the events that unfolded before, during and after the event's approval, as well as the parallel communication activities that were undertaken to finally bridge the knowledge gap. In so doing, we highlight how the different concerns of a diverse group of stakeholders were considered and came into a convergence that allowed the approval of Bt maize in the country.

## Science Communication and the Need for 'Socially Robust Knowledge'

More and more literature points to a concrete discipline called science communication. As defined by Gregory and Miller (1998), it is a process of generating new, mutually acceptable knowledge, attitudes and practices. It is a dynamic exchange as disparate groups find a way of sharing a single message. It is a process of negotiation based on trust that leads to mutual understanding, rather than through statements of authorities or of facts. Hence, communication is necessary to enable stakeholders to participate in the social processes of debate and decision-making.

In a sense, science communication is breaking down the barrier between science and society, aiming to offer the public knowledge on what scientists do, while allowing this same public to answer back and offer its opinion. Precisely because society has participated in knowledge generation and validation, such knowledge is less likely to be contested than if it were produced by a specific sector alone. Issues can be answered by conventional science, but it is not 'socially robust' until the peer group is broadened to include the perspectives and concerns of a wider section of the community. The reliability of scientific knowledge is thus complemented and strengthened by the participatory involvement of various stakeholders. Thus, we see an emerging pattern where science and society are transforming each other (Gibbons, 1999).

Such a process can be undertaken through various communication modalities, which include interpersonal and multi-media channels. The means with the widest reach however, are still the mass media, which include radio, television, print and, now, the Internet.

### The role of the media

The role and influence of the mass media in focusing public attention on key public issues is well documented. Elements prominent on the media agenda become significant in the public mind. The news media can set the agenda for the public's attention to issues around which public opinion forms. Hence, the agenda-setting influence of the news media does not stop at merely focusing public attention on a particular topic. It pervades the communication process, the understanding and perspective on the topics in the news (McCombs, 2002).

Mass media's ability to effect cognitive change among individuals as well as to structure their thinking stresses the power to selectively choose what people see or hear in the media. That is, if the media broadcast negative images of a certain technology well enough for them to be remembered, the public will most likely believe it and choose to shun the technology. On the other hand, if the media broadcast positive images, then the public may be urged to accept the technology and its benefits. The knowledge that the media presents, however, may not necessarily be socially robust nor, at the very least, science-based. This is where science communicators (to include not only scientists but all those engaged in the communication of science) play a major role: in finding ways and means to get the truth out while still garnering listeners and viewers.

## The Case of Bt Maize in the Philippines

The approval of Bt maize in the Philippines in December 2002 was not without controversy. It was the first genetically modified food/feed product ever to be allowed for commercial planting in Asia, and therefore attracted enormous amounts of media and public attention both locally and internationally.

During the 7 years of the local evaluation of the technology, there was a continuous communication tug-of-war among the technology developers, the scientists, scientific organizations, advocacy groups/non-government organizations, the farmers involved in the trials and the governmental sector. The debate in the Philippines continued from 1996 to 2002, and well after Bt maize was approved for planting and commercialization. The debate also saw a plethora of stake-holders, who included even the religious community, all trying to win the hearts and minds of the public and the government agencies assigned to assess the technology.

Some cause-oriented groups uprooted a field trial, sued the technology developers and lobbied for a moratorium on GM crops. A group of Catholic priests and nuns pleaded with local government units to refrain from giving support to GM activities in the community. Even politicians, including two senators, joined the fray by alleging that GM products could cause cancer and that it was a crime to do GM research. Filipino scientists battled it out with various groups in order to clarify the various concerns regarding the Bt maize technology.

Addressing the different concerns of such a diverse group of stakeholders became a real challenge, but was critical to the eventual commercial approval of Bt maize in the country.

## Timeline of Bt maize approval and parallel communication activities

Although the furore and clamour surrounding Bt maize came as early as 1998, when the technology developers filed for a limited field trial, its players had already been established nearly a decade beforehand. The National Committee on Biosafety of the Philippines (NCBP) was established in October 1990, by virtue of Executive Order No. 430. A year later, the first edition of the country's Biosafety Guidelines was issued, detailing procedures for work involving genetic engineering, as well as activities requiring importation, introduction, field release and breeding of exotic organisms that could be potentially harmful to people and to the environment, even though such organisms were not genetically modified.

In 1996, a congressional hearing on Bt rice was conducted in response to growing interest in GM technology. Soon after, two activist groups exposed the importation of the crop for research by a scientific institution, with one group initiating a 'rice-napping' action in Zurich and the other leading the local media campaign.

At the same time, however, Bt maize was already being evaluated in a greenhouse at the same scientific institute and was beginning to attract protests by some activist groups. It was against this backdrop that academic and government institutions decided to collaborate on an information and communication campaign on biotechnology in 1998.

The early institutional players in biotechnology outreach activities included: (i) the different institutes of the University of the Philippines (UP), Los Banos, such as the Institute of Plant Breeding (IPB) and the National Institute of Molecular Biology and Biotechnology (BIOTECH); (ii) the sectoral planning councils of the Department of Science and Technology (DOST), namely the Philippine Council for Advanced Science and Technology Research and Development (PCASTRD) and the Philippine Council for Agriculture, Forestry ad Natural Resources Research and Development (PCARRD); and (iii) The National Academy of Science and Technology (NAST).

During the same period, several public, congressional and senate hearings were being conducted, and bills regarding GMOs and Bt maize were drafted. Some of these supported the use of GMOs as a tool to improve food productivity, while others proposed a moratorium.

Despite the heated environment, scientists continued to file their field trial applications and regulatory review proceeded. By late 1999, the first limited field trial of Bt maize Mon 810 was approved by the NCBP and was located in General Santos City. However, in less than a year, the city had approved a resolution imposing a 5-year moratorium on Bt maize. Protests by Catholic groups and other NGOs followed in other sites across the country. When multi-location trials began in 2001, one of the field trials was even destroyed by militant groups.

It was during this turbulent period that communication and scientific outreach activities gained significant momentum. In September 2000, the SEAMEO Regional Center for Graduate Study and Research in Agriculture–Biotechnology Information Center (SEARCA BIC) was established, and was tasked to play a major role in looking after communication activities regarding crop biotechnology. Its efforts complemented the communication outreach conducted by line government agencies such as: (i) the Department of Agriculture and the Department of Science and Technology; (ii) state universities such as UP's National Institute of Molecular Biology and Biotechnology (NIMBB) and Isabela State University (ISU); (iii) the consortia of agencies in agriculture and natural resources such as the Cagayan Valley Agriculture and Resources Research and Development Consortium (CVARRD); and (iv) the NAST.

Not too long after, a coalition of academic and government institutions, private sector and non-government organizations was established to promote the safe use of biotechnology. This group, first launched as the Biotechnology Conference of the Philippines, later became known as the Biotechnology Coalition of the Philippines (BCP).

The BIC, together with the Coalition and the other institutional players mentioned above, embarked on a deliberate communication programme that focused on crop biotechnology, with emphasis on GM crops. Initially, communication outreach activities were targeted to areas where GM field trials were conducted and to specific groups of stakeholders. They each organized seminars for more general audiences including students, teachers, researchers, government workers, extension workers, local government units, media and farmers. The BIC, in particular, developed a website and created an e-group where news and other developments in biotechnology were shared with its community of users. Numerous communication materials like brochures, posters and radio plugs were developed. They even organized study visits to biotech. facilities and farms for various stakeholders, which proved to be extremely effective.

Finally, in April 2002, significant headway was made with the issuance of the Department of Agriculture's (DA) Administrative No. 8 (AO 8), which governs the importation and release into the environment of plants and plant products derived from the use of modern biotechnology. However, AO 8 only came into existence after several nationwide public consultations were conducted by the DA. It was not long after this, in December 2002, when Bt maize was finally approved by the DA Bureau of Plant Industry for propagation (ISAAA, 2004).

Still, public debates and protests continued. In May 2003, anti-GM proponents held a hunger strike in front of the DA. They asked for the withdrawal of the approval of Bt maize and the imposition of a moratorium on GMOs. The DA stood firm on its decision to allow planting of Bt maize because of insufficient scientific evidence to warrant a moratorium on the commercialization of the said product. Subsequent news stories focused on various reactions to this event, both in support of and in opposition to biotechnology. Statements of support were issued by food groups, the science community and legislators. The month was capped by a story on the first GM maize harvest in the province of Ilocos Norte.

Since its commercialization, about 70,000 ha have been planted to Bt maize. Aside from Mon 810, three events have been approved for propagation, namely Maize Mon Bt 11, Maize NK 603 and Maize NK 603 × Mon 810. In addition to these, more than 30 events have been approved for direct use as food and/or feed. These events are deployed in crops like maize, soybean, canola, cotton, potato and sugarbeet (Bureau of Plant Industry, Philippines, 2006, personal communication).

## Importance of numerous stakeholders in overall communication strategy

In order for any biosafety measures to be effective, they must be science based, flexible and transparent. This first allows for objective risk evaluation, the second for new knowledge and the third for public participation.

Public awareness and understanding at various critical stages of the R & D process is crucial, as well as an openness of the scientific community and other stakeholders to the need to answer to public opinion. Diverse opinions from various sectors serve as a check and balance to refine technology so that it is acceptable by the ultimate beneficiaries. The knowledge gap is partly bridged by providing the public with the appropriate seminars, conferences, symposia and information, education and communication materials (IECs), in order for the public to make an informed decision.

On the other hand, scientists and extension workers are trained in risk communication in order for them to be able to communicate effectively to various audiences, many of whom do not have a technical grasp of issues and concerns. In like manner, workshops and media encounters with different stakeholders are offered to media practitioners so they are updated on scientific concerns as well as have a better understanding of issues that may affect public opinion.

## Roles of major stakeholders in the entire process leading up to Bt maize approval and parallel communication activities

Unlike with previous technologies, the introduction of Bt maize has seen unprecedented involvement of various stakeholders in the debate and discussion of its merits and limitations, as well as its eventual approval. Initially, scientists kept to their laboratories and field trials. There were not many scientists who actively participated in the public debates on biotechnology or communication campaigns. Soon, however, they could no longer remain in the background. Both technology developers and scientists in both public and private sectors found themselves suddenly accountable for their work, not only to their peers but also to cause-oriented and consumer groups, as well as policymakers. They not only had to talk about their research or biotechnology, but had to actually participate in communication activities. Still another dimension was the scientists themselves not having consensus with one another about aspects of the technology.

There were farmers on both sides of the GM divide. Some of them supported the technology and joined scientists at symposia and gatherings. Others joined in protests and hunger strikes organized by activist groups. Many remained non-committal about the technology. Nevertheless, there were quite a few farmers leaders who joined scientists in various government hearings at both the local and national levels. Their involvement and perspectives gave a pragmatic, real-case scenario to the debate.

The support of the local government units (LGUs) was critical, not only during field trials but more so when the technology had been adopted by farmers. Because the regulatory process required public consultation about where trials were to be conducted, it was important for local governments to have a fair hearing on GM technology. In fact, many of the passionate debates occurred at the local level. Some LGUs supported the transgenic crop field trials through local resolutions, while others opted not to have field trials in their community.

The private sector, particularly technology developers, also contributed to public awareness activities by tapping resource persons in the academe to help explain and clarify the issues surrounding Bt maize, especially during the hearings at the local level. International organizations, such as the International Service for the Acquisition of Agri-Biotech Applications (ISAAA), likewise assisted in capacity-building activities for researchers and scientists involved in crop biotechnology.

The media operated on what information came most prominently into the press. That is, if some groups created attractive anti-GM sound bytes, then the news highlighted these sentiments. Conversely, if scientists chanced to answer news or opinion columns, then the media promptly took up the baton and reported pro-GM news. Since national newspapers have sections on agriculture, science, environment and news from provinces, developments in agriculture and crop biotechnology receive regular coverage. The presence of media associations for science, agriculture and environment also helps in assuring media mileage for agriculture.

In order to quantify how media had presented the issues of crop biotechnology in general, and Bt maize in particular, ISAAA analysed data from a media-monitoring study of nine national daily newspapers in the Philippines from January 2001 to May 2003. A total of 446 articles were featured, over half of which were positive in tone. The articles clustered around specific events such as the events following the signing of AO 8 in April 2002 and the approval of Bt maize (Navarro and Villena, 2004).

One aspect of the Philippine scenario quite unique to the GM environment is the influence of the Church. Some religious leaders, such as priests and nuns, took their turn to protest and organized their own anti-GM rallies and meetings, putting banners in churches calling for the end of Bt maize planting. On the other hand, there were also members of the religious sector that espoused the religious stand of the Vatican regarding the safe use of biotechnology by citing the Compendium of the Social Doctrine of the Church, released in 2004 by the Pontifical Council for Justice and Peace.

Policymakers were likewise divided on the issue. A number of congressional resolutions and bills were drafted, with the aim of directing the appropriate committees to: 'conduct an inquiry, in aid of legislation, into the reported field testing of genetically modified organisms, including the possibility that they may have already entered the country, to determine the adequacy of any safety measures that may have been adopted to prevent unnecessary accidents, mishaps and health impacts to the consuming public and the environment and for other purposes', as in Resolution 282; or: 'requiring mandatory labelling of genetically engineered food products and providing penalties for violation thereof', as in House Bill No. 7793; or even requesting that the appropriate government committee 'conduct an inquiry, in aid of legislation, into the entry in the country of GMOs, their use in our food products and the extent of danger, if any, that they pose to the public', as in House Resolution No. 552.

Policymakers from government at the national and regional levels were given briefings regarding GM technology, especially its application and extent of adoption in other countries that have grown GM crops.

After much discussion and debate, the Senate Committee on Agriculture and Food concluded its deliberation by rejecting the proposed moratorium on the entry of GM crops. According to the statement issued by the same committee, the Philippine Senate found that: 'the proposed moratorium on activities related to biotechnology and GMOs would, in effect, stifle the impetus of human innovation and inventiveness and exclude the Philippines from the tide of technological advances now prevalent elsewhere in the world' (Philippine Senate Committee on Agriculture and Food (2000), available at: http://www.gene.ch).

The committee recommended, among other things, that the powers of the NCBP be strengthened; that a law be enacted to penalize the unauthorized entry, use or release of GM products; and that better information dissemination should be undertaken by appropriate agencies in coordination with local government units.

## Conclusion

The case of Bt maize commercialization in the Philippines has clearly demonstrated the crucial role of the involvement of stakeholders and communication. On hindsight, it can be said that the confluence of many factors – beyond just the technology itself – has led to a perceived complex, if not circuitous, route to the eventual approval of Bt maize in the country. Despite diverse interests and ideological viewpoints, each stakeholder contributed to a better understanding of the technology. Such diversity was tempered by communication initiatives and stakeholder outreach through informed debate, media reach and opportunities for interaction and eventual convergence of ideas and opinions. The process also enabled stakeholders to appreciate the crucial role of communication in fostering awareness and appreciation for science and technology.

Indeed, communication is an ability that must extend to and be shared amongst the various stakeholders of the technology. The commercialization of Bt maize in the country enabled a concrete, tangible biotechnology issue to be discussed in the mass media and between various stakeholders. Thus, for any technology to be introduced and commercialized, it must first be accepted by the public, and must be accompanied by a campaign involving socially robust knowledge.

The story of Bt maize in the Philippines did not end with its commercial planting. It continues to unfold in spite of current government support for the technology, farmers' continuing acceptance of Bt maize as an alternative crop and favourable public opinion. Cause-oriented groups and other sectors continue to espouse their respective viewpoints and call for a moratorium. What the Bt maize experience has proved, however, is that by recognizing the role of various stakeholders, and an openness to communicate and respect each other's perspectives, the development of socially robust knowledge and acceptance of technology is possible.

## References

Gibbons, M. (1999) Science's new social contract with society. *Nature* 402 (Suppl.), C81–C84.

Global Biodiversity Institute and the International Institute for Tropical Agriculture (2000) *Biodiversity, Biotechnology, and Law: West Africa* (available at: http://www.aaas.org/international/africa/gbdi/mod1a.html).

Gregory, J. and Miller, S. (1998) *Science in public: Communication, Culture, and Credibility.* Perseus Publishing, Cambridge, Massachusetts.

International Service for the Acquisition of Agri-Biotech Applications (ISAAA) (2004) *Bt Maize Videos: Asia's First – the Bt Maize Story in the Philippines.* ISAAA, the Philippines.

McCombs, M. (2002) Democracy, citizens and the news: the agenda-setting role of the news media and the formation of public opinion. *The London School of Economics Conference on Mass Media.* London, June 2002.

Navarro, M. and Villena, M. (2004) Media monitoring of agri-biotechnology in the Philippines: understanding the biotech debate. *The Philippine Agricultural Scientist* 87, 439–451.

Philippine Senate Committee on Agriculture and Food (2000) *Report on GMO field testing* (available at: http://www.gene.ch).

# 15

# Food Aid Crisis and Communication about GM foods: Experiences from Southern Africa

L.E. Mumba

*NEPAD S & T, SANBIO, Pretoria, Republic of South Africa; e-mail: Mumba@sanbio.co.za; lemumba2004@yahoo.co.uk*

## Introduction

This chapter reviews the food aid crisis of 2001–2002 in southern Africa and examines how communication about genetically modified (GM) foods was handled in Zambia. However, much of the content will be applicable to other countries in the region. As a result of two consecutive poor cereal harvests due to drought, an estimated 14 million people needed emergency food assistance to survive. Food donations from the international community, which had some GM events, were in some cases rejected outright, whereas in other cases it was accepted only after protracted negotiations with national governments. There is much to be learned in southern African countries from the GM food aid crisis experience. This chapter makes reference to Malawi, Lesotho, Swaziland, Mozambique and Zimbabwe, and covers in depth the GMO debate in Zambia.

Southern Africa is considered a poor, highly indebted and most disadvantaged region of Africa, with 50% of countries being land-locked. The population growth in these countries is unsustainable and the inhabitants are generally: (i) income poor and undernourished, especially the young and aged; (ii) immunocompromised, especially from HIV/AIDS; (iii) literacy levels are low; and (iv) cultural values have been lost. Frequently, there are also political challenges.

Agriculture is the main occupation, consisting of both small and commercial farmers. This sector, however, has problems of low net production resulting from, among other factors, pests, disease and natural production failures – like drought and wastage during storage.

Of late, the sector is facing seed security pressures arising from intellectual property rights of commercial seed, and marginalization of traditional crops.

The above factors, which can be divided into: (i) geographical/ecological; (ii) socio-cultural; and (iii) economic and agricultural, affect the extent to which food can be produced, distributed and accessed, and these ultimately affect food security, nutrition and health. This often leads to net food deficits from time to time, resulting in the need for food imports which, in turn, raise issues of biosecurity and other food safety concerns. Food crisis might occur when countries in the region are unable to produce enough staple food (maize) at affordable prices to feed both the growing urban and rural populations.

In addition, food crises are also due to the fact that, even in situations where both locally produced and imported food may be available, the local people may be unable to purchase it due to high prices caused by inflation, low incomes and the liberalization of the African economies caused by restructuring (Kajoba, 1993).

This chapter examines: (i) the extent of the famine; (ii) the offer of humanitarian assistance in the form of GM maize by the WFP and other international non-governmental organizations (NGOs) as an intervention in the crisis; (iii) the reasons for limited acceptance of the offer; and (iv) the key players that participated in communicating the pros and cons of GM foods in the region. Was the decision-making process by national governments transparent and inclusive of the rural communities who where often the most acutely affected by the famine? Could the communication on GM foods have been handled differently?

This chapter is not meant to apportion blame for actions taken or not taken, but rather to share experiences from the region with others upon which lessons can be drawn for the future as regards the handling of GM food aid. Admittedly, food aid is one of the most controversial instruments of international cooperation, and the GM dimension to it only exacerbates the magnitude of the controversy. Therefore, this chapter is not exhaustive. The author merely scratches the surface of the controversy, so to say! The staple food in the region is maize; as such, all reference to GM foods here will refer to maize.

## Extent of Famine in Southern Africa in 2001–2002 and Response by the International Community

Several countries in southern Africa, including Malawi, Lesotho, Swaziland, Mozambique, Zimbabwe and Zambia, experienced a food crisis in 2001–2002. At its peak in 2002, the crisis had reached such proportions that the governments of some of these countries declared that they were facing national disasters due to actual and anticipated food shortages: Malawi, Lesotho, Zimbabwe and Zambia.

The food crisis was precipitated by drought. This seriously compounded existing structural and other weaknesses within the agricultural sector in the region, which had themselves been brought about by inappropriate and expedient government policies, trade liberalization, structural adjustment programmes and the impact of HIV/AIDS (Greenpeace International, 2003, unpublished report).

Following weeks of concern over low summer rainfall and the development of moderate El Niño conditions in southern Africa, a drought warning was issued by the Southern African Development Community's (SADC) Drought Monitoring Centre (DMC). The worst affected areas were southern Malawi, Swaziland, southern Mozambique, Lesotho, eastern South Africa, eastern Botswana and part of north-western Zambia, while the situation in western and north-western Zimbabwe was deemed as critical. As a result of two consecutive poor cereal harvests, an estimated 14 million people in Zimbabwe, Zambia, Malawi, Mozambique, Lesotho and Swaziland needed emergency food assistance to survive (Greenpeace International, 2003, unpublished report).

In response to the emerging crisis, on 18 July 2002, the United Nations launched an appeal for US$508,745,176 to fund a 9-month (July 2002–March 2003) food aid distribution solution. At the time, the World Food Programme (WFP) estimated that 993,050 metric tonnes of food would be needed for emergency relief. By December 2002, 662,945 t of food aid (66.76% of the original estimate) had been pledged. In cash terms, US$298,602,381 of confirmed contributions had been pledged (58.69% of the total originally requested by the WFP) (WFP, 2002b).

As conditions in the region worsened, the estimates of required food aid increased to a total of 1.2 million t for those countries in the region where food shortages were the most acute (see Table 15.1). Furthermore, due to a lack of viable seeds during the 2002 planting season (WFP, 2002a), and with adverse weather conditions likely to continue, a favourable harvest in certain regions in 2003 was unlikely, thus exacerbating regional food shortages.

It is within this context that food aid and, more specifically, food aid containing GM maize, was introduced to provide famine relief for affected countries in the region. This in turn heightened the global debate on health, socio-economic and environmental impacts of GM crops as governments in southern Africa continued to resist or seriously question GM technology as a whole. The often acrimonious

**Table 15.1.** 2002 Food AID Needs in southern Africa and proposed US assistance (from Greenpeace International, 2003, unpublished report).

| Country | Peak number of people in need of food aid | Cereal food aid needs (t) | Total US food assistance (t) | Food aid needs met by US assistance with GM maize (%) |
|---|---|---|---|---|
| Zimbabwe | 2,075,000 | 705,000 | 217,000 | 30.8 |
| Malawi | 3,188,000 | 208,000 | 141,895 | 68.2 |
| Zambia | 2,329,000 | 174,000 | 74,000 | 43.0 |
| Mozambique | 515,000 | 62,000 | 19,790 | 32.0 |
| Lesotho | 445,000 | 50,000 | 27,760 | 55.5 |
| Total | 8,552,000 | 1,199,000 | 480,445 | 40.1 |

debate, despite its validity, also obscured many traditional factors that needed to be considered in the broader debate of food aid. These included the freedom to choose, the nature of food aid, the role of the international trade regime and so forth (Greenpeace International, 2003, unpublished report).

In the summer of 2002, Zambia, Zimbabwe, Mozambique, Malawi, Lesotho and Swaziland faced dramatic food shortages which threatened more than 10 million people with starvation. Their governments decided to refuse maize donations from the USA on the grounds that the cereal was genetically modified. In the autumn of 2002, Malawi, Mozambique and Zimbabwe requested that all US-imported GM maize be milled prior to distribution in order to prevent its inadvertent use as seed.

Lesotho and Swaziland authorized the distribution of non-milled GE food aid, but warned the public that the grain should be used strictly for consumption and not for cultivation. The Zambian government, on the other hand, remained unconvinced and rejected 63,000 t of maize from the USA, despite the threat of more than 2 million Zambians facing starvation. The government refused to accept the food aid and effectively took a decision to ban the distribution of GM food aid within Zambian borders.

## Focus on the Zambian Debate

The food crisis in Zambia was most acute in the Southern Province of Zambia and, to a varying degree, in parts of Eastern, Central, Western and Lusaka Provinces. North-western and Northern Provinces were also affected to varying degrees. Equally afflicted by the hunger situation were the refugees Zambia was hosting in the various camps around the country. The estimated total population in need of food assistance was 2.4 million, which was expected to swell to 3 million by March 2003. The country as a whole did not have adequate maize stocks to supply the affected areas. The country required 224,200 t of cereal to meet the shortfall and to supply the affected population adequately (GRZ, 2002a). To mitigate the adverse effects of the food shortage on the affected population, the Zambian government declared the southern parts of the country a national disaster and appealed for food donations (GRZ, 2002a).

In response to this food crisis, the WFP of the United Nations offered Zambia food aid in the form of maize containing GM events. One of the criticisms levelled against the WFP was that the government was only informed that the maize had GM events after a large consignment of it had already been shipped into the country. The WFP did not obtain Prior Informed Consent from the recipient country, as required by Article 8 of the Cartagena Protocol on Biosafety. It is a requirement under this principle that recipient countries of food aid should be fully informed by donors about the food being offered to them and should have the choice of deciding whether to accept the food or not. This omission by the WFP was to be later advanced as one of the reasons for ordering that GM food aid be removed from Zambia.

Three institutes – the Soils and Crop Research Branch (SCRB) of the Ministry of Agriculture and Cooperatives, the National Science and Technology Council (NSTC) and the National Institute for Scientific and Industrial Research (NISIR) – advised their respective Government Ministries against the acceptance of the GM food aid. The government advised the WFP not to distribute the GM food aid until further notice.

First, it was argued that circulation of GM maize in southern Africa might lead to its uncontrolled spread, if kernels were used for planting rather than for consumption. It was pointed out that planting GM maize might have unpredictable consequences in terms of gene flow and, in particular, that genetic materials might eventually spread to fields on which non-GM maize could be grown for export. Given the *de facto* moratorium in the EU and its reluctance to accept imports of GM foods, it was feared that a major future export market might be lost.

Second, although the governments of Zimbabwe, Malawi, Mozambique, Lesotho and Swaziland had eventually decided to accept GM food aid, the government of Zambia did not wish to take any risks and was sceptical about whether GM food was safe to eat. While acknowledging that GM maize might be safe to eat for the US population, where this crop forms a relatively small proportion of the diet, it was noted that maize accounted for as much as 90% of the typical Zambian diet. It was also feared that the high prevalence of HIV/AIDS in Zambia could bias the transferability of studies on food safety undertaken in developed countries. Thus, it was argued that GM maize might be unsafe for consumption by Zambians.

The policy dilemma that confronted Zambia over whether or not to accept GM maize offered as food aid by the USA drew questions over the way and the extent to which the debate over the issue was allowed in the country.

When the government rejected the US offer in August 2002, many commentators described the move as a bold step aimed at asserting the country's national pride. But with the UN WFP estimating that nearly 3 million people faced starvation in Zambia, the rejection was perceived by some Western observers as unreasonable – the UK newspaper the *Financial Times* called it 'absurd'. The Zambian government's refusal of the aid was 'a crime against humanity', claimed Tony Hall, the American ambassador to the United Nations Food and Agriculture Organization (FAO), with indignation. The ambassador emphasized that representatives of governments who refused food for their people and let them starve should be called to account before the highest courts in the world.

On 12 August 2002, the government organized a public debate in order to gauge the scientific evidence and other views. The debate was convened by the office of the Vice President and was chaired by the Secretary to the Cabinet. It was attended by over 200 participants from a broad spectrum of stakeholders and interested parties. It highlighted deep divisions among Zambian scientists on the benefits of biotechnology. The two Ministers of Agriculture and Cooperatives and Science, Technology and Vocational Training gave speeches. An overview paper on GMOs, potential benefits and risks, prepared by the three government institutes (SCRB, NSTC, NISIR) was presented by a representative from NISIR. Following this paper, four invited papers – from WFP, Organic Producers and Processors Association of Zambia (OPPAZ), Zambia National farmers Union (ZNFU) and Kasisi Agricultural Training College (KATC) – were presented (GRZ, 2002b).

It should be noted that, except for the presenter from WFP, the other four speakers – including the Chair and the two ministers – were all senior government officers hitherto strongly opposed to GM food aid. Given the backgrounds of the key speakers, it would appear as though the direction of the discussions and probably the conclusion to the debate had already been predetermined. In plenary session, seven institutions/individuals spoke in favour of accepting GM maize in milled form against eleven who were against GM maize in any form. This brought the total number of those who spoke in favour of GM food aid to 8 against, 15 opposed.

Focus group discussions organized by an organization called Panos Southern Africa in 2001, in conjunction with the Zambia National Farmers' Union (ZNFU), showed that farmers, too, were divided on the issue (Banda, 2004). While most small-scale farmers wanted more information on the subject, commercial farmers were opposed to GMOs, citing as their main reason the possibility of losing European markets for their existing non-GM exports.

The European Union accounts for 53% of Zambian exports – mostly made up of processed and refined foods, primary agricultural commodities and floricultural, horticultural, animal and leather products. Largely based on this trade-related rationale, ZNFU was among those organizations that welcomed the government's rejection of US food consignments (Banda, 2004). Others included the Organic Producers and Processors Association of Zambia (OPPAZ), the Jesuit Commission for Justice and Peace (JCJP), Kasisi Agricultural Training Centre (KATC), Zambia Medical Association (ZMA) and Zambia Consumers Association (ZACA).

These discussions were held against a backdrop of little media coverage of GMOs. According to one media content analysis, only four newspaper articles appeared on the issue throughout 2002. Almost all articles covered biotechnology in a general way, with little or no contextualization (Banda, 2004). The section on media coverage will discuss this point further.

The scientific case for rejection took the view that there was compelling evidence that GMOs would have a negative impact on the local breeds such as millet, sorghum and traditional maize, with the possibility of causing an ecological problem that would affect farming. It was further emphasized that these fears were borne out of a peer-reviewed study which suggested that gene flow had occurred from GM maize to native Mexican maize (FAO, 2003).

A number of non-governmental organizations (NGOs) interpreted this as an instance of 'genetic pollution', claiming that the 'well had been poisoned' (GRAIN, 2003). While it was not clear how the GM maize might have been introduced in Mexico – where a ban on GM crops was in place – subsequent debate about the scientific validity of the research led the journal *Nature* to disavow the published paper. The editors of *Nature* admitted in the 4 April 2002 issue of the journal that the said paper by Quist and Chapela was riddled with methodological errors and should not have been published in the first place (Metz and Fütterer, 2002). Nevertheless, Zambian scientists opposed to GM food still advanced this paper as a strong case for rejecting GM food aid.

It was further argued that GM foods might contain new food toxins or new allergens, and might increase antibiotic resistance because of the widespread use of antibiotic resistance marker genes in GM products. It was noted that the millions of Americans who consumed GM maize did so mostly in processed foods such as maize flakes and taco chips, and the new genetic formations that might cause heath problems would be rendered harmless during processing of these products.

In contrast, Zambians eat unprocessed maize as the staple food and usually as the only source of carbohydrates, so its impact would be different. Zambians consume it for breakfast, lunch, supper and as a snack between meals. Another consideration was that the likely recipients of the food aid were the most vulnerable members of the society – the old, women and children – some of whom were in a poor state of health and immunocompromised. A representative from the Medical Association of Zambia submitted:

> The absence of explicit health effects does not imply that they do not exist. Zambians are generally malnourished and thus very vulnerable to foreign substances. Results of experiments done elsewhere can not directly apply in Zambia. We need to be cautious, as maize in Zambia is consumed as a staple food in large quantities and on a daily basis. The susceptibility to reactions from GM foods is thus very high. Allergies and deaths of people had been reported in other countries like Japan after consumption of certain GM foods. In Medicine, for the family, any person who dies is a 100% loss to them. Ethical issues should also be considered where GMOs are concerned.

In view of these strong sentiments against GM food, the government was advised to err on the side of caution by invoking the 'precautionary principle' clause of the Cartagena Protocol on Biosafety. According to the precautionary principle, even if there is no clear evidence that a seed type is dangerous, the government can decide to take the precaution of refusing it if there is likelihood that it might be harmful.

Two preconditions were laid down for allowing GMOs into the country: First, the need to develop a national biosafety framework to regulate biotechnology and GMOs. Secondly, the government must build the capacity to detect and monitor GM substances in foodstuffs coming into Zambia.

Another view advanced by critics of the WFP was that non-GM maize was available in some parts of Zambia, in the region and elsewhere in the world. Suggestions were made to the effect that the northern parts of Zambia had a surplus of maize (a position disputed by some officials within the Ministry of Agriculture). According to this point of view, what were required were resources to transport the maize to areas of Zambia that had food deficits. It was also argued that a number of African countries had available surplus maize that was non-GM.

Furthermore, non-GM maize was available at the global level, even in the USA.

To render support to this viewpoint, the Greenpeace Briefing of 2003 claimed that USAID had chosen to supply genetically engineered (GE) maize as food aid, even though there were numerous grain companies in the USA from which they could have supplied certified non-GE grain. Greenpeace quoted a survey by the American Maize Growers Association suggesting that over 50% of US elevators segregated GM and non-GM grains (American Maize Growers Association, 2001). Greenpeace also quoted a survey conducted in 2000 by seed giant Pioneer Hi-Bred that found that nearly 20% of US maize elevators were dedicated to using only non-GM varieties (Pioneer Hi-Bred International, 2000).

These arguments were strongly rejected by the US Ambassador, William Farish, who argued that it was impossible for the USA to provide non-GM maize because more than 95% of US maize was GM (Farish, 2003). Guy Scott, former Zambian Agricultural Minister argued that: 'If the aid agencies had cash rather than maize they could resolve the crisis without touching GM. But it is the official policy of USAID to promote GM' (Carroll, 2002). It was also noted that the Zambian government had made a decision not to accept GM food aid as far back as July and August 2002, and the impact of the food crisis was going to be critical 7 months later in March/April 2003. This, therefore, gave well-wishers enough time to source for non-GM food aid (Lewanika, 2003).

There was also a strong response from the faith-based organizations, particularly the Jesuit Centre for Justice and Peace who had, in fact, prior to the debate commissioned a study on the impact of GMOs on sustainable agriculture in Zambia. This study was an attempt to serve small-scale farmers, promote sustainable agriculture, promote social justice and encourage poverty eradication (JCJP, 2002).

JCJP views were that the push for the adoption of GM crops in Zambia was posing a serious challenge to the current agricultural infrastructure of Zambia, with consequent danger to the viability of food production to meet the needs of over 10 million Zambians. They noted that GM crops posed a particular threat to the survival of the powerless majority (small-scale farmers) of the farming population.

They further noted that, in their quest to increase their profit margins in the agricultural business, a powerful minority group of the farming population was attempting to use the small-scale farmer to persuade the government to allow them to bring GM crops into the country. They observed that the current commercial GM crops had, in fact, little if anything to offer to the small-scale farmers on the one hand while, on the other hand, these crops were likely to exacerbate the rural household food insecurity and further erode the little cash income which might be there.

Simply stated, the critical point of debate was that the very serious problem of food consumption (the presence of hunger) must not be dealt with in ways that created even more serious problems of food production (the destruction of agricultural infrastructure) (JCJP, 2002).

Although most of the debate was confined to scientific polemics, there was also some ideological/nationalistic opposition to GM foods. The argument from this group suggested that the US government, represented by huge seed trans-national corporations, had an interest in establishing future markets on the African continent for its GM food exports. The asserted, further, that the USA was just not willing to offer non-GM maize in place of GM food aid.

The Zambian public was reminded that particular care needed to be given to the way in which the precautionary principle was being applied in making decisions about the use of GM crops. Highly restrictive interpretations invoke the fallacy of thinking that the option of 'doing nothing' is itself without risk. However, in some cases the use of a GM crop variety may well pose fewer risks than the agricultural system already in operation. Therefore, in applying the precautionary principle, risks arising from the option of inaction must also be considered. It was further asserted that research focusing on 'second generation' transgenic crops – those more to do with increased nutritional and/or industrial traits – had led to such beneficial products as iron- and vitamin-enriched rice, potatoes with higher starch content, edible vaccines in bananas and maize varieties able to grow in poor conditions. In drought-prone Zambia, hardy, GM maize would have been a useful contribution to ensuring food security.

The proponents of GM food aid acknowledge that there have been reports of gene flow from GM crops to other cultivars or wild relatives. However, this phenomenon is not specific to GM crops. It also frequently occurs in the case of organic and conventionally bred crops and from improved crops that have been changed in their genetic structure by exposure to radiation or chemical substances. The possibility of gene flow, as such, cannot justify the prohibition of the planting of a crop, but only the specific possible adverse consequences which result from it.

However, while measurable gene flow from GM crops has been reported, it is not clear that this has posed environmental hazards (Nuffield Council on Bioethics, 2003). Furthermore, this school of thought is not persuaded that the possibility of gene flow should be sufficient to deny hungry people the chance to consume GM food aid in situations of famine. Taking into account the gravity of the food crisis in the country, and the concerns raised about gene flow, the proponents of GM food aid appealed to government that the GM maize grain be milled so as to ensure that it was consumed by the starving masses without there being the possibility of storing any of it for the next farming season (Mumba, 2002).

As regards food safety concerns, it was noted that the WHO had affirmed that GM foods currently available on the international market had passed risk assessments and were not likely to present risks for human health. In addition, no effects on human health had been shown as a result of consumption of such foods by the general population in the countries where they had been approved. It was emphasized that all foods derived from biotechnology had been thoroughly evaluated by independent assessors to ensure that they were safe to eat. They had all undergone extensive regulatory processes. And, while many common foods such as wheat, cows' milk, eggs and soybeans cause allergies in some people, allergenic proteins can be removed (or silenced) in GM varieties. When it comes to toxicity, the effects of GM foods are expected to be no greater than for conventional or regular foods.

The proponents took the view that a reasonable interpretation of the precautionary principle should also be applied when assessing the safety of GM crops intended for human consumption. In this context, they recommended the use of the concept of 'substantial equivalence' that has been endorsed by the World Health Organization (WHO), which involves comparing the GM crop in question to its closest conventional counterpart (FAO and WHO, 2000). They explained that the purpose of the procedure was to identify similarities and differences between a GM crop and a comparator that had a history of safe use. The approach does not aim to establish absolute safety, which is impossible to attain for any type of food. Rather, it aims to assure that a new type of food, such as a GM crop, is as safe to eat as its closest traditional counterpart. This procedure is very useful for identifying intended or unintended differences between a GM crop and its comparator, which might require further safety assessments.

Other concerns raised were related to the fact that some forms of genetic modification involve genetic material that is foreign to the organism that is modified. Often, viral sequences are used to facilitate the insertion of a specific gene sequence. For example, the cauliflower mosaic plant virus is used as a promoter, which means that a short sequence of the genetic materials of the virus is inserted together with a particular gene, to facilitate its expression (this function is known as 'switching on' the gene) (Royal Society, 2002). Some people regard this as a threshold that should not be breached because, in their view, an organism has been created that has not previously existed in nature.

Fears that viral promoters could produce new viruses that would affect humans were allied in that there are a number of difficulties with this speculation: first, only a small part of such viruses is used (usually the 35S promoter from the cauliflower mosaic virus). Secondly, viruses usually infect a very narrowly defined range of species. It is therefore unlikely that viruses that are adapted to infect cauliflower would infect humans (Royal Society, 2002).

Another possibility is that plant viruses may produce new viruses in humans by recombination with remnants of viral DNA sequences that exist in human DNA. However, research has shown that there are considerable natural barriers to such a process (Aaziz and Tepfer, 1999; Worobey and Holmes, 1999). Indeed, humans have eaten virally infected plants for millennia and there is no evidence that new viruses have been created as a consequence (Royal Society, 2002).

There were also questions with regard to how other foreign genetic materials that have been introduced in a GM crop will be taken up by the body. It was explained that when humans eat plants or animals, they also eat the DNA of these organisms. Similarly, this is the case with GM crops. However, the fact that such crops have been genetically altered does not mean that this necessarily creates new health risks.

According to a recent FAO/WHO document, the amount of DNA which is ingested varies widely, but it is estimated to be in the area of 0.1–1.0 g/day. Novel DNA from a GM crop would represent less than 1/250,000 of the total amount consumed (FAO and WHO, 2000). This means that the possibility of transfer of genes that were introduced through genetic modification is extremely low. Thus, the DNA of the modified crop will usually be processed and broken down by the digestive system in the same way as that of conventionally bred, or otherwise modified, crops (FAO and WHO, 2000; Royal Society, 2002).

Finally, we pointed out that while we were not aware of any studies documenting proven damage arising from the consumption of GM crop products, the use of some conventional varieties of crops could have grave health consequences. For, example, most varieties of *Lathyrus sativus*, a lentil formerly grown widely in north India and now most widespread in Ethiopia, are known to cause the crippling disease of lathyrism. Plant scientists have bred varieties that do not cause this disease; they have also produced cassava varieties that, following processing into food, do not endanger the consumer with high levels of hydrocyanic acid as do many traditional cassava varieties in Nigeria.

Research on GM crops could well aim to create safer varieties of these crops that could replace harmful traditional varieties. In our judgement, there was no empirical or theoretical evidence that GM crops posed greater hazards to health than plants resulting from conventional plant breeding. However, we welcome the fact that concerns about GM have focused attention on issues of safety with regard to new crops and varieties. We recommended that the same standards should be applied to the assessment of risks from GM and from non-GM plants and foods, and that the risks of inaction be given the same careful analysis as risks of action, in a responsible application of the precautionary principle.

What was clear from the debate up to this point, though, was the absence of the voices of the most affected people in rural areas. Bishop Peter Ndhlovu, the Head of the Bible Gospel Church in Africa, who had visited hunger-stricken villagers, informed Dr Banda of Panos Southern Africa that: ‘The food crisis in rural Zambia is more grave than can be imagined from an urban perspective.’

This echoed many concerns that the debate had been so urban-centred and elite-based that it had largely ignored the concerns and urgent needs of the rural poor. The emphasis on scientific evidence as a basis for policy-making rendered the GM public debate in Zambia elitist. Those who were not schooled in science had largely been on the sidelines, apart from some vocal civil societies. Those with full bellies and having three full meals per day were making decisions on behalf of those who had nothing to eat.

Participants in the national debate were invited to react to the arguments above. An overwhelming majority of participants spoke against accepting the GE food aid. I do recall one prominent politician warning participants that: ‘If Zambia accepts GE maize, the country risks having children in future with eyes at the back of their heads.’

A representative from KATC thanked the US government for the donation, but wondered why the donation has been tied to GM maize by a country like the USA, which has 70% of its surplus grain stocks consisting of non-GM maize. ‘America champions freedom but has denied Zambia the freedom to choose between GM and non-GM maize’, he said. He mentioned that the European Union (EU) regulations required all GM foods to be labelled and wondered why the GM maize brought to Zambia was not labelled as such.

Moreover, the EU has zero tolerance for non-approved GM foods. He informed the gathering that, in June 2002, the Bolivian Forum on Environment and Development announced that a sample of USAID food aid had tested positive for StarLink® maize, a GM variety not approved for human consumption due to health concerns over possible allergic effects. They criticized USAID and WFP and demanded that GM Crops should not be sent as food aid to countries that had not formulated and developed biosafety regulations and capacity.

A female representative of the Women's Movement said that during the Second World War, scientists were used to commit crimes against humanity and hoped that Zambian scientists were not being used in a similar manner. She passionately argued: 'The women's movement in Zambia is best placed to know the hunger situation in the country, but are not accepting food that is questionable. It is unethical for America to link the donation with GMOs. People in Zambia are not able to access medical facilities; therefore, it would be very difficult to monitor the impacts on people consuming GM maize.' She supported the government in rejecting GM maize.

At this point it was becoming apparent that those opposed to GM food aid had succeeded in convincing the audience about the perceived dangers of GMOs. Sighting largely what have become popularly known as 'myths busters' in biotechnology circles because of their lack in scientific merit, enough fear had been instilled among Zambians.

Based on the proceedings of the meeting and the discussions, it was concluded as follows:

- Genetically engineered products are relatively new creations and their long-term effects in both humans and environment are not known and cannot be predicted. In this regard, all GM products can be considered experimental.
- Safety assurances from producers of GMOs cannot be realistically relied upon due to vested commercial, political and social interests.
- The greater majority of Zambians consume maize in greater quantity and frequency than in countries where limited evaluation of GMO has been done.
- No data on the bio-engineering has been availed with regard to the GM maize that the US government has procured and dispatched to Zambia.
- There is worldwide uncertainty on the use of GMOs in food, particularly certain GM maize (StarLink® developed by Aventis) in the USA.
- Zambia has no national biosafety framework to regulate the importation and application of biotechnology and GMOs.
- Zambia has no capacity to detect GMOs and manage unplanned or unanticipated entry of GMOs into the Zambian environment.
- Zambia has not ratified the internationally agreed instruments governing the use of GMO and biotechnology (Cartagena Protocol).
- Zambia should exploit international norms (mainly Precautionary

Principle and Advanced Informed Agreement) during procurements or importation of GMOs.

- The Zambian government must immediately source GMO-free maize and other foods from both within and outside Zambia.
- The USA does have segregated GMO-free maize. If Zambia is to source maize from the USA, the latter must supply only GMO-free maize.
- Many other countries have rejected GM maize, even in the face of high-handed methods employed by the USA.
- Embracing GMOs will undermine the productivity, marketability and profitability on non-GM crops such as organic farming crops.

Following the deliberations and upon scrutinizing the record of the discussions, the organizers of the national debate recommended that:

- The Zambian Government should not accept the GM food but should source food from non-GM areas.
- In the absence of a national biosafety framework, the Zambian government should establish an interim administrative structure to address issues pertaining to GMOs.
- As a matter of urgency, the Zambian government should ensure that the National Biotechnology and Biosafety Policy Framework are put in place.
- The Zambian government should develop national capacity to deal with GMOs.
- The government imposes an interim moratorium on GMOs until the framework is in place.
- The government sensitizes Zambians not to import as seed any unapproved GM maize.
- The government creates a platform for ongoing debate and input on biotechnology development in Zambia.
- Zambia should ratify the Cartagena Protocol.
- The government should immediately review relevant legislation to cover GMO issues.
- Zambia needs to address issues of trans-boundary movement of GM maize, considering that some of our neighbours have accepted it.

In the final analysis, a report of the national public debate on GM foods was presented to government, recommending that the GM food aid should be rejected. The Zambian Government deliberated on the report and the recommendations that emanated from it. In August, the Minister of Information and Broadcasting announced to the nation and the world at large the decision by the Zambian government not to accept GM food aid. The Minister explained that the decision by the government was not an indication of lack of appreciation of assistance that was offered to Zambia. He went on to urge all well-wishers to source for non-GM food aid.

Although 18,000 t of US-produced grain had originally been imported into the country, the Zambian government demanded that the WFP remove it from Zambia. The consignments were subsequently shipped to countries within the region which had accepted the donation. Although around 3 million people in his country were suffering from hunger, President Levy Mwanawasa was adamant that he would not expose his people to the risks associated with the GM maize.

## International community reactions to Zambia's decision

Pressure mounted on the Zambian government to rescind its decision, and this forced President Mwanawasa to commission a team of Zambian scientists who travelled to the USA and some European countries with the aim of obtaining further information regarding the safety of GM food crops on the environment and human health, including ethical issues and impact on trade.

Through financial support from the US Agency for International Development (USAID), Department for International Development (DFID), Norwegian Agency for Development (NORAD) and the Netherlands government, the team visited the USA, South Africa, the UK, Norway and the Netherlands during the period 10 September–2 October 2002. They also met EU officials in Brussels, and met and held discussions with representatives from government and non-governmental organizations. The required information was gathered through open discussions and both unpublished and published reports (GRZ, 2002a).

In October 2002, the scientists submitted their report to the government. The report was consistent with the advice given to the government earlier in August that: 'There were a lack of long-term studies capable of demonstrating that GM maize carried no risks, particularly in view of the fact that maize formed the staple diet of most Zambians.' Moreover, according to the report, there were ecological concerns. If maize were sown as seed, it might cross with native varieties – as had already happened in Mexico – and pose a threat to biodiversity and future grain exports to the EU, where it is important for products to be GM-free. In rendering support to this report, the Zambian Minister of Agriculture and Cooperatives, Mr Mundia Sikatana said that: 'Accepting GM maize would be like lighting a fire, which could perhaps burn out of control.'

The apparent uncompromising stance taken by the Zambian government was universally condemned. The heaviest pressure on the Zambian Government to accept GE food aid had come from agencies of the United Nations (UN) – especially the World Food Programme (WFP), World Health Organization (WHO) and the Food and Agriculture Organization. These three agencies issued a joint statement to the effect that there was no reason for African countries not to accept GM food aid, since GM foods were consumed by millions of people globally and no adverse effects had been observed thus far. The statement further made a plea that: 'In the current crisis, governments of southern African countries must consider carefully the severe and immediate consequences of limiting the food aid that is made available for millions of people so desperately in need' (FAO/WHO/WTO/Codex, 2002).

The USA put pressure on the Zambian government through statements of senior officials. In his address to the World Summit on Sustainable Development, the US Secretary of State, Colin Powell, stated that there was no reason for African countries not to accept GM food, since Americans consumed it. The Zambian government's refusal of the aid was 'a crime against humanity', claimed with indignation Tony Hall, the American ambassador to the UN Food and Agriculture Organization. The ambassador said that representatives of governments who refused food for their people and let them starve should be called to account before the highest courts in the world. The US Secretary of Agriculture, Anne Veneman, blamed the anti-biotech forces for scaring Zambians into believing that GM maize would harm them.

The USA did not confine itself to putting pressure on the African countries; it also criticized the attitude of the EU. The Zambian incident had escalated into a full-blown diplomatic row. The US accused the 'anti-science-thinking' Europeans of persuading the Africans into believing that GM foods might be unsafe. In turn, Europeans suggested that Americans were cynically trying to shove GM maize they could not sell elsewhere down the throats of starving Africans, then calling it charity. The US government claimed that the European policy had stirred up anti-GM feeling in a few African states, undermining American efforts. President George W. Bush said: 'European governments should join – not hinder – the great cause of ending hunger in Africa.'

However, the EU maintains the policy that it is up to recipient countries alone to decide whether they wish to accept aid consignments. The EU provides money to partners rather than agricultural surpluses. It is EC policy 'to buy as much as possible from markets in the region. We generally look at the situation country by country – in southern African food crisis, the EU has a clause in its contract with WFP that the money should be used to purchase the food locally – the idea is that the WFP should try as hard as possible to buy food in southern Africa' (Franco Viault, European Aid Cooperation Office, 18 September 2002).

The EU would not persuade African governments to accept donations of GM food, EU trade commissioner Pascal Lamy said on 2 December 2002, rejecting a US complaint that its stance was worsening starvation. 'Our policy is very different from US policy', Lamy said; 'there is no way we are going to change it just for the sake of being nice to the Americans' (Bloomberg Report, 2002). The EU had questioned the use of food aid donations as surplus disposal measures, stating that: 'Some WTO members have used food aid donations more as a production and commercial tool to dispose of surpluses and promote sales in foreign markets than as a development tool tailored to the needs of the recipient countries.' The UK Environmental Minister, Michael Meacher, said in response to the US government's insistence on supplying large quantities of GM grain to Africa that: 'It is wicked, when there is such an excess of non-GM food available, for GM to be forced on countries for reasons of GM politics' (Oxfam, 2002).

With the coming into force of the Cartagena Protocol on Biosafety in September 2003, one analyst argued that: 'One is inclined to think that, with hindsight, the policy of the Zambian government was justified – every country can decide for itself whether or not it wants to import genetically modified foods.' This international regulatory framework controlled the cross-border movement of GM organisms for the first time. It originates from the Biodiversity Convention and has been ratified by more than 50 countries, including Zambia and the EU countries, but not the USA. The Protocol affirms the precautionary principle: accordingly, signatory countries may impose import bans even if there is no definitive evidence of possible risks.

On this particular issue, however, the USA and EU member states still hold divergent views. According to the analysis of Tewolde Egziabher, the Director of Ethiopia's Ministry of Environment, a genetically modified organism (GMO) is the result of a combination of genes that would never have occurred in nature. Consequently, a transgenic product can only be viewed as harmless when its safety has been proved. The USA, on the other hand, takes the view that a GMO is harmless until such time as there is evidence to the contrary.

The rejected aid had one serious repercussion: it inflamed the debate between the USA and the EU on genetic engineering. In May 2003, the USA brought a case before the World Trade Organization, challenging the EU's unofficial moratorium on GMO. According to

Tamas Nagy of the European Institute of Food and Nutrition Sciences, for a good many years this European approach had been a thorn in the flesh of the USA because it prevented exports of GM maize and soy from the USA to Europe.

Ulrike Brendels, a genetic engineering expert at the environmental organization Greenpeace, was more inclined to see US aid to Africa as a self-interested act on behalf of the US economy: 'The USA could not export production surpluses to other countries, and hence tried to dispose of them by diverting them to famine relief.' Critics of GM technology also saw these aid consignments as an attempt on the part of multinational agricultural concerns to mix their products in with conventional seed, in order to gain a foothold in further segments of the worldwide seed market. Genetically modified seed is protected by patents, and farmers re-sowing it would have to pay royalty fees to the seed companies. Critics protested that this would make agriculture in the south dependent on corporations in the north.

The consensus amongst aid agencies working on the ground was that financial donations were a more effective means of combating hunger and that it was wrong to suggest that the distribution of GE food was necessary. Moreover, according to Donald Mavunduse from ActionAid, there was enough non-GM on the world market which meant: 'We [were] not at the point where we should be saying to starving countries: "Take GE or nothing"' (Greenpeace International, 2003, unpublished report). This line of reasoning was shared by Oxfam, who also demanded a moratorium on GMOs and the improved enforcement of monitoring systems to stop GMOs from entering vulnerable populations through food aid (Oxfam, 2002).

NGOs from 39 countries signed the Statement of Solidarity with Southern African Nations over GM food and Crops, condemning the use of food aid as a tool of propaganda to force acceptance of GE food and crops by Southern nations (26 August 2002). One hundred and thirty five African NGOs signed a statement: 'In Support of the Zambian and Zimbabwean Government position to reject food aid contaminated by genetic engineering' (31 August 2002).

A team of African scientists, set up by 14 nations of the Southern African Development Community (SADC) to investigate the effects of GM foods, concluded that they posed no immediate risk to humans and animals. The approximately 20 scientists, who were sent on a fact-finding mission to the USA and Europe in 2003, advised that southern African nations should embrace the technology because of its potential to increase agricultural yields. However, the team also warned that: 'Potential environmental risks remain a challenge, especially in Africa because of its rich plant and animal genetic resources.'

The team therefore recommended that genetic modification technologies be evaluated in African environments and called for African nations to develop their own capacity to regulate and test GM products. According to the report, scientific evidence suggested that GM crops did not pose any health risks different from non-GM crops.

On account of the findings from their mission the SADC team recommended the following:

- Donors should be advised that, where possible, food aid should be sourced from within the region.
- The donors of food aid containing GM events should comply with the Prior Informed Consent Principle and with the notification requirements in accordance with Article 8 of the Cartagena Protocol on Biosafety.
- SADC should develop and adopt a harmonized transit policy, as well as information and management systems for food aid that should be designed to facilitate trans-boundary movement in a safe and expeditious manner.
- The delegation also reaffirmed the position previously taken by the SADC Council of Ministers on GMOs with respect to handling of emergency food aid, the establishment of national legislations, a regional Advisory Committee and the milling of food aid grain before distribution.

### Donors bow to Zambia's demand for non-GM food

After prolonged wrangling, Zambia finally received food aid that was GM-free. Neighbouring African countries, along with China and, finally, even the USA, donated conventional food, while the EU assisted with financial resources. The USA bowed to international pressure and promised Zambia 30,000 t of GM-free grain. 'I am pleased to announce that the USA has secured 15,000 t of sorghum and 15,000 t of wheat to help Zambia in this time of need', Martin Brennan, US ambassador in Zambia told Reuters on 9 December 2002.

On 18 September 2002, the Zambia National Vulnerability Assessment (VAC) identified the need for 224,000 t of food aid for Zambia from September 2002 through to March 2003 – 64,000 more than had been forecast by the FAO's Crop Food Supply Assessment Mission (CFSAM) in May 2002.

The WFP in Zambia appealled for US$61 million to feed almost 3 million starving people. Since the start of the emergency operation on 1 July 2002, the WFP in Zambia had received donations from the EU, Japan, France, New Zealand, the UK, the USA, Switzerland and the

Netherlands. The WFP had pledged to provide Zambia with 82,000 t, but the country was still faced with a deficit of 120,000 t.

Against maize import requirements in marketing year 2002/03 (May/April) of 575,000 t, planned commercial imports by private millers amounted to 150,000 t and the government granted tenders to private traders for an additional 300,000 t. However, imports received in the country were much lower, and by early November amounted to only about 50,000 t. While WFP monthly food aid requirements were 21,000 t, only half of the people in need could be reached in October due to shortages in the food aid pipeline. Pledges of food aid by early December amounted to 106,000 t.

WFP was using cash contributions received to purchase maize in the region. Zambia was still looking to other sources of non-GM maize and finance to supplement the shortfall. The WFP eventually reacted to protests and critique of its purchase of GM maize from the USA, and eventually decided to use cash donations to purchase GM-free maize regionally.

December pipeline: In December 2002, Zambia needed 33,000 t of food aid. The WFP purchased 10,000 t of this aid locally, for distribution. The Zambian government agreed to make up the 23,000 t shortfall and the WFP assumed responsibility for its distribution (WFP, 13 December 2002).

January pipeline: In December 2002 the EU donated 20 million euros, which the WFP used to purchase and deliver non-GE grain (WFP, 17 December 2002).

The second pipeline: a consortium of NGOs, comprising World Vision International (WVI), CARE and Catholic Relief Services, requested donations for a second pipeline for victims in order to make up the shortfall of 64,000 t, which had not been anticipated by the CFSAM. This second pipeline was expected to arrive in January 2003.

## Media coverage by Zambia's leading print media

As stated earlier, the discussions on GM foods in Zambia were held against a backdrop of little media coverage of GMOs. The study summarized below will illustrate this point further. At the height of the GM debate the National Agriculture Information Service (NAIS), a department under the Ministry of Agriculture and Cooperatives, commissioned a study covering the period August 2001–September 2002. This study involved the three leading print media in the country, namely *Times of Zambia*, *Daily Mail* and *The Post* newspapers.

The objectives of the study were to gain some understanding of the coverage of biotechnology in the media over time in terms of both quality and quantity, and to gain some basic understanding of the knowledge of biotechnology among Zambian's journalists and editors (Kakunta, 2002).

The study came up with the following summary conclusions:

- There was very limited coverage of biotechnology.
- Media concentrated on breaking news about biotechnology more than on information to educate the population.
- There were hardly any editorials, suggesting that biotechnology is not a priority area for the editors, also proving the paucity of knowledge of the subject among editors.
- There were no photographs, suggesting that journalists 'did not see' biotechnology on the ground: it was like an abstract concept.
- No newspaper covered the issue of biotechnology adequately until the GMO 'fever and government rejection of GM maize' in 2002, due to the hunger situation and the food aid, suggesting that the media had been passive followers rather than proactive in the debate. This was also confirmed by the high concentration of articles after conferences and discussions on whether government should accept GM maize. The issue was widely discussed, mainly as a result of the hunger in the country.
- No newspaper had biotechnology articles in their business columns, suggesting that they did not see biotechnology from a business perspective.
- Over 95% of articles featured were inside pages, indicating the relegated position of biotechnology in the order of prominence.
- Only one article was specifically on GMOs. This confirmed the limited understanding of the other forms of biotechnology and application of biotech. – in areas other than plant genetic resources – by journalists. Also, media failed to report that biotech. was not a new science.
- There was limited depth of coverage (in terms of number of sources quoted), and this applied to features as well. Either journalists have limited time or simply do not know where to go for more information.
- The media tended to be merely speculative. For instance, whereas many articles claimed that GMOs could damage human health, they did not explain how.

On the debate about GM maize and food aid, and whether or not the country should accept the offer of such food from the USA, the study came up with the following findings:

- The media were more interested in knowing how much maize had been sourced or donated, without analysing the implications in terms of health, the environment and the economy.
- Politicians or leaders in general have an important role in determining how news are generated, and it is journalists' responsibility to analyse it but it would appear that, in the Zambian situation, journalists did not analyse the news as such. Also, journalists were ambivalent about science reporting, due to a lack of general understanding of science issues and/or lack of interest.
- No stories covered farmers or expressed farmers' opinions.
- In almost all the articles, the sources of information were politicians or donor agencies: very few were scientists and yet they were the experts regarding biotechnology.

As regards to access to information, the study concluded that there was little information in the public domain regarding GM technology.

## Concluding Remarks

Given the outcome of the GM debate in some countries in the region, one cannot avoid introspecting and concluding that the media are partially to blame. They failed to provide timely and balanced information on the debate and thus denied the citizenry the opportunity to make an informed decision. Although there is no concrete evidence to suggest that the media were directed to influence the debate in any one direction, there were instances when articles in favour of GM foods were not published by either private or public newspapers.

In some cases, those in favour of GM foods were denied the opportunity to air their views on national radio and television stations. It was not uncommon to see discussion programmes (so-called debates) with a one-sided panel opposed to GM foods. It is also possible that the individuals – especially editors – in these media houses took sides in the debate and, as such, abrogated their role of disseminating balanced information. They became emotionally involved in the debate themselves and began to spearhead the anti-GM food campaign.

Public attitude towards GM foods in Africa often smacks of a victim mentality. Fears of corporate control of an agricultural system that traditionally was communally owned, coupled with apprehensions of marginalization and the memory of colonial domination, lead to distrust of solutions that appear imposed externally. At the OAU Workshop on an African Model Law on Biosafety held in Addis Ababa, Ethiopia (May 2001), the text of a draft model law was tabled for adoption as an African initiative until the following OAU Council of Ministers meeting.

This model law was described by some as 'preventative' and aimed at depriving Africa from deriving the benefits of biotechnology. The model contained numerous provisions, inconsistent with the Cartagena Protocol on Biosafety (CPB) that member states had already signed. The bill was opposed by African scientists and NGOs, such as Africabio and African Biotechnology Stakeholders Forum (ABSF).

Whatever the arguments for or against GM technology in Africa, it is evident that the nature of international economic interdependence means that the freedom of developing countries to choose technologies that they judge to be to their own advantage has implications for the behaviour of developed countries. This is especially the case in the EU, where agricultural protectionism already poses considerable barriers to the economic growth of poor countries. The complexities to which this interdependence can give rise have been illustrated in this chapter, in the context of the food aid to southern Africa case.

The issues raised by food aid are complex. For example, it is noteworthy that the USA donates food aid in kind, whereas the three other major donors worldwide – the WFP, the EU and the UK – donate in cash. The latter group argues that financial assistance allows for the quickest and most effective form of aid, which also supports local economies of countries close to the recipient country.

The USA, on the other hand, has provided aid in southern African countries entirely in the form of shipments of US maize. Indeed, the US Agency for International Development (USAID) website even stresses that, in buying cereals from US farmers rather than from the world market or markets in developing countries, it actively seeks to subsidize US farmers and the US economy (USAID, 2002). Furthermore, the question may be asked why the USA did not offer to provide milled maize once it had become apparent that several African countries would have preferred the donation in that form.

Many, therefore, suspected that USAID was seeking to play a role in a US-led marketing campaign designed to introduce GM food to developing countries (Greenpeace International, 2002; Friends of the Earth International, 2003). There have also been reports that donations through the WEP have previously included GMOs, and that the recipient countries have not been informed accordingly (Friends of the Earth International, 2003).

While these events were quoted as evidence that food aid was being used to promote the marketing of GM crops, there were also reports that pressure had been put on developing countries from the opposite end of the spectrum. For example, it has been alleged that African leaders were advised by EU officials not to accept GM maize, as this would jeopardize current and future trade relations. However, this claim was refuted vehemently by, among others, EU Development Commissioner Poul Nielson (Nuffield Council on Bioethics, 2003). It was also reported that both local and international NGOs were very active in persuading the Zambian government to refuse GM maize.

It is recognized that long-term reliance on food aid, whether provided in the form of GM or non-GM cereals, is highly undesirable. It is therefore necessary to assist developing countries in becoming self-sufficient in food production. This is a complex process and GM crops could play a substantial role in it. However, the question remains as to how developed countries can comply with their ethical obligations in the case of food aid in emergencies that may continue to arise in a number of developing countries.

In view of the current evidence relating to assessments of food safety of GM crops, or products produced from GM crops, I am not convinced that these pose significant risks to humans who eat them. I therefore take a critical stance towards activities of some NGOs – and, indeed, African scientists who fail to provide a rational assessment of the risks and benefits which GM crops may have in the context of agriculture in developing countries. At the same time, I take the view that preferences of developing countries dependent on emergency food aid must be taken seriously. A genuine choice between GM and non-GM food must be offered wherever possible. It is therefore necessary to provide full information about whether or not donated food is derived wholly or in part from GM crops.

On this very complicated issue of GMFs and food aid much more could be said, but let me conclude by saying that, where developing countries prefer to receive non-GM grain, the WFP and other food aid organizations should purchase it as was the case in Zambia. However, this should be subject to such grain being available in sufficient amounts, with reasonable financial and logistical costs, and where it can be provided quickly enough to address emergency situations. Where only donations of GM varieties are available and developing countries object to their import solely on the basis of environmental risks, I recommend that food aid be provided in milled form. Grain from food aid donations is likely to be planted in developing countries. It would be unacceptable to introduce a GM crop into a country in this way. Although milling increases the costs of providing food aid, it allows for fortification of the milled grain with micronutrients.

Finally, whereas it is acknowledged that we all need to respect the right of people to hold different views, scientists have an obligation to counteract misinformation. Concerns about GMFs ought to be addressed by sourcing factual information from credible institutions. The media should be partners with scientists to disseminate information and thus create awareness among consumers and policymakers.

## References

Aaziz, R. and Tepfer, M. (1999) Recombination in RNA viruses and in virus-resistant transgenic plants. *Journal of General Virology* 80, 1339–1346.

American Corn Growers Association (2001) *Corn Growers' Third Annual Survey Shows More Elevators Requiring GMO Segregation.* American Corn Growers Association, Washington, DC (http://www.acga.org/news/2001/121801.html, accessed 18 December 2001).

Banda F. (2004) *Can Beggars be Choosers?* (http://allafrica.com/stories/200408170844.html, accessed 17 August 2004).

Bloomberg Report (2002) *Bloomberg Business Report, South Africa* (http://www.busrep.co.za, accessed 2 December 2002).

Carroll, R. (2002) Zambians starve as food aid lies rejected. *The Guardian,* 17 October 2002 (http://www.guardian.co.uk/gmdebate/story/0,2763,813220,00.html).

FAO (2003) Discussion at a *FAO Electronic Forum on Biotechnology in Food and Agriculture,* 28 April–25 May 2003 (http://www.fao.org/biotech/logs/c9logs.htm).

FAO/WHO (2000) Safety aspects of genetically modified foods of plant origin. *Report of a Joint FAO/WHO Expert Consultation on Foods Derived from Biotechnology,* World Health Organization, Geneva, Switzerland 29 May–2 June 2000. WHO, Geneva, p. 11.

FAO/WHO/WTO/Codex (2002) *Food Laws News: GM Foods – UN Statement on the Use of GM food aid in Southern Africa.* FAO Press Release, 27 August 2002 (http://www.foodlaw.rdg.ac.uk/news/in-02027.htm).

Farish, W. (2003) *US Aid for Africa's Famine* (http://www.guardian.co.uk/letters/story/0,3604,859029.html).

Friends of the Earth International (2003) *Playing with Hunger: the Reality behind the Shipment of GMOs as Food Aid.* FoEI, Amsterdam.

Government of the Republic of Zambia (GRZ) (2002a) *Report of the Fact-finding mission on genetically modified foods* by a team of Zambian scientists who visited the USA, the Republic of South Africa, the UK, Belgium, Norway and the Netherlands from 10 September to 2 October 2002, Lusaka, Zambia, pp. 1–56.

Government of the Republic of Zambia (GRZ) (2002b) *Report of the National Debate on Genetically Modified Organisms (GMOs) and Foods (GMFs).* Mulungushi International Conference Centre, Zambia, 12 August 2002, pp. 1–27.

GRAIN (2003) Poisoning the well: the genetic pollution of maize. *Seedling,* 20 January.

Greenpeace International (2002) *USAID and GM Food Aid* (http://www.greenpeace.org.uk/multimediafiless/Live/FullReports/5243.pdf).

Jesuit Centre for Justice and Peace (JCJP) (2002) *What is the Impact of GMOs on Sustainable Agriculture in Zambia?* A research study sponsored by Kasisi Agricultural Training Centre and the Jesuit Centre for Theological Reflection. JCTR Publications, Lusaka, Zambia, pp. 1–17.

Kajoba, G. (1993) *Food Crisis in Zambia.* ZPC publications, Lusaka, Zambia, pp. 1–2.

Kakunta, C. (2002) An analysis of media coverage of biotechnology issues in Zambia's leading print media (August 2001 to September 2002) vs reality – the case of genetically modified maize and food aid in Zambia. Paper presented at the *Biotechnology Awareness Workshop for Journalists and Scientists from Malawi and Zambia,* organized by MBERU in Blantyre, Malawi, 25–31 December 2002.

Lewanika, M.M. (2003) *The Real Story behind the Food Crisis in Zambia* (http://www.biotech_activists@iatp.org, accessed 22 September 2003).

Metz, M. and Fütterer, J. (2002) Biodiversity: suspect evidence of transgenic contamination. *Nature* 416, 600–601.

Mumba, L.E. (2002) GM maize food aid imports: what are the options for Zambia? In: *GRZ Report of the National Debate on Genetically Modified Organisms (GMOs) and Foods (GMFs).* Mulungushi International Conference Centre, Zambia, 12 August 2002, pp. 13–14.

Nuffield Council on Bioethics (2003) *The Use of Genetically Modified Crops in Developing Countries.* A follow-up discussion paper to the 1999 report: *Genetically Modified Crops: the Ethical and Social Issues,* pp. 1–89 (http://www.nuffieldbioethics.org).

Oxfam (2002) Oxfam condemns the distribution of food aid contaminated with Genetically Modified Organisms. Press Release, Oxfam International.

Pioneer Hi-Bred International (2000) *Elevator Biotech Grain Acceptance Survey – 2000* (http://www.pioneer.com/biotec/asta/accep%SFsurvey.htm).

Royal Society (2002) *Genetically Modified Plants for Food Use and Human Health – an Update.* Royal Society, London, pp. 8–9.

USAID (2002) *Direct Economic Benefits of US Assistance by State* (http://www.usaid.gov/procurement_bus-opp/state/).

Worobey, M. and Holmes, E. (1999) Evolutionary aspects of recombination in RNA viruses. *Journal of General Virology* 80, 2535–2544.

World Food Programme (WFP) (2002a) More than 70% of households have no cereal seed in Zambia and Malawi while in Zimbabwe, more than 94 per cent of farmers were without Seeds as of September, 2002. Press release, 29 October, WFP, Rome.

World Food Programme (WFP) (2002b) Confirmed Contributions to Southern African Region through the World Food Programme of the United Nations, 9 December.

16

# Approval Process and Adoption of Bollgard Cotton in India: a Private Company Perspective

R.B. Barwale,* M. Char,** S. Deshpande,*** M.K. Sharma**** and U.B. Zehr*****

*Maharashtra Hybrid Seeds Co. Ltd. (Mahyco), Mumbai, India; e-mail: *raju.barwale@mahyco.com; **madhavi.char@mahyco.com; ***sanjay.deshpande@mahyco.com; ****mahedra.sharma@mahyco.com; *****usha.zehr@mahyco.com*

## Introduction

Cotton cultivation is important to India from many aspects: (i) It has the largest land area under cotton in the world: 9 million ha against 34 million ha total area; (ii) a textile industry which is cotton dependent; and (iii) is the second largest employment generator, providing direct employment to 60 million people, accounting for 20% of industrial production – 7.5% of the GDP (Basu, 2004), contributing approximately one-third of the total export. In spite of the importance of cotton, the average productivity is only 300 kg/ha against a global average of 650 kg/ha.

This lower productivity can be attributed to many factors; however, one of the main reasons is the damage caused by insect pests, particularly the bollworm complex. The significant efforts made to control the pests by Indian farmers through spending Rs 12 billion (US$266 million) per year have proved ineffective due to the pest developing high levels of resistance for a number of pesticides.

Several approaches are used by farmers for control of crop pests, including biological control, improved crop husbandry, development of resistant cultivars, etc. However, no known sources of resistance are available against the bollworm complex of cotton. Thus, development of resistant lines through conventional breeding has not been feasible. Hence, an alternate method to develop insect resistant plants was required.

Using plant transformation methods, a gene from the native soil bacterium *Bacillus thuriengiensis* subsp. kurstaki (Bt) was introduced

 *The Media, the Public and Agricultural Biotechnology* (ed. D. Brossard *et al.*)

into the cotton plant. The gene commonly known as cry1Ac produces an endotoxin which has specific activity against the lepidopteran pest complex, which includes the bollworm complex. This endotoxin, due to its action specificity, does not work against non-lepidopteran pests. Plants carrying this gene were then put into the breeding programme for incorporation of the Bollgard, or cry1Ac, gene (Event Mon 531) to produce Bollgard cotton.

Bollgard cotton, being a transgenic crop, requires clearance under The Environment (Protection) Act (1986) for large-scale release into the environment. Due to the transgenic nature of the crop, there are concerns about the potential risks associated with their use to human health, environment and biological diversity. To address these concerns, biosafety regulations have been developed in India to monitor transgenic research, evaluation and large-scale release.

A strong regulatory system for assessing the biosafety of transgenic plants and food items before release into the environment is mandatory. A three-tier regulatory system for field testing of transgenic plants, structured on the basis of guidelines issued by the Department of Biotechnology (DBT) under the Environment (Protection) Act 1986, is in place. The Rules for the Manufacture, Use, Import, Export and Storage of Hazardous micro-organisms and Genetically Engineered Organisms or Cells (1989) were framed under the Environment (Protection) Act 1986 in the year 1989.

Biosafety guidelines were formulated by the RDAC (Recombinant DNA Advisory Committee) in 1990 and were adopted by the Government (Ministry of Science and Technology, 1990). They were revised in 1994 and 1998, incorporating allergenicity and toxicity evaluation of transgenic material (Ministry of Science and Technology, 1998). The guidelines incorporating changes up to September 1999 have been published recently. These guidelines prescribe the codes for experimentation and field testing of transgenic material for the assessment of safety. It is pertinent that no testing of transgenics can be done without permit under the Environment Protection Act. The Act provides penalties, including prosecution for violation of the Act.

The Indian regulatory system is a three-tier structure. It comprises the following:

- Institutional Biosafety Committee (IBSC), set up at each institution for monitoring institute-level research in genetically modified organisms (GMOs).
- Review Committee on Genetic Manipulation (RCGM), set up at DBT to monitor the ongoing research activities in GMOs. The Monitoring and Evaluation Committee (MEC), comprising agricultural scientists, was constituted in July 1998 by the RCGM to monitor and supervise the field trials permitted by the government.

- Genetic Engineering Approval Committee (GEAC) in the Ministry of Environment and Forests has been set up to authorize large-scale trials and environmental release of GMOs (http://www.envfor.nic.in).
- State Biotechnology Coordination Committee (SBCC).
- District Level Coordination Committee (DLCC).

The SBCC and DLCC are basically the law-enforcing committees formed under the Environment (Protection) Act, 1989.

Bollgard cotton became the first transgenic agriculture crop approved for large-scale release in the country by the GEAC, the apex-competent authority, under the local Act, in April 2002.

When the work on Bollgard cotton was initiated in the early 1990s, the regulatory system was formulated; however no precedence of any approval of transgenic crop was in place. The existing procedures were required to be tested and implemented in coordination with the various authorities, namely the Ministries of: (i) Agriculture; (ii) Science and Technology; (iii) Health and Environment; and (iv) Forests.

Prior to the large-scale environmental release of Bollgard cotton-specific data and information were required to demonstrate that Bollgard cotton is similar to the commonly grown non-Bollgard cotton in composition and agronomic performance, and that the Bt protein expressed by the inserted genes causes no adverse effect when consumed by domesticated or wild animals and non-target organisms, including beneficial insects. As yet not established was the actual process of taking this evaluation forward, which included the development of appropriate protocols, experimental protocols, data collection and, finally, interpretation of the results. These were carried out by using the expertise of the Mahyco Research Centre and various premier, specialized research institutes around the country.

With the approval process for Bollgard cotton, these issues were addressed and the regulatory approval path for transgenic crops in the country was firmly established. We present here the various aspects involved in seeking the approval for large scale environmental release of Bollgard cotton under the Environment Protection Act.

## The Introduction of Bollgard Cotton

### Biosafety aspects

Indian law requires that any transgenic product developed is safe to the environment, man and cattle. The safety assessment is performed to evaluate whether there is incremental risk or not. In the case of

Bollgard cotton, the process evolved over time to address biosafety issues such as pollen transfer, aggressiveness, weediness, soil residue, impact on non-target organisms – including but not limited to beneficial insects, soil micro-organisms and invertebrates. Food safety aspects were evaluated based on the principle of substantial equivalence.

In 1995, the IBSC of Maharashtra Hybrid Seeds Company Ltd. (Mahyco) obtained permission for the import and testing of Bollgard cotton seed in India from the appropriate authorities. The seed was imported from the Monsanto Company, St Louis (USA). The imported seed of the line containing cry1Ac gene event Mon 531 was grown in greenhouses approved by the regulatory authorities for carrying out various experiments and incorporating the Bt trait into locally adapted lines by the traditional breeding method of backcrossing.

The converted Indian cotton lines were tested for germination and vigour in the laboratory and greenhouse. The results indicated no significant differences between Bollgard and non-Bt cotton in germination and/or vigour characteristics. Subsequently, experiments were designed in the contained, open environment to confirm the environmental safety of Bollgard cotton. This included experiments on gene flow (pollen flow), persistence, weediness characteristics, crossability with non-transgenic cotton as well as near relatives, effect on non-target organisms and changes in soil microflora and fauna – such as nematodes and earthworms. All experiments were designed with appropriate controls.

To understand the level of resistance to cry1Ac protein expressed in Bollgard cotton plants and for future monitoring purposes, it was necessary to develop baseline susceptibility data on the bollworm complex on an annual basis. The resistance development to cry1Ac protein in cotton bollworms, if any, can be determined by such studies.

To assess the efficacy of the cry1Ac gene, the presence of cry1Ac protein was quantified in various plant parts at different time points in Bollgard cotton. The composition of cotton seed and oil from Bollgard cotton hybrids was compared to that of its non-Bt counterparts and other conventional cotton varieties. The components measured in the cotton seed included protein, fat, fibre, moisture, ash, amino acids, fatty acids and the anti-nutrients like gossypol, cyclopropenoids and aflatoxin.

These analyses were carried out to demonstrate that Bollgard cotton varieties are substantially equivalent and as safe and nutritious as non-Bt and other conventional cotton varieties. Furthermore, nutritional studies in cows and buffaloes showed no difference in feed intake, milk yield and composition, confirming the food and feed safety of Bollgard cotton.

To comply with the DBT guidelines, a 90-day toxicological feeding study on ruminants was conducted at the Industrial Toxicology Research Centre, Lucknow, India. The study showed that Bollgard cotton seed was as nutritious as its non-Bt counterpart, and similar to the non-Bt material in its effect on ruminants. Similarly, toxicological studies were also conducted using fish and chicken as animal models, which confirmed the safety and wholesomeness of cotton seed meal derived from Bollgard cotton.

The presence of cry1Ac protein was also determined in cotton lint and oil by the Central Cotton Research Institute, Nagpur, India. No cry1Ac protein was detected in lint. However, it was detected in crude seed samples of Bollgard cotton because of the seed debris that is usually present in crude extracts.

Assessment of any allergenic effect by transgenic plants and derived products is essential as per the guidelines, and the same was carried out using the Brown Norway Rat model. The result of this study led to the clear conclusion that there is no difference in endogenous allergens of Bollgard cotton feed compared to non-Bt cotton seed.

The rigorous and extensive scientific studies conducted in India – in addition to the global data – in compliance with the regulatory guidelines, have established that Bollgard cotton and its products are as safe for the environment, humans and animals as is non-Bt cotton.

## Technical aspects

Bollgard Cotton was developed by transforming the parental cotton cultivar Coker 312, using the *Agrobacterium*-mediated method for gene transfer. cry1Ac, npt-II and aad genes were transferred. Of these three genes, only cry1Ac encodes for a protein that has insecticidal properties; this gene is derived from the common soil microbe *Bacillus thuriengiensis* subsp. kurstaki (Bt.k). The other two genes encode for bacterial-selectable marker enzymes and have no insecticidal properties.

To comply with the regulatory guidelines, various issues related to the gene had to be addressed. These were:

- Molecular characterization of insertion pattern of cry1Ac gene in Bollgard cotton hybrid.
- Segregation ratio observed in the transgenic cotton varieties.
- Gene stability.
- Absence of the cre-recombinase gene that is an integral component of the so-called 'terminator technology'.

- Copy number.
- Quantification of cry1Ac protein expressed in various plant parts and tissues during the growth cycle of the plant.

Molecular analysis of Bollgard cotton event Mon 531 was carried out using various techniques, including Southern blot analysis, Western blot analysis and PCR based methods. It was shown that the event MON 531 contains a single functional T-DNA insert in the cotton plant genome. The experiment was conducted to demonstrate the site of integration of the cry1Ac gene in the genome of tested Bollgard cotton hybrids and the donor Bollgard cotton line. This also indicated the stability of the incorporated transgene in the cotton genome at the molecular level.

To observe the segregation ratio in the transgenic cotton, in each generation the cry1Ac protein was detected by both ELISA and insect bioassay techniques. From the data generated, it was deduced that the trait behaves as a single dominant Mendelian locus.

Gene stability was further ascertained by the analysing F-2 segregation pattern of cry1Ac gene. Seeds produced as F-2 population by growing F-1 hybrids segregated into the dominant (ELISA positive) and recessive (ELISA negative) phenotypes, in the typical Mendelian monohybrid ratio of 3:1.

The absence of cre gene(s) in Bollgard cotton plants was determined by PCR analysis of DNA samples isolated from individual seedlings grown from Bollgard cotton hybrid seeds. The PCR analysis results showed that Bollgard cotton hybrid seedlings positive for cry1Ac gene amplification did not show any amplification product for cre primers, indicating the absence of the cre gene or 'terminator technology' in the tested hybrids (Pental *et al.*, 2001).

It was important to determine the level of protein expression in different plant parts in order to evaluate the efficacy against the target pest. To achieve this objective, samples of terminal leaves, squares and bolls were collected at different time points and were used to quantify the expression of cry1Ac protein by a sensitive quantitative bioassay method (Greenplate, 1999). The results obtained confirmed that the expression of cry1Ac protein in various plant parts at different time intervals was adequate for efficacious control of the target pests. The terminal leaf had highest expression, followed by the squares and bolls. This observation carries significance because the Cotton Bollworm (*Helicoverpa armigera*) and the Cotton Spotted Bollworm (*Erias vittella*) predominantly lay eggs on these plant parts, and their neonates feed on these tissues.

During the process of Bollgard cotton development at Mahyco, extensive testing of cotton lines was carried out to ascertain the presence of cry1Ac gene. This was required in connection with the aforesaid studies, as well as for the purpose of breeding Bollgard hybrids. During this process, the quality control laboratory tested more than 133,000 plant samples for the presence/absence of cry1Ac gene through ELISA. In the post-release period of Bollgard cotton, the number of such tests has further risen to 554,888 samples in one season. This includes the testing of each seed lot produced by the company for the purpose of commercial sale. Those seed lots meeting the said standards of purity for the trait, along with other quality parameters, were the only ones released for sale by the company.

## Agronomic benefits

Bollgard cotton hybrids were extensively tested in the field prior to their large-scale release into the environment. Various types of field trials were conducted at several locations during the period 1998–2002. These trials were conducted by Mahyco and agricultural universities, as well as by Indian Council of Agricultural Research under the All India Co-ordinated Cotton Improvement Project, to assess the agronomic superiority of Bollgard cotton over its non-Bt counterparts. The trials conducted were monitored by the Monitoring-cum-Evaluation Committee (MEC), nominated for this purpose by RCGM.

On the basis of result of these trials it was concluded that Bollgard cotton hybrids provided:

- Effective control against three bollworm species in different agro-climatic zones of the country.
- Significantly higher boll retention resulting in higher yield when compared with non-Bt cotton hybrids – due to yield savings.
- Reduction in pesticide use and associated costs.
- Additional income, ranging from Rs 2610 to 4185 (US$58–93) per acre to farmers using Bollgard cotton as compared to non-Bt cotton.
- No adverse effect on non-target and beneficial insects or adjacent non-Bt cotton crops.

A socio-economic study was conducted by the Indian Institute of Management, Ahmedabad. The results of this study supported the above findings and further pointed out that adoption of such technologies would play a key role in developing national competitiveness in cotton cultivation, besides generating additional employment.

The Bollgard farmer is expected to gain more than two-thirds of the incremental benefits generated by using Bollgard cotton (Pray and Danmeng, 2001), and its use will enhance competitiveness of cotton grown in India and therefore employment opportunities for the people dependent on this sector (Naik, 2001; Qaim, 2003; Qaim and David, 2003; Bennett *et al.*, 2004).

After a thorough biosafety assessment of Bollgard cotton through various scientific studies conducted by national institutes and extensive field trials to establish the agronomic, social and economic benefits which mught be derived by use of this technology, the GEAC considered the proposal – for the large-scale environmental release of Bollgard cotton – in its meeting held on 26 March 2002 and, after a careful and in-depth examination of the findings, accorded approval for the large-scale environmental release of three Bollgard cotton hybrids developed by the Maharashtra Hybrid Seeds Co. Ltd.

The permission letter was issued by the Government of India, Ministry of Environment and Forests, on 5 April 2002. The permission order was issued subject to the compliance of certain conditions:

- A validity of order for 3 years.
- Bollgard cotton should be surrounded by five rows, or 20% of the total sown area, of non-Bt counterpart as a refuge crop.
- Packs of seeds should contain both Bt and non-Bt counterpart seeds.
- Labelling of pack should clearly mention the transgenic details.
- A list of farmers planting Bollgard cotton should be maintained and submitted to the regulatory authority.
- The development of Bt-based integrated pest management.
- The Monitoring of the susceptibility of bollworm to Bt protein by Mahyco, as well as by the designated Government Institute.
- The undertaking of an awareness programme of Bollgard cotton for farmers.

This commendable, bold decision of GEAC was well received by the cotton-farming community. However, it was opposed by certain environmentalists/NGOs who oppose the use of transgenic technologies in agriculture. They raised their concerns on the process of release, the impact of technology and the violation of law, challenging the result of the scientific experiments and the benefits of the technology. These issues were overcome by interaction with the various stakeholders directly and by the Government, through regulatory bodies, scientists and companies.

With the passage of time, the Bollgard cotton farmers, the end-users of the technology, have started realising the benefits of the technology and finally accepted it, irrespective of the continued opposition from certain NGOs. In addition, the companies have now become more proactive in communicating the scientific facts and benefits thereof to farmers, which has resulted in the creation of favourable opinion about the technology. The textile industry and trade also extended their full support for adoption of Bollgard cotton, because of its many benefits. This extensive awareness-creating process has helped in placing the facts before the farmers.

From a company perspective, the goal was to make every Bollgard farmer a satisfied farmer. To make it happen, the company developed a comprehensive field communication plan and successfully implemented it in every planting season. Under this programme, the state government officials, dealers, distributors, sales representatives and farmers were trained in technology application and discussions were held on the modalities for the implementation of conditions laid down in the approval letter, besides other related issues such as agronomic practices, management of non-lepidopteran pests and nutrient management. These meetings proved to be very fruitful in understanding the technicalities related to Bollgard cotton.

To disseminate the information related to Bollgard cotton, mass media methods were used extensively. These included use of educational advertisements, audio cassettes, flip charts, multilingual product leaflets, banners, posters and video films. All these conveyed the message of the importance of refugia planting, spraying decisions based on economic threshold levels and integrated pest management. The application of this approach has greatly helped Bollgard cotton farmers in reducing their expenses involved in chemical spraying.

## Pre-sale activities

To educate farmers on this new technology for its beneficial utilization, a massive and detailed communication programme for various stakeholders was organized by Mahyco, prior to the distribution of Bt cotton seeds in the various states. A summary of these activities organized under this programme follows.

### *Selection of Bollgard cotton dealers*

To create a better distribution channel and communication with Bollgard users in selected districts/areas, it was decided to sell Bollgard cotton hybrids only through the selected outlets designated as Bollgard seed outlets. The final selection of 549 Bollgard dealers was made on the fulfilment of certain qualifications, including the signing of a memorandum of understanding that stated various conditions for implemention by Bollgard dealers for the cultivation of the cotton by their customer farmers.

To comply with the conditions of approval, the dealers were asked to assist the government agencies, regulatory bodies and authorities – as well as the farmers – in ensuring compliance with the refugia requirements and in maintaining a complete database of customer farmers.

All dealers were asked to appoint trained person(s) for the provision of technical support – on a full-time basis – to the farmers during the entire cotton-growing and -harvesting season. The dealers and the technical personnel appointed by each dealer were trained in communication and technical aspects such as sowing method, spray decision, insect identification and their specific economic threshold level and integrated pest management etc. This group of technical field staff assisted farmers by providing technical guidance during the entire crop season and helped them reap the fruits of Bt technology.

### *Packing of Bollgard cotton hybrids*

In compliance with the condition of the approval letter, Bollgard cotton hybrid seed was supplied in special and innovative packing. This pack contained a composite can of 450 g Bollgard cotton hybrid seed and 120 g non-Bt hybrid counterpart. The seed was treated with Imidacloprid (to provide protection from sucking pests) during the early age of the crop. For the guidance and convenience of the user farmers, also enclosed within the pack was a product leaflet in eight Indian languages providing detailed information on cultivation of Bollgard cotton, IPM and precautions such as planting of five surrounding rows as refugia, etc.

The Bollgard cotton container was labelled as per the Seed Act and GEAC approval letter. In compliance with the release order, the label clearly stated that: 'These transgenic seeds contain cry1ac gene and nptII and aad marker genes. GEAC approval DO No.10/1/2002-CS dated April 5, 2002.' Bollgard cotton seeds were supplied in a composite can containing transgenic Bollgard and non-Bollgard cotton hybrid seeds, along with a multilingual instruction leaflet providing full details about the nature of the Bollgard hybrid, its agronomy and the planting procedure. For a clear distinction between the hybrids, a hybrid-specific colour scheme was used for the composite can and outer cardboard pack.

### *Pre-sale training programmes*

The company focused on communication about the technology through education and training programmes 2 years prior to the approval of the product, and it continued to pursue the same prior to the sales of Bollgard cotton (Kharif, 2002). A comprehensive training plan was developed, which included various components like introduction of Bollgard technology, its functions and benefits, regulatory conditions and their implementation and, finally, the management of Bollgard cotton for maximization of benefits.

The participants were trained in the technical aspects of transgenic cotton, its method of cultivation with special reference to the planting mechanism, commonly asked questions and important communication channels in order to manage the expectations of the stakeholders. The main area of focus was on the following aspects of Bt cotton:

- Bollgard cotton requires normal insecticidal spray/s for management of sucking pests.
- Economic threshold (ETL)-based decisions for insecticidal spray for bollworm.
- Regular and timely assessment of various ETLs.

The training programmes were successfully implemented. The participants in these programmes were Department of Agriculture officials, dealers, distributors, sales representatives and farmers. The Government officials who participated in the training programmes were those from the State Agriculture Department holding the positions of Joint Director, Suprintending Agriculture Officer, Development Officer, etc. Table 16.1 gives the details of various training programmes conducted by the Company.

### *Positioning*

Bollgard was positioned as a type which provided 24-hour, non-stop protection against the bollworm complex and, in addition, the dual

**Table 16.1.** Numbers of individuals trained and training sessions with regard to Bt cotton.[a]

| | Numbers of persons trained (sessions) | | | | | |
|---|---|---|---|---|---|---|
| State | Bollgard outlets | Field executives | Field assistants | Counter boys | Dealers | Govt officials |
| Andhra Pradesh | 54 | 20 (1) | 15 (1) | 55 (8) | 55 (2) | 47(1) |
| Karnataka | 100 | 24 (1) | 20 (1) | 100 (12) | 100 (3) | 27(1) |
| Tamilnadu | 45 | 22 (1) | 10 (1) | 45 (10) | 45 (2) | 58 (3) |
| Gujarat | 100 | 16 (1) | 10 (1) | 200 (15) | 200 (5) | 16 (1) |
| Maharashtra | 200 | 35 (2) | 35 (2) | 50 (7) | 50 (2) | 49 (1) |
| Madhya Pradesh | 50 | 15 (2) | 10 (2) | 100 (7) | 100 (2) | 114 (1) |
| Total | 549 | 132 (7) | 100 (7) | 550 (59) | 550 (19) | 311 (8) |

[a] Figures in parentheses represent the numbers of training sessions carried out.

advantage of improved production due to yield savings and reduced expenses from application of chemical sprays for control of the bollworm complex.

### *Critical communication*

Communication was provided on the management of other pests of cotton such as sucking pests, namely aphids, jassids and thrips. The seed treatment was allocated to chemicals that could provide inbuilt protection against sucking pest for a period of 30–45 days, followed by additional sprays. The concept of Economic Threshold Level determination for these pests was also communicated to the farmers for implementation, and the use of solely effective chemical sprays.

### *Effective pest management: keys to success*

Bollgard cotton appraisal and awareness sessions for officials of the State Agriculture Department in all the six states were organized. The respective Agriculture Commisioner/Director chaired these sessions and senior officials of the ranks Deputy Director, Agriculture Development Officer and Suprintending Agricultural Officer attended these sessions. In these sessions, discussions were held on the modalities for the implementation of conditions laid down in the permission letter, besides the technical discussion on issues related to Bt cotton. These meeting sessions proved to be really fruitful in understanding the technicalities related to Bt cotton commercialization.

### *Publicity and educational inputs on Bollgard cotton*

Bollgard cotton, being the first transgenic product approved in India, required preparation of various materials which enabled efficient and accurate dissemination of the information related to the technology, cultivation practices, concept of refugia, its necessity and various other agronomical aspects. As this was to be achieved in a short span of time, the company took a multifarious approach. These included the following:

- Booklet and handbill providing information on Bt cotton.
- Multilingual product leaflet on cultivation practices with Bollgard cotton pack (see Box 16.1).
- Audio cassette describing Bollgard cotton in question-and-answer form in different Indian languages.
- Direct mailers; these letters were sent to those farmers who had seen the Bollgard cotton trials in the earlier years and were aware

**Box 16.1.**

**CROP MANAGEMENT PRACTICES FOR BOLLGARD™ (Bt) COTTON**

**What is Bollgard cotton?**

•Bollgard cotton has in-built strength to control bollworms (American bollworm, Pink bollworm and Spotted bollworm). The whole cotton plant fights bollworm day and night non-stop. •When bollworm larvae feed on the Bollgard cotton plant, they become inactive and stop feeding further. Larvae die due to starvation within two days. •Bollgard cotton may require supplemental sprays to control bollworms, in case Economic Threshold Level (ETL) is reached. • Bollgard cotton needs to be sprayed for control of sucking pests like aphids, jassids, whiteflies, thrips and mites. •Bollgard cotton fits in very well as an important component of Integrated Pest Management (IPM). •Bollgard cotton cultivation practices are similar to cultivation of other cotton hybrids.

The following general guidelines can help to get maximum benefits from Bollgard cotton cultivation.

**Land Preparation :**

Remove and burn debris of previous crop for field sanitation. Give two deep summer ploughings to reduce soil borne diseases, insects and weeds.

Prepare the field by repeated ploughings to bring the soil to suitable tilth.

**Manures and Fertilisers:**

Organic manure: 3-4 weeks before sowing, apply organic manure @ 10 to 12.5 tonnes per hectare(ha) and incorporate into soil.

Fertilisers : The fertiliser schedule should be based on the soil test and local recommended practices. However, general recommendations are as under:

| **Fertiliser nutrient ( kg/ha)** | **N (kg)** | **P (kg)** | **K (kg)** |
|---|---|---|---|
| Total requirement | 120 | 60 | 60 |
| Basal dose | 20 | 60 | 60 |
| 1st top dressing | 40 | -- | – |
| 2nd top dressing | 40 | -- | – |
| 3rd top dressing | 20 | -- | – |

Note: For MECH- 184, add 20 kg/ha K at 2nd top dressing

**Sowing:**

Sowing time : Follow the normal regional sowing schedule.

Spacing : 90 x 90cm between rows and plants

**Planting method:**(in accordance with conditions laid down in the Genetic Engineering Approval Committee (GEAC) vide letter D O No.10/1/2002-CSdated 5th April, 02 ).

- Plant one seed per hill. Bollgard cotton should be planted in the centre of the plot. For one acre area plant 5 rows of non-Bollgard cotton seeds ( as a refuge belt ) surrounding the Bollgard cotton plot.
- For more than one acre area, The field where Bt cotton (Bollgard cotton) is planted shall be fully surrounded by a belt of land in which the non Bt cotton (non-Bollgard cotton)variety shall be sown. The size of the refuge(the belt of land surrounding the Bt cotton field) should be such as to take atleast 5 rows of non-Bt cotton or shall be 20% of the total sown area whichever is more.
- Planting of non-Bollgard cotton rows (refuge belt) is suggested by the Government as part of insect resistance management strategy. This will help growers to manage bollworms effectively for many years with less chance/risk of resistance development. To meet this requirement non-Bollgard cotton seeds of the same hybrid are also provided along with Bollgard cotton seeds.

In no case, should the Bollgard and non-Bt seeds be mixed before planting.

**For planting, follow the diagram given below:**

**PLANTING LAYOUT:**

For planting in one acre maintain at least five rows of non-Bollgard cotton surrounding the Bollgard sown area. "Farmers are advised to maintain the refuge conditions as specified heretofore, laid down by the GEAC under the Environment (Protection) Act and the Rules framed thereunder".

Sowing depth : Plant seeds at 4-5 cm deep, cover with soil and press.

Gap filling : To achieve optimum plant stand, if necessary go for gap filling in the respective Bollgard and non-Bollgard areas. Non-Bollgard seed should not be used for gap filling in Bollgard area. Similarly, Bollgard seeds should not be used for sowing in non-Bollgard area.

**Inter-cultivation and weed control:**

Follow need based hand weeding and inter cultivation practices to check the growth of the weeds. It helps in better soil aeration and soil moisture conservation.

Note: Follow recommended herbicide application, if it is in practice.

**Irrigation:**

Ensure moisture availability during the critical stages of crop growth i.e, germination, seedling growth, flowering, boll formation and development. Cotton is susceptible to water stagnation for long duration and therefore care should be taken to drain the excess water from the field.

**INTEGRATED PEST MANAGEMENT RECOMMENDATION**

The following integrated practices are recommended :

1.Follow summer ploughing and field sanitation as mentioned earlier. 2.Seed treatment: Bollgard and Non Bollgard seeds are pre treated for control of sucking pests.3. If the infestation of sucking pests crosses the economic threshold level (ETL) at any stage of the crop growth, spray recommended insecticides for control.

| Pest | Economic Threshold level |
|---|---|
| Jassid | 1-2 nymphs or adults per leaf |
| Whitefly | 8-10 nymphs or adults per leaf |
| Aphid | 10 % plants infested |
| Thrips | 10 nymphs or adults / leaf |
| Mites | 10 adults or 20 nymphs / leaf. |

4. Bollworm control: For control of Bollworms in Bollgard and non-Bollgard cotton areas, regular scouting should be done and take spraying decision based on ( ETL).

Scouting method to find out ETL:

- Scouting should be followed twice a week, in the morning hours.
- In one acre area, select at least 20 plants at random ( excluding border rows ).
- Count the number of live bollworm larvae on each of the above selected plants
- If the total number of larvae exceeds the total number of plants counted, then there is a need for spray. Use the insecticides as per local recommendations.
- In case of any doubt, count bollworm larvae again on another set of 20 plants and take spraying decisions.

***Bollgard cotton: It has inbuilt strength to fight against bollworms.***

Based on scouting, if ETL is reached, spray recommended insecticides for bollworm control.

Non-Bollgard cotton plot: Non-Bollgard plot should be sprayed based on recommended practices for bollworm control based on ETL.

Note: For other non-target insects like '*Spodoptera*', stem weevil etc. follow recommended practices.

5.Other recommended IPM practices including trap/barrier crops, pheromone/light traps, bird perches, use of natural enemies, bio-rational insecticides (Ha NPV, Neem) etc. can also be followed for Bollgard cultivation.

6.Avoid use of synthetic pyrethroids upto 110 days after sowing.

7.Avoid repeated use of same class of insecticides.

8.Follow need based recommended practices for the control of diseases.

9. Harvesting: Bollgard need to be Harvested as per the normal harvesting practices.

**Characteristics of Hybrids:**

| **Characteristics** | **Bollgard MECH-12** | **Bollgard MECH-162** | **Bollgard MECH-184** |
|---|---|---|---|
| Leaf shape | Semi-okra | Normal | Semi-okra |
| Leaf texture | Smooth | Slightly hairy | Hairy |
| Boll size | Big | Medium | Big |
| Maturity (days) | 150-160, medium early | 160-170, medium | 160-170, medium |
| Suitability | Normal and early sowing | Normal and Early sowing | Early sowing |
| Fibre quality, Staple length(mm) | Superior, 28-29 | Medium, 26-27 | Superior,28-29 |
| Recommended for | All soil types, rainfed and irrigated | All soil types, rainfed and irrigateo | Light soil, irrigated |

Cotton plants fight Bollworms

What is Bollgard Cotton ?
The strength to fight bollworms is inside the Bollgard cotton plant.
The whole plant of Bollgard Cotton fights bollworms day & night, non-stop.
It fights only 3 types of bollworms i.e. American, spotted & pink bollworms.

Bollgard cotton - Dual benefit.
More bolls on the plant
Less number of insecticide sprayings for bollworm control
The strength to fight bollworms is now within the plant and it fights against bollworms day & night, non-stop. Hence more bolls retained on the plant at less bollworm control cost.
It is being grown in lakhs of acres in USA, Australia, China, Argentina, Mexico, South Africa & Indonesia.

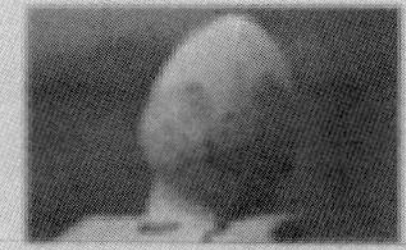
How does it work?
When a bollworm larva feeds on the Bollgard cotton plant, it becomes inactive, and dies within 2 days.
As the larva does not die immediately, it may remain on the plant. But, the dying larva can not do any further damage to the crop.
Hence, the damage caused to the Bollgard cotton plant, by the bollworms, will be much less.

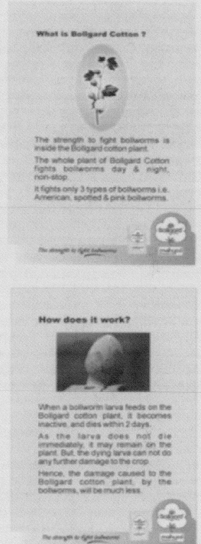

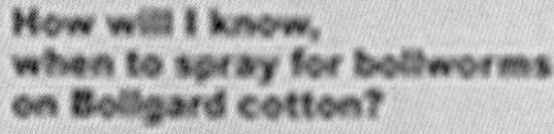
How will I know,
when to spray for bollworms
on Bollgard cotton?

If bollworm infestation is high then you will have to spray insecticides.
To know the level of infestation you will have to count the bollworm larvae in the Bollgard cotton plot.

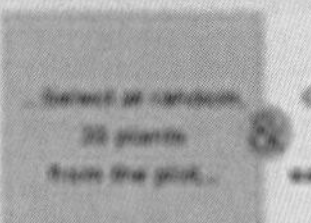
Counting of bollworms?
How will I do it?
If you have grown Bollgard cotton in 1 acre of land...
...Select at random 20 plants from the plot...
&
Count the number of live larvae on each of these plants
If the total number of larvae on 20 plants is 20 or more, then you need to spray.
In case of any doubt, count again.
Note :

How to select 20 plants
in 1 acre of Bollgard cotton?
Bollgard cotton area
Non Bollgard cotton in 5 rows
Selected
Not Selected
- Select plants at random
-Don't select plants from the border row
In each of these plants, search for live bollworm larvae on the whole plant especially on the upper portion and on the growing tips including flowers, squares and bolls.
Don't count the dead bollworms.

Planting / Sowing Instructions.
Every pack of Bollgard cotton contains 2 units:
1. Bollgard Cotton seeds in a can - 450g
2. Non - Bollgard Cotton seeds in a pouch - 120g
Bollgard cotton should be planted in the centre of the plot. For 1 acre area, plant 5 rows of non-Bollgard cotton seeds surrounding the Bollgard plot, as shown in the following diagram.
For gapfilling, use Bollgard Cotton Seeds only in the Bollgard Cotton plot.

of the technology; the database consists, in total, of 24,000 farmers.

- Farmer meetings: conducted in the villages for those cotton farmers wishing to grow Bollgard cotton; the meetings were organized over a period of time (see Fig. 16.1).
- Mass media support: advertisements were published in the interests of the farming community at large (in the reputable newspapers),

**Fig. 16.1.** Farmers' meetings-cum-training programmes.

cautioning against the (so-called illegal) use of Bt seed; this was as per the decision taken at the meeting of 17 April 2002.

- Primary farmers: farmers on whose fields Bollgard cotton trials were conducted in previous years were designated as 'Bollgard cotton leaders', and their help was sought in guiding their fellow farmers by sharing their experience of Bollgard cotton.
- Telephone helplines: to provide immediate replies to farmers' questions on Bollgard cotton, two mobile helpline numbers were made available in all the six states in which Bollgard cotton was to be distributed.
- Flip charts: these multi-page flip charts were used by the company's field staff to educate farmers on important matters related to Bollgard cotton, such as ETL, sowing mechanisms, spray decision, etc.
- Video film: this was also used to show the benefits of Bollgard cotton, ETL counting mechanisms, sowing of peripheral non-Bt rows, etc. to farmers in a simpler and more practical manner.

## Sale and post-sale activities

Bollgard cotton seed was sold through specified Bollgard dealers to ensure compliance with the conditions laid down. Besides this, the dealers were provided with extensive training on the features of the products, etc. for correct and timely communication to achieve the desired results from the product.

Similarly, Bollgard cotton seed was sold only to those farmers who underwent training on the specific cultivation practices required for Bollgard cultivation, including the planting of five rows of non-Bt cotton around the core area of Bollgard cotton.

### *Post-sowing education programme*

This was organized from July to December 2002 in all six states where Bollgard planting had been carried out by farmers. The company organized field days (i.e. farmers' meetings beside Bollgard plots) to continue education on the management of Bollgard cotton, in order to achieve the anticipated results, as well as to show farmers the benefits of Bollgard in comparison to other cotton plots. This activity was carried out across the six states, and 143,507 farmers were contacted through 1144 communication meetings. A typical farmers' meeting is shown in Fig. 16.1.

Mahyco also successfully developed and implemented an after-sales service plan for Bollgard cotton. This was achieved through direct mailers, personal visits to farmers' plots, helpline phone numbers, educational advertisements in local language print media, farmers' meetings at regular intervals and routine performance surveys in different areas. The company also developed an online complaint-resolving system for Bollgard cotton farmers. The farmers using Bollgard cotton could register their complaints and obtain resolution in the minimum possible time. A schematic presentation of this is shown in Fig. 16.2.

## Intellectual property (IP) rights

India did not have Plant Variety Protection legislation in place when the initial work on Bollgard cotton was started. However, over a

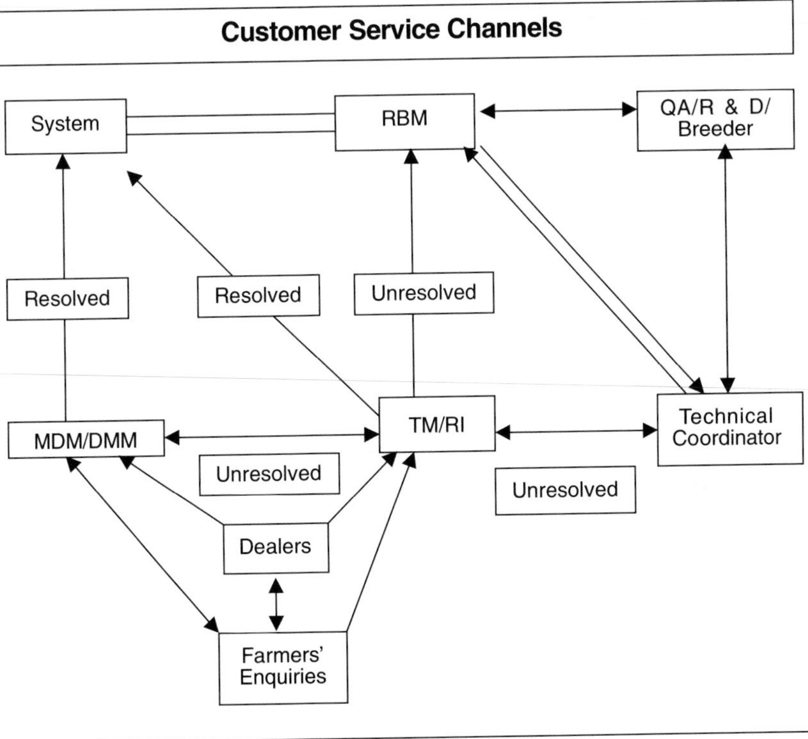


**Fig. 16.2.** Schematic representation of the complaint-resolving mechanism for Bollgard cotton at Mahyco.

period of time, the Protection of Plant Varieties and Farmers' Rights Act (PPV&FR Act) was enacted in 2001 and is now being operationalized. It is anticipated that the PPV&FR Act, 2001 will enable protection of intellectual property rights in plants, and transgenic varieties/hybrids per se will also be protected. An example of the unauthorized use of transgenic cotton is well known: in this case the government was able to act against the concerned company under the provision of the Environment (Protection) Act, 1986, and no action was contemplated on account of intellectual property rights.

The Indian legislation did not recognize patenting of life forms and products under its then-existing norms. However, changes have been made by the Indian government by issuance of ordinance taken out in December 2004 and the subsequent amendments to the Patents Act, enabling the patenting of unique DNA sequences, man-made DNA sequences, etc. This was carried out to make IP laws in India compliant with Trips.

## Conclusion

To conclude, Bollgard cotton approval has been an interesting experience for us. The challenges thrown downduring the course of the process have been numerous. However, all these have set the stage for adoption of transgenic varieties in various field crops in future years.

The adoption of Bollgard cotton by more than 1 million farmers over an area of 3.25 million ha in the year 2005 is a living example of the benefits provided to them by using this new technology. Numerous farmers' surveys and extensive post-release investigations have proved the significant benefits provided by the use of Bollgard technology, and it is expected to be adopted more extensively in the coming years as new hybrids are approved and stacked gene products are made available, providing superior control.

## References

Basu, A.K. (2004) Indian cotton scenario and future perspective. Souvenir of All India *Coordinated Cotton Improvement Programme*, Annual Group Meeting, pp. 19–21.

Bennett, R.M., Ismael, Y., Kambhampati, U. and Morse, S. (2004) Economic impact of genetically modified cotton in India. *AgBioForum* 7 (3), 96–100.

Greenplate, J.T. (1999) Quantification of *Bacillus thuringiensis* insect control protein cry1Ac over time in Bollgard cotton fruit and terminals. *Journal of Economic Entomology* 92, 1377–1383.

Ministry of Science and Technology (1990) *Recombinant DNA Safety Guidelines.* Government of India, New Delhi, p. 75.

Ministry of Science and Technology (1998) *Revised Guidelines for Research in Transgenic Plants and Guidelines for Toxicity and Allergenicity Evaluation of Transgenic Seeds, Plants and Plant Parts.* Government of India, New Delhi.

Naik, G. (2001) *An Analysis of Socio-economic Impact of Bt Technology on Indian Cotton Farmers.* Indian Institute of Management, Centre for Management in Agriculture, Ahmedabad, India.

Pental, D. *et al.* (2001) *Confirmation of the Absence of 'Terminator Gene', i.e. a Patented Embryogenesis Deactivation System in Mahyco Bt Cotton Hybrid Seeds.* University of Delhi, South Campus, New Delhi, India.

Pray, C.E. and Danmeng, M. (2001) Impact of Bt cotton in China. *World Development* 29 (5), 813–825.

Rules for the Manufacture, Use/Import/Export and Storage of Hazardous Micro-organisms/Genetically Engineered Organisms or Cells (1989) Notification No. GSR1037(E), 5 December.

The Environment (Protection) Act (1986) *Gazette of India* Ext., Pt. II, S. 1, 26–5–1986. Bare Act. Professional Book Publishers.

The Patents (Amendment) Act (2005) *Patents Act, 1970.* Universal Law Publishing Co. Pvt. Ltd., New Delhi, India, sections 2j, 3c, 3j.

The Protection of Plant Varieties and Farmers' Rights Act (2001) Universal Law Publishing Co. Pvt. Ltd., New Delhi, India.

Qaim, M. (2003) Bt cotton in India: field trial results and economic projections. World Table 1: Training Programs

Qaim, M. and Zilberman, D. (2003) Yield effects of genetically modified crops in developing countries. *Science* 7 (299), 900–902.

# Index